ALGEBRA

ALGEBRA

Editor

Philomena Rosenberger

scitus
academics

Algebra
Edited by **Philomena Rosenberger**

This Book is Printed on Acid-Free Paper. ♾

ISBN: 9781681179551
Library of Congress Control Number: 2017949084

SCITUS Academics LLC,
616, Corporate Way, Suite 2, 4766,
Valley Cottage, NY 10989

www.scitusacademics.com

Printed in United States of America.

Preface

Algebra, a huge area in mathematics, is one of the broad parts of mathematics, together with number theory, geometry and analysis. In its most universal form, algebra is the study of mathematical symbols and the rules for manipulating these symbols; it is a unifying thread of almost all of mathematics. The roots of algebra can be traced to the ancient Babylonians, who developed an advanced arithmetical system with which they were able to do calculations in an algorithmic fashion. The Babylonians developed formulas to calculate solutions for problems typically solved today by using linear equations, quadratic equations, and indeterminate linear equations. As such, it includes everything from elementary equation solving to the study of abstractions such as groups, rings, and fields. Over the last 25 years, worldwide research trends in algebra have increasingly emphasized the subject's connections with, and applications to, other areas of mathematics and science. It's easy to think of algebra as an abstract notion that has no use in real life. Understanding the history and the practical applications of algebra that are put into use every day might make it a little differently.

This volume 'Algebra' demonstrates high quality research results in all areas of algebra, including algebraic geometry or algebraic number theory. It covers the mathematical orientation in various aspects of algebra and discrete structures along with their applications to other fields, e.g. mathematical, physical and biological sciences; engineering, information technology etc. It also publishes original research papers of mathematical orientation on various aspects of computational mathematics and optimization including numerical analysis, approximation theory, number theory, Probability theory and Statistics, computational aspects of geometry and algebra, mechanics, design of efficient numerical/qualitative methods for solving differential equations.

Table of Contents

CHAPTER 1

A New Type of Restarted Krylov Methods

Achiya Dax

Hydrological Service, Jerusalem, Israel

ABSTRACT

In this paper we present a new type of Restarted Krylov methods for calculating peripheral eigenvalues of symmetric matrices. The new framework avoids the Lanczos tridiagonalization process, and the use of polynomial filtering. This simplifies the restarting mechanism and allows the introduction of several modifications. Convergence is assured by a monotonicity property that pushes the eigenvalues toward their limits. The Krylov matrices that we use lead to fast rate of convergence. Numerical experiments illustrate the usefulness of the proposed approach.

Keywords: Restarted Krylov Methods, Exterior Eigenvalues, Symmetric Matrices, Monotonicity, Starting Vectors

1. INTRODUCTION

In this paper we present a new type of Restarted Krylov methods. Given a symmetric matrix $G \in \mathbb{R}^{n \times n}$, the method is aimed at calculating a cluster of k exterior eigenvalues of G. Other names for such eigenvalues are "peripheral eigenvalues" and "extreme eigenvalues". The method is best suited for handling large sparse matrices in which a matrix-vector product needs only $0(n)$ flops. Another underlying assumption is that k^2 is considerably smaller than n. The need for computing a few extreme eigenvalues of such a matrix arises in many applications, see [1] - [21].

The traditional restarted Krylov methods are classified into three types of restarts: "Explicit restart" [1] , [9] , [11] , "Implicit restart" [1] , [3] , [12] , and "Thick restart" [18] , [19]. See also [7] , [11] , [14] , [16]. When solving symmetric eigenvalue problems all these methods are carried out by repeated use of the Lanczos tridiagonalization process, and the use of polynomial filtering to determine the related starting vectors. This way each iteration generates a new tridiagonal matrix and computes its eigensystem. The method proposed in this paper is based on a different framework, one that avoids these techniques. The basic iteration of the new method have recently been presented by this author in [4]. The driving force is a monotonicity property that pushes the estimated eigenvalues toward their limits. The rate of convergence depends on the quality of the Krylov matrix that we use. In this paper we introduce a modified scheme for generating the Krylov matrix. This leads to dramatic improvement in the rate of convergence, and turns the method into a powerful tool.

Let the eigenvalues of G be ordered to satisfy

$$\left|\hat{\lambda}_1\right| \geq \left|\hat{\lambda}_2\right| \geq \cdots \geq \left|\hat{\lambda}_n\right| \geq 0 \tag{1.1}$$

or

$$\tilde{\lambda}_1 \geq \tilde{\lambda}_2 \geq \cdots \geq \tilde{\lambda}_n \tag{1.2}$$

Then the new algorithm is built to compute one of the following four types of target clusters that contain *k* extreme eigenvalues.

A dominant cluster

$$\left\{\hat{\lambda}_1, \cdots, \hat{\lambda}_k\right\} \tag{1.3}$$

A right-side cluster

$$\left\{\tilde{\lambda}_1, \cdots, \tilde{\lambda}_k\right\} \tag{1.4}$$

A left-side cluster

$$\left\{\tilde{\lambda}_{n+1-k}, \cdots, \tilde{\lambda}_{n-1}, \tilde{\lambda}_n\right\} \tag{1.5}$$

A two-sides cluster is a union of a right-side cluster and a left-side cluster.

For example, $\left\{\tilde{\lambda}_1, \tilde{\lambda}_n\right\}$ or $\left\{\tilde{\lambda}_1, \tilde{\lambda}_2, \tilde{\lambda}_n\right\}$.

Note that although the above definitions refer to clusters of eigenvalues, the algorithm is carried out by computing the corresponding *k* eigenvectors of

G. The subspace that is spanned by these eigenvectors is called the target space.

The basic iteration

The qth iteration, $q = 0, 1, 2, \cdots$, is composed of the following five steps. The first step starts with a matrix $V_q \in \mathbb{R}^{n\times k}$ that contains "old" information on the target space, a matrix $Y_q \in \mathbb{R}^{n\times \ell}$ that contains "new" information, and a matrix $X_q = [V_q, Y_q] \in \mathbb{R}^{n\times(k+\ell)}$ that includes all the known information. The matrix X_q has $p = k + \ell$ orthonormal columns. That is

$$X_q^{\mathrm{T}} X_q = I \in \mathbb{R}^{p\times p}$$

(Typical values for ℓ lie between k to $2k$.)

Step 1: Eigenvalues extraction. First compute the Rayleigh quotient matrix

$$S_q = X_q^{\mathrm{T}} G X_q$$

Then compute k eigenpairs of S_q which correspond to the target cluster. (For example, if it is desired to compute a right-side cluster of G, then compute a right-side cluster of S_q.) The corresponding k eigenvectors of S_q are assembled into a matrix

$$U_q \in \mathbb{R}^{p\times k}, \quad U_q^{\mathrm{T}} U_q = I \in \mathbb{R}^{k\times k}$$

which is used to compute the related matrix of Ritz vectors,

$$V_{q+1} = X_q U_q$$

Note that both X_q and U_q have orthonormal columns, and V_{q+1} inherits this property.

Step 2: Collecting new information. Compute a Krylov matrix $B_q \in \mathbb{R}^{n\times \ell}$ that contains new information on the target space.

Step 3: Orthogonalize the columns of B_q against the columns of V_{q+1}. There are several ways to achieve this task. In exact arithmetic the resulting matrix, Z_q, satisfies the Gram-Schmidt formula

$$Z_q = B_q - V_{q+1}\left(V_{q+1}^{\mathrm{T}} B_q\right)$$

Step 4: Build an orthonormal basis of Range (Z_q). Compute a matrix,

$$Y_{q+1} \in \mathbb{R}^{n\times\ell}, \quad Y_{q+1}^{\mathrm{T}} Y_{q+1} = I \in \mathbb{R}^{\ell\times\ell}$$

whose columns form an orthonormal basis of Range (Z_q). This can be done by a QR factorization of. Z_q (If rank (Z_q) is smaller than ℓ, then ℓ is temporarily reduced to be rank (Z_q).)

Step 5: Define X_{q+1} by the rule

$$X_{q+1} = \left[V_{q+1}, Y_{q+1}\right]$$

which ensures that

$$X_{q+1}^{\mathrm{T}} X_{q+1} = I \in \mathbb{R}^{p\times p}$$

At this point we are not concerned with efficiency issues, and the above description is mainly aimed to clarify the purpose of each step. Hence there might be better ways to carry out the basic iteration.

The plan of the paper is as follows. The monotonicity property that motivates the new method is established in the next section. Let $\lambda_j^{(q)}, j = 1,\cdots,k$, denote the Ritz values which are computed at Step 1 of the qth iteration. Then it is shown that each iteration gives a better approximation of the target cluster. Moreover, for each $j, j = 1,\cdots,k$, the sequence $\lambda_j^{(q)}, q = 1,2,\cdots$, proceeds monotonously toward the desired eigenvalue of G. The rate of convergence depends on the information matrix B_q. The method proposed in Section 3 is based on a three-term recurrence relation that leads to rapid convergence. A further improvement can be gained by Power acceleration, see Section 4. The paper ends with numerical experiments that illustrate the usefulness of the proposed methods.

2. THE MONOTONICITY PROPERTY

The monotonicity property is an important feature of the new iteration, whose proof is given in [4]. Yet, in order to make this paper self-contained, we provide the proof. We start by stating two well-known interlacing theorems, e.g., [8] , [10] and [20].

Theorem 1 (Cauchy interlace theorem) *Let* $G \in \mathbb{R}^{n\times n}$ *be a symmetric matrix with eigenvalues*

$$\lambda_1 \geq \lambda_2 \geq \cdots \geq \lambda_n \qquad (2.1)$$

Let the symmetric matrix $H \in \mathbb{R}^{k\times k}$ *be obtained from G by deleting n-k rows and the corresponding n-k columns. Let*

$$\eta_1 \geq \eta_2 \geq \cdots \geq \eta_k \qquad (2.2)$$

denote the eigenvalues of H. Then

$$\lambda_j \geq \eta_j \quad \text{for } j = 1, \cdots, k \qquad (2.3)$$

and

$$\eta_{k+1-i} \geq \lambda_{n+1-i} \quad \text{for } i = 1, \cdots, k \qquad (2.4)$$

In particular, for k=n-1 we have the interlacing relations

$$\lambda_1 \geq \eta_1 \geq \lambda_2 \geq \eta_2 \geq \lambda_3 \geq \cdots \geq \lambda_{n-1} \geq \eta_{n-1} \geq \lambda_n \qquad (2.5)$$

Corollary 2 (Poincarà separation theorem) *Let the matrix* $V \in \mathbb{R}^{n\times k}$ *have k orthonormal columns. That is* $V^{\mathrm{T}}V = I \in \mathbb{R}^{k\times k}$. *Let the matrix* $H=V^{T}GV$ *have the eigenvalues (2.2). Then the eigenvalues of H and G satisfy (2.3) and (2.4).*

Let us turn now to consider the qth iteration of the new method,q=1,2,3,.... Assume first that the algorithm is aimed at computing a cluster of k right-side eigenvalues of G,

$$\left\{\tilde{\lambda}_1, \tilde{\lambda}_2, \cdots, \tilde{\lambda}_k\right\}$$

and let the eigenvalues of the matrix

$$S_q = X_q^{\mathrm{T}} G X_q = \left[V_q, Y_q\right]^{\mathrm{T}} G \left[V_q, Y_q\right]$$

be denoted as

$$\lambda_1^{(q)} \geq \lambda_2^{(q)} \geq \cdots \geq \lambda_k^{(q)} \geq \cdots \geq \lambda_p^{(q)}$$

Then the Ritz values which are computed at Step 1 are

$$\lambda_1^{(q)} \geq \lambda_2^{(q)} \geq \cdots \geq \lambda_k^{(q)}$$

and these values are the eigenvalues of the matrix

$$V_{q+1}^{\mathrm{T}} G V_{q+1}$$

Similarly,

$$\lambda_1^{(q-1)} \geq \lambda_2^{(q-1)} \geq \cdots \geq \lambda_k^{(q-1)}$$

are the eigenvalues of the matrix

$$V_q^{\mathrm{T}} G V_q$$

Therefore, since the columns of V_q are the first k columns of X_q,

$$\lambda_j^{(q)} \geq \lambda_j^{(q-1)} \qquad \text{for } j = 1, \cdots, k$$

On the other hand from Corollary 2 we obtain that

$$\tilde{\lambda}_j \geq \lambda_j^{(q)} \qquad \text{for } j = 1, \cdots, k$$

Hence by combining these relations we see that

$$\tilde{\lambda}_j \geq \lambda_j^{(q)} \geq \lambda_j^{(q-1)} \tag{2.6}$$

for j=1,...,k and q=1,2,3,....

The treatment of a left-side cluster is done in a similar way. Assume that the algorithm is aimed at computing a cluster of k left-side eigenvalues of G,

$$\left\{\tilde{\lambda}_{n+1-k}, \cdots, \tilde{\lambda}_{n-1}, \tilde{\lambda}_n\right\}$$

Then similar arguments show that

$$\lambda_{p+1-i}^{(q-1)} \geq \lambda_{p+1-i}^{(q)} \geq \tilde{\lambda}_{n+1-i} \tag{2.7}$$

For i=1,...,k, and q=1,2,3,....

Recall that a two-sides cluster is the union of a right-side cluster and a left-side one. In this case the eigenvalues of S_q that correspond to the right-side satisfy (2.6) while eigenvalues of S_q that correspond to the left-side satisfy (2.7). A similar situation occurs in the computation of a dominant cluster, since a dominant cluster is either a right-side cluster, a left-side cluster, or a two-sides cluster.

3. THE BASIC KRYLOV MATRIX

The basic Krylov information matrix has the form

$$B_q = \left[\boldsymbol{b}_1, \boldsymbol{b}_2, \cdots, \boldsymbol{b}_\ell \right] \in \mathbb{R}^{n \times \ell} \tag{3.1}$$

where the sequence $\boldsymbol{b}_1, \boldsymbol{b}_2, \ldots$, is initialized by the starting vector $\boldsymbol{b}_0$. The ability of a Krylov subspace to approximate a dominant subspace is characterized by the Kaniel-Paige-Saad bounds. See, for example, ([7] , pp. 552-554), ([10] , pp. 242-247), ([11] , pp. 147-151), ([14] , pp. 272-274), and the references therein. One consequence of these bounds regards the angle between $\boldsymbol{b}_1$ and the dominant subspace: The smaller the angle, the better approximation we get. This suggests that b_0 should be defined as the sum of the current Ritz vectors. That is,

$$\boldsymbol{b}_0 = V_{q+1} \boldsymbol{e}$$

where $\boldsymbol{e} = (1, 1, \cdots, 1)^{\mathrm{T}} \in \mathbb{R}^k$ is a vector of ones. Note that there is no point in setting $\boldsymbol{b}_1 = V_{q+1} \boldsymbol{e}$, since in the next step B_q is orthogonalized against V_{q+1}.

The proof of the Kaniel-Paige-Saad bounds relies on the properties of Chebyshev polynomials, while the building of these polynomials is carried out by using a three term recurrence relation, e.g. [10] , [11]. This observation suggests that in order to achieve these bounds the algorithm for generating our Krylov sequence should use a "similar" three term recurrence relation. Indeed this feature is one of the reasons that make the Lanczos recurrence so successful, see ([11] , p. 138). Below we describe an alternative three term recurrence relation, which is based on the Modified Gram-Schmidt (MGS) orthogonaliza- tion process.

Let $\boldsymbol{r} \in \mathbb{R}^n$ be a given vector and let $\boldsymbol{q} \in \mathbb{R}^n$ be a unit length vector. That is $\|\boldsymbol{q}\|_2 = 1$ where $\|\cdot\|_2$ denotes the Euclidean vector norm. Then the statement "orthogonalize r against q" is carried out by replacing r with $\boldsymbol{r} - (\boldsymbol{r}^{\mathrm{T}} \boldsymbol{q}) \boldsymbol{q}$. Similarly, the statement "normalize r" is carried out by replacing r with $\boldsymbol{r} / \|\boldsymbol{r}\|_2$. With these conventions at hand the construction of the vectors $\boldsymbol{b}_0$-, $\boldsymbol{b}_1, \ldots, \boldsymbol{b}_\ell$, is carried out as follows.

The preparations part

a) Compute the starting vector:

$$V_{q+1}\boldsymbol{e} \Big/ \left\| V_{q+1}\boldsymbol{e} \right\|_2 \tag{3.2}$$

b) Compute $\boldsymbol{b}_1$: Set $\boldsymbol{b}_1 = G b_0$.

Orthogonalize $\boldsymbol{b}_1$ against $\boldsymbol{b}_0$.

Normalize $\boldsymbol{b}_1$.

c) Compute $\boldsymbol{b}_2$: Set $\boldsymbol{b}_2 = G b_1$.

Orthogonalize $\boldsymbol{b}_2$ against $\boldsymbol{b}_0$.

Orthogonalize $\boldsymbol{b}_2$ against $\boldsymbol{b}_1$.

Normalize $\boldsymbol{b}_2$.

The iterative part

For $j=3,\ldots,\ell$, compute $\boldsymbol{b}_j$ as follows:

a) Set $\boldsymbol{b}_j = G\boldsymbol{b}_{j-1}$.

b) Orthogonalize $\boldsymbol{b}_j$ againstb $\boldsymbol{b}_{j-2}$.

c) Orthogonalize $\boldsymbol{b}_j$ against. $\boldsymbol{b}_{j-1.}$

d) Reorthogonalization: For $\boldsymbol{i=1,\ldots,j-1}$, orthogonalize $\boldsymbol{b}_j$ against $\boldsymbol{b}_i$.

e) Normalize $\boldsymbol{b}_j$.

The reorthogonalization step is aimed to ensure that the numerical rank of $\boldsymbol{B}_q$ will stay close to ℓ. Yet for small values of ℓ it is not essential.

4. THE POWER-KRYLOV MATRIX

Assume for a moment that the algorithm is aimed at calculating a cluster of k dominant eigenvalues. Then the Kaniel-Paige-Saad bounds suggest that slow rate of convergence is expected when these eigenvalues are poorly separated from the other eigenvalues. Indeed, this difficulty is seen in Table 2, when the basic Krylov matrix is applied on problems like "Very slow geometric" and "Dense equispaced". Let $v \geq 2$ be a small integer. Then the larger eigenvalues of the powered matrix, $\boldsymbol{G}^v$, are better separated than those of G. This suggests that a faster rate of convergence can be gained by replacing G with $\boldsymbol{G}^v$.

The implementation of this idea is carried out by introducing a small modification in the construction of $\boldsymbol{B_q}$: Here the computation of $\boldsymbol{b}_j$, $\boldsymbol{j}$=**1**,..., $\boldsymbol{\ell}$, starts with

$$\boldsymbol{b}_j = G^{\nu}\boldsymbol{b}_{j-1} \tag{4.1}$$

(In our experiments ν=4.) It is important to stress that this is the only part of the algorithm that uses $\boldsymbol{G}^{\nu}$. All the other steps of the basic iteration remain unchanged. In particular, the Ritz values which are computed in Step 1 are those of G (not $\boldsymbol{G}^{\nu}$). Of course, in practice $\boldsymbol{G}^{\nu}$ is never computed. Instead $\boldsymbol{b}_j$ is computed by a sequence of ν matrix-vector multiplications.

The usefulness of the Power-Krylov approach depends on two factors: The cost of a matrix-vector product and the distribution of the eigenvalues. As noted above, it is expected to reduce the number of iterations when the k largest eigenvalues of G are poorly separated from the rest of the spectrum. See Table 3. Another advantage of this approach is that $\boldsymbol{\ell}$ is kept small. Note that although the computational effort per iteration increases, a smaller portion of time is spent on orthogonalizations and on the Rayleigh-Ritz procedure.

5. THE INITIAL ORTHONORMAL MATRIX

To start the algorithm we need to supply an "initial" orthonormal matrix, $X_0 \in \mathbb{R}^{n\times p}$, where.$p > k$ This task can be done in the following way. Define p=k+ $\boldsymbol{\ell}$ and let n×k matrix

$$B_0 = \left[\boldsymbol{b}_1, \boldsymbol{b}_2, \cdots, \boldsymbol{b}_p\right] \tag{5.1}$$

be generated as in Section 3, using some arbitrary starting vector $\boldsymbol{b_0}$. Then $\boldsymbol{X_0}$ is obtained by computing an orthonormal basis of Range(B_0). A similar procedure is used in the Power-Krylov method.

In our experiments B_0 is initiated by the vector $\boldsymbol{b}_0 = \boldsymbol{e}/\|\boldsymbol{e}\|_2$ where $\boldsymbol{e} = (1,1,\cdots,1)^T \in \mathbb{R}^n$. Yet a random starting vector is equally good. In the next section we shall see that the above choice of $\boldsymbol{X_0}$ is often sufficient to provide accurate estimates of the desired eigenpairs.

6. NUMERICAL EXPERIMENTS

In this section we describe some experiments with the proposed methods. The test matrices have the form

$$G = \text{diag}\{\lambda_1, \lambda_2, \cdots, \lambda_n\} \in \mathbb{R}^{n \times n} \tag{6.1}$$

where

$$\lambda_1 \geq \lambda_2 \geq \cdots \geq \lambda_n \geq 0 \tag{6.2}$$

Recall that in Krylov methods there is no loss of generality in experimenting with diagonal matrices, e.g., ([6] , p. 367). The diagonal matrices that we have used are displayed in Table 1. All the experiments were carried out with **n = 12,000**. The experiments that we have done are aimed at computing a cluster of k dominant eigenvalues. The figures in Table 2 and Table 3 provide the number of iterations that were needed to satisfy the inequality

$$\left(\sum_{j=1}^{k} \left|\lambda_j - \lambda_j^{(q)}\right|\right) \Big/ \left(k\left|\lambda_1\right|\right) \leq 10^{-14} \tag{6.3}$$

Thus, for example, from Table 2 we see that only 4 iterations are needed when the algorithm computes k=200 eigenvalues of the Equispaced test matrix.

The ability of the basic Krylov matrix to achieve accurate computation of the eigenvalues is illustrated in Table 2. We see that often the algorithm terminates within a remarkably small number of iterations. Observe that the method is highly successful in handling low-rank matrices, matrices which are nearly low rank, like "Harmonic" or "Geometric", and matrices with gap in the spectrum. In such matrices the initial orthonormal matrix, $\boldsymbol{X}_0$, is sufficient to provide accurate eigenvalues. Note also that the method has no difficulty in computing multiple eigenvalues.

As expected, a slower rate of convergence occurs when the dominant eigen-alues that we seek are poorly separated from the other eigenvalues. This situation is demonstrated in matrices like "Dense equispaced" or "Very slow geometric". Yet, as Table 3 shows, in such cases the Power-Krylov method leads to considerable reduction in the number of iterations.

Table 1. Types of test matrices, n=200.

Matrix type	Matrix eigenvalues λ_j, $j=1,\cdots,n$
Harmonic squares	$\lambda_j=(1/j)^2$
Harmonic	$\lambda_j=1/j$
Harmonic roots	$\lambda_j=(1/j)^{1/2}$
Very fast geometric decay	$\lambda_j=(0.5)^j$
Fast geometric decay	$\lambda_j=(0.9)^j$
Geometric decay	$\lambda_j=(0.95)^j$
Moderate geometric decay	$\lambda_j=(0.99)^j$
Slow geometric decay	$\lambda_j=(0.999)^j$
Very slow geometric decay	$\lambda_j=(0.9999)^j$
Dense equispaced	$\lambda_j=n+1-j$
Equispaced	$\lambda_j=1001-j$ for $j=1,\cdots,1000$ and $\lambda_j=0$ for $j=1001,\cdots,n$
Low-Rank-100	$\lambda_j=101-j$ for $j=1,\cdots,100$ and $\lambda_j=0$ for $j=101,\cdots,n$
Low-Rank-50	$\lambda_j=51-j$ for $j=1,\cdots,50$ and $\lambda_j=0$ for $j=51,\cdots,n$
Low-Rank-10	$\lambda_j=n+1-j$ for $j=1,\cdots,10$ and $\lambda_j=0$ for $j=11,\cdots,n$
Harmonic Triples	$\lambda_{3j-2}=\lambda_{3j-1}=\lambda_{3j}=1/j$ for $j=1,\cdots,n/3$
Multiple-Harmonic	$\lambda_j=1$ for $j=1,\cdots,10$ and $\lambda_j=1/j$ for $j=11,\cdots,n$
Multiple-Geometric	$\lambda_j=1$ for $j=1,\cdots,10$ and $\lambda_j=(0.95)^j$ for $j=11,\cdots,n$
Equispaced-Geometric Gap	$\lambda_j=101-j$ for $j=1,\cdots,100$ and $\lambda_j=(0.95)^j$ for $j=101,\cdots,n$

Table 2. Computing k dominant eigenvalues with the basic Krylov matrix, $\ell=k+40$.

Matrix type	Number of iterations					
	$k=6$	$k=10$	$k=20$	$k=40$	$k=100$	$k=200$
Harmonic squares	0	0	0	0	0	0
Harmonic	0	0	0	0	0	0
Harmonic roots	0	0	0	1	2	3
Very fast Geometric	0	0	0	0	0	0
Fast Geometric	0	0	0	0	0	0
Geometric	0	0	0	0	0	0
Moderate Geometric	1	1	2	1	0	0
Slow Geometric	7	7	8	9	10	7
Very slow Geometric	28	27	28	23	26	23
Dense Equispaced	43	43	39	33	31	32
Equispaced	6	7	7	9	7	4
Low-Rank-100	1	1	1	0	0	0
Low-Rank-50	0	0	0	0	0	0
Low-Rank-10	0	0	0	0	0	0
Harmonic triples	2	2	2	3	3	4
Multiple-Harmonic	3	2	2	2	2	1
Multiple-Geometric	3	4	3	2	2	1
Equispaced-Geometric Gap	1	1	1	0	0	0

7. CONCLUDING REMARKS

The new type of Restarted Krylov methods avoids the use of Lanczos algorithm. This simplifies the basic iteration, and clarifies the main ideas behind the method. The driving force that ensures convergence is the monotonicity proper- ty, which is easily concluded from the Cauchy-Poincaré interlacing theorems. The proof indicates too important points. First, there is a lot of freedom in choosing the information matrix, and that monotonicity is guaranteed as long as we achieve proper orthogonalizations. Second, the rate of convergence depends on the "quality" of the information matrix. This raises the question of how to define this matrix. Since the algorithm is aimed at computing a cluster of exterior eigenvalues, a Krylov information matrix is a good choice. In [4] we

Table 3. Computing k dominant eigenvalues with the Power-Krylov matrix, $\ell=k+40$.

Matrix type	Number of iterations					
	$k=6$	$k=10$	$k=20$	$k=40$	$k=100$	$k=200$
Harmonic squares	0	0	0	0	0	1
Harmonic	0	0	0	0	0	0
Harmonic roots	0	0	0	0	0	0
Very fast Geometric	0	0	0	0	0	0
Fast Geometric	0	0	0	0	0	0
Geometric	0	0	0	0	0	1
Moderate Geometric	0	0	0	0	0	0
Slow Geometric	3	4	4	4	4	2
Very Slow Geometric	14	17	18	15	15	12
Dense Equispaced	16	17	19	18	16	15
Equispaced-1000	3	4	3	4	3	2
Low-Rank-100	0	0	0	0	0	0
Low-Rank-50	0	0	0	0	0	0
Low-Rank-10	0	0	0	0	0	0
Harmonic triples	1	2	2	2	2	2
Multiple-Harmonic	2	2	2	1	1	1
Multiple-Geometric	1	3	1	2	0	0
Equispaced-Geometric Gap	0	0	0	0	0	0

have tested the classical Krylov matrix where $\boldsymbol{b}_j$ is obtained by normalizing $\boldsymbol{G}\boldsymbol{b}_{j-1}$. However, as shown in [5] , this matrix suffers from certain deficiencies. In this paper the Krylov basis is built by a three term recurrence relation, which leads to dramatic reduction in the number of iterations.

Indeed, the results of our experiments are quite encouraging. We see that the algorithm requires a remarkably small number of iterations. In particular, it efficiently handles various kinds of low-rank matrices. In these matrices the initial orthonormal matrix is often sufficient for accurate computation of the desired eigenpairs. The algorithm is also successful in computing eigenvalues of “difficult” matrices like “Dense equispaced” or “Very slow geometric decay”.

REFERENCES

1. Bai, A., Demmel, J., Dongarra, J., Ruhe, A. and van der Vorst, H. (1999) Templates for the Solution of Algebraic Eigenvalue Problems: A Practical Guide. SIAM, Philadelphia, PA.
2. Bjorck, A. (1996) Numerical Methods for Least-Squares Problems. SIAM, Philadelphia. https://doi.org/10.1137/1.9781611971484
3. Calvetti, D., Reichel, L. and Sorenson, D.C. (1994) An Implicitly Restarted Lanczos Method for Large Symmetric Eigenvalue Problems. Electronic Transactions on Numerical Analysis, 2, 1-21.
4. Dax, A. (2015) A Subspace Iteration for Calculating a Cluster of Exterior Eigenvalues. Advances in Linear Algebra and Matrix Theory, 5, 76-89. https://doi.org/10.4236/alamt.2015.53008
5. Dax, A. (2016) The Numerical Rank of Krylov Matrices. In: Linear Algebra and Its Applications.
6. Demmel, J.W. (1997) Applied Numerical Linear Algebra. SIAM, Philadelphia. https://doi.org/10.1137/1.9781611971446
7. G.H. Golub and C.F. Van Loan (2013) Matrix Computations. 4th Edition, Johns Hopkins University Press, Baltimore.
8. Horn, R.A. and Johnson, C.R. (1985) Matrix Analysis. Cambridge University Press, Cambridge. https://doi.org/10.1017/CBO9780511810817
9. Morgan, R.B. (1996) On Restarting the Arnoldi Method for Large Non-Symmetric Eigenvalues Problems. Mathematics of Computation, 65, 1213-1230.
10. Parlett, B.N. (1980) The Symmetric Eigenvalue Problem. Prentice-Hall, Englewood Cliffs, NJ.
11. Saad, Y. (2011) Numerical Methods for Large Eigenvalue Problems: Revised Edition. SIAM, Philadelphia. https://doi.org/10.1137/1.9781611970739
12. Sorensen, D.C. (1992) Implicit Application of Polynomial Filters in a k-Step Arnoldi Method. SIAM Journal on Matrix Analysis and Applications, 13, 357-385. https://doi.org/10.1137/0613025
13. Stewart, G.W. (1998) Matrix Algorithms, Vol. I: Basic Decompositions. SIAM, Philadelphia.
14. Stewart, G.W. (2001) Matrix Algorithms, Vol. II: Eigensystems. SIAM, Philadelphia.
15. Trefethen, L.N. and Bau III, D. (1997) Numerical Linear Algebra. SIAM, Philadelphia.
16. Watkins, D.S. (2007) The Matrix Eigenvalue Problem: GR and Krylov Subspace Methods. SIAM, Philadelphia. https://doi.org/10.1137/1.9780898717808
17. Wilkinson, J.H. (1965) The Algebraic Eigenvalue Problem. Clarendon Press, Oxford.
18. Wu, K. and Simon, H. (2000) Thick-Restarted Lanczos Method for Large Symmetric Eigenvalue Problems. SIAM Journal on Matrix Analysis and Applications, 22, 602-616. https://doi.org/10.1137/S0895479898334605
19. Yamazaki, I., Bai, Z., Simon, H., Wang, L. and Wu, K. (2010) Adaptive Projection Subspace Dimension for the Thick-Restart Lanczos Method. ACM Transactions on Mathematical Software, 37, Article No. 27. https://doi.org /10.1145 /1824801.1824805
20. Zhang, F. (1999) Matrix Theory: Basic Results and Techniques. Springer-Verlag, New York. https://doi.org/10.1007/978-1-4757-5797-2
21. Zhou, Y. and Saad, Y. (2008) Block Krylov-Schur Method for Large Symmetric Eigenvalue Problems. Numerical Algorithms, 47, 341-359. https://doi.org/10.1007/s11075-008-9192-9

CHAPTER 2

A Generalization of Cramer's Rule

Hugo Leiva

Department of Mathematics, Louisiana State University, Baton Rouge, LA, USA

ABSTRACT

In this paper, we find two formulas for the solutions of the following linear equation*Ax=b*, $x \in IR^m, b \in IR^n, m \geq n$, where $A = (a_{j,i})_{n \times m}$ is a $n \times m$ real matrix. This system has been well studied since the 1970s. It is known and simple proven that there is a solution for all $b \in IR^n$ if, and only if, the rows of A are linearly independent, and the minimum norm solution is given by the Moore-Penrose inverse formula, which is often denoted by $A^{\dagger}$;in this case, this solution is given by $A^{\dagger}b = A^{*}(AA^{*})^{-1}b$ Using this formula, Cramer's Rule and Burgstahler's Theorem (Theorem 2), we prove the following representation for this solution

$$x_i = \frac{\begin{vmatrix} \|l_1\|^2 + a_{1i}b_1 & \langle l_1, l_2 \rangle + a_{2i}b_1 & \cdots & \langle l_1, l_n \rangle + a_{ni}b_1 \\ \langle l_2, l_1 \rangle + a_{1i}b_2 & \|l_2\|^2 + a_{2i}b_2 & \cdots & \langle l_2, l_n \rangle + a_{ni}b_2 \\ \vdots & \vdots & \ddots & \vdots \\ \langle l_n, l_1 \rangle + a_{1i}b_n & \langle l_n, l_2 \rangle + a_{2i}b_n & \cdots & \|l_n\|^2 + a_{ni}b_n \end{vmatrix}}{\begin{vmatrix} \|l_1\|^2 & \langle l_1, l_2 \rangle & \cdots & \langle l_1, l_n \rangle \\ \langle l_2, l_1 \rangle & \|l_2\|^2 & \cdots & \langle l_2, l_n \rangle \\ \vdots & \vdots & \ddots & \vdots \\ \langle l_n, l_1 \rangle & \langle l_n, l_2 \rangle & \cdots & \|l_n\|^2 \end{vmatrix}} - 1,$$

i=1,2,3---,m, where $l_1, l_2 --- l_n$ are the row vectors of the matrix *A*. To the best of our knowledge and looking in to many Linear Algebra books, there is not formula for this solution depending on determinants. Of course, this formula coincides with the one given by Cramer's Rule when *m=n*.

Keywords: Linear Equation, Cramer's Rule, Generalized Formula

1. INTRODUCTION

In this paper, we find a formula depending on determinants for the solutions of the following linear equation

$$Ax = b,\ x \in IR^m,\ b \in IR^n,\ m \geq n \tag{1}$$

or

$$\begin{cases} a_{1,1}x_1 + a_{1,2}x_2 + \cdots + a_{1,m}x_m = b_1 \\ a_{2,1}x_1 + a_{2,2}x_2 + \cdots + a_{2,m}x_m = b_2 \\ \vdots \\ a_{n,1}x_1 + a_{n,2}x_2 + \cdots + a_{n,m}x_m = b_n \end{cases}. \tag{2}$$

Now, if we define the column vectors

$$l_1 = \begin{bmatrix} a_{1,1} \\ a_{1,2} \\ \vdots \\ a_{1,m} \end{bmatrix}, l_2 = \begin{bmatrix} a_{2,1} \\ a_{2,2} \\ \vdots \\ a_{2,m} \end{bmatrix}, \cdots, l_n = \begin{bmatrix} a_{n,1} \\ a_{n,2} \\ \vdots \\ a_{n,m} \end{bmatrix},$$

then the system (2) also can be written as follows:

$$\langle l_i, x \rangle = b_i, \quad i = 1, 2, \cdots, n, \tag{3}$$

where $\langle \cdot, \cdot \rangle$ denotes the innerproduct in IR^m and A is $n \times m$

real matrix. Usually, one can apply Gauss Elimination Method to find some solutions of this system, and this method is a systematic procedure for solving systems like (1); it is based on the idea of reducing the augmented matrix

$$\begin{bmatrix} a_{1,1} & a_{1,2} & \cdots & a_{1,m} & b_1 \\ a_{2,1} & a_{2,2} & \cdots & a_{2,m} & b_2 \\ \vdots & \vdots & \ddots & \vdots & \vdots \\ a_{n,1} & a_{n,2} & \cdots & a_{n,m} & b_n \end{bmatrix}, \tag{4}$$

to the form that is simple enough such that the system of equations can be solved by inspection. But, to my knowledge, in general there is not formula for the solutions of (1) in terms of determinants if $m \neq n$

When $m=n$ and det(A)≠0, the system (1) admits only one solution given by$x=A^{-1}b$, and from here one can deduce the well known Cramer Rule which says:

Theorem 1.1. (Cramer Rule 1704-1752) If *A is n×n matrix with det(A) ≠0, then the solution of the system (1) is given by the formula*:

$$x_i = \frac{\det\left(\left(A\right)_i\right)}{\det\left(A\right)}, \quad i = 1, 2, 3, \cdots, n, \tag{5}$$

where $(A)_i$ is the matrix obtained by replacing the entries in the ith column of A by the entries in the matrix

$$\begin{bmatrix} b_1 \\ b_2 \\ \vdots \\ b_n \end{bmatrix}.$$

A simple and interested generalization of Cramer Rule is done by Prof. Dr. Sylvan Burgstahler ([1]) from University of Minnesota, Duluth, where he taught for 20 years. This result is given by the following Theorem:

Theorem 1.2. (Burgstahler 1983) If the system of equations

$$\begin{cases} a_{1,1}x_1 + a_{1,2}x_2 + \cdots + a_{1,n}x_n = b_1 \\ a_{2,1}x_1 + a_{2,2}x_2 + \cdots + a_{2,n}x_n = b_2 \\ \vdots \\ a_{n,1}x_1 + a_{n,2}x_2 + \cdots + a_{n,n}x_n = b_n \end{cases}$$

(6)

has(unique) solution*x1,x2---,xn,*, then for all $\lambda_i \in IR, i = 1, 2, \cdots, n$ one has

$$\lambda_1 x_1 + \lambda_2 x_2 + \cdots + \lambda_n x_n = \frac{\begin{vmatrix} a_{1,1} + \lambda_1 b_1 & a_{1,2} + \lambda_2 b_1 & \cdots & a_{1,n} + \lambda_n b_1 \\ a_{2,1} + \lambda_1 b_2 & a_{2,2} + \lambda_2 b_2 & \cdots & a_{2,n} + \lambda_n b_2 \\ \vdots & \vdots & \ddots & \vdots \\ a_{n,1} + \lambda_1 b_n & a_{n,2} + \lambda_2 b_n & \cdots & a_{n,n} + \lambda_n b_n \end{vmatrix}}{\begin{vmatrix} a_{1,1} & a_{1,2} & \cdots & a_{1,n} \\ a_{2,1} & a_{2,2} & \cdots & a_{2,n} \\ \vdots & \vdots & \ddots & \vdots \\ a_{n,1} & a_{n,2} & \cdots & a_{n,n} \end{vmatrix}} - 1.$$

(7)

Using Moore-Penrose Inverse Formula and Cramer's Rule, one can prove the following Theorem. But, for better understanding of the reader, we will include here a direct proof of it.

Theorem 1.3. For all $b \in IR^n$, the system (1) is solvable if, and only if,

$$\det\left(AA^*\right) \neq 0.$$

(8)

Moreover, one solution for this equation is given by the following formula:

$$x = A^*\left(AA^*\right)^{-1} b,$$

(9)

where A^* is the transpose of A (or the conjugate transpose of A in the complex case).

Also, this solution coincides with the Cramer formula whe n=m. In fact, this formula is given as follows:

$$x_i = \sum_{j=1}^{n} a_{j,i} \frac{\det\left(\left(AA^*\right)_j\right)}{\det\left(AA^*\right)}, \quad i = 1,2,3,\cdots,m,$$

(10)

where $(AA^*)_j$ is the matrix obtained by replacing the entries in the jth column of AA^* by the entries in the matrix

$$\begin{bmatrix} b_1 \\ b_2 \\ \vdots \\ b_n \end{bmatrix}.$$

In addition, this solution has minimum norm, i.e.,

$$\|x\| = \inf\left\{\|w\| : Aw = b,\ w \in IR^m\right\}$$

(11)

and $\|x\| = \|w\|_{\text{with}} Aw = b \Leftrightarrow x = w$.

The main results of this work are the following Theorems.

Theorem 1.4. The solutions of (1)-(3) given by (9) can be written as follows:

$$x_i = \frac{\begin{vmatrix} \|l_1\|^2 + a_{1i}b_1 & \langle l_1, l_2\rangle + a_{2i}b_1 & \cdots & \langle l_1, l_n\rangle + a_{ni}b_1 \\ \langle l_2, l_1\rangle + a_{1i}b_2 & \|l_2\|^2 + a_{2i}b_2 & \cdots & \langle l_2, l_n\rangle + a_{ni}b_2 \\ \vdots & \vdots & \ddots & \vdots \\ \langle l_n, l_1\rangle + a_{1i}b_n & \langle l_n, l_2\rangle + a_{2i}b_n & \cdots & \|l_n\|^2 + a_{ni}b_n \end{vmatrix}}{\begin{vmatrix} \|l_1\|^2 & \langle l_1, l_2\rangle & \cdots & \langle l_1, l_n\rangle \\ \langle l_2, l_1\rangle & \|l_2\|^2 & \cdots & \langle l_2, l_n\rangle \\ \vdots & \vdots & \ddots & \vdots \\ \langle l_n, l_1\rangle & \langle l_n, l_2\rangle & \cdots & \|l_n\|^2 \end{vmatrix}} - 1, \quad i = 1,2,3,\cdots,m.$$

(12)

Theorem 1.5. The system (1) is solvable for each $b \in IR^n$, if, and only if, the set of vectors $\{l_1, l_2 --- l_n\}$ formed by the rows of the matrix *A* is lineally independent in IR^m.

Moreover, a solution for the system (1) is given by the following formula:

$$x_i = \upsilon_{1i} \|\upsilon_1\|^2 b_1 + \upsilon_{2i} \|\upsilon_2\|^2 \left(b_2 - \langle l_2, \upsilon_1 \rangle \|\upsilon_1\|^2 c_1\right), \tag{13}$$

$$+ \cdots + \upsilon_{ni} \|\upsilon_n\|^2 \left(b_n - \sum_{i=1}^{n-1} \langle l_n, \upsilon_i \rangle \|\upsilon_i\|^2 c_i \right), \quad i = 1, 2, \cdots, m, \tag{14}$$

where the set of vectors $\{v_1, v_2 --- v_n\}$ is obtain by the Gram-Schmidt process and the numbers $c_1, c_2 --- c_n$ are given by

$$\begin{aligned} c_1 &= b_1 \\ c_2 &= b_2 - \langle l_2, \upsilon_1 \rangle \|\upsilon_1\|^2 c_1 \\ c_3 &= b_3 - \langle l_3, \upsilon_1 \rangle \|\upsilon_1\|^2 c_1 - \langle l_3, \upsilon_2 \rangle \|\upsilon_2\|^2 c_2 \\ &\vdots \\ c_n &= b_n - \sum_{i=1}^{n-1} \langle l_n, \upsilon_i \rangle \|\upsilon_i\|^2 c_i \end{aligned} \tag{15}$$

and $\upsilon_i = [\upsilon_{i1}, \upsilon_{i2}, \upsilon_{i3}, \cdots, \upsilon_{im}]^T, i = 1, 2, \cdots, n$.

2. PROOF OF THE MAIN THEOREMS

In this section we shall prove Theorems 1.3, 1.4, 1.5 and more. To this end, we shall denote by ⟨x,y⟩ the Euclidian innerproduct in IR^k and the associated norm by $\|x\| = \sqrt{\langle x, x \rangle}$. Also, we shall use some ideas from [2] and the following result from [3] , pp 55.

Lemma 2.1. Let *W* and *Z* be Hilbert space, $G \in L(W, Z)$ and $G^* \in L(Z, W)$ the adjoint operator, then the following statements holds, [(i)] $\text{Rang}(G) = Z \Leftrightarrow \exists \gamma > 0$ such that

$\|G^{*}z\|_{W} \geq \gamma \|z\|_{Z}, \; z \in Z.$

[(ii)] $\overline{\text{Rang}(G)} = Z \Leftrightarrow Ker(G^{*}) = \{0\} \Leftrightarrow G^{*}$ is $1-1$.

We will include here a direct proof of Theorem 1.3 just for better understanding of the reader.

Proof of Theorem 1.3. The matrix A may also viewed as a linear operator $A: IR^{m} \to IR^{n}$; therefore $A \in L(IR^{m}, IR^{n})$ and its adjoint operator A^* is the transpose of A and $A^*: IR^{n}$ ▯

IR^{m}.

Then, system (1) is solvable for all $b \in IR^{n}$ if, and only if, the operator A is surjective. Hence, from the Lemma 2.1 there exists $\gamma > 0$ such that

$$\|A^{*}z\|_{IR^{m}} \geq \gamma \|z\|_{IR^{n}}, \; z \in IR^{n}.$$

Therefore,

$$\langle AA^{*}z, z \rangle \geq \gamma^{2} \|z\|_{IR^{n}}^{2}, \; z \in IR^{n}.$$

This implies that AA^* is one to one. Since AA^* is a n▯n

matrix, then $\det(AA^{*}) \neq 0$.

Suppose now that $\det(AA^{*}) \neq 0$. Then $(AA^*)^{-1}$ exists and given $b \in IR^{n}$ we can see that $x=A^*(AA^*)^{-1}{}_{b}$ is a solution of$Az=b$.

Now, since $z=(AA^*)^{-1}{}_{b}$ is the only solution of the equation

$(AA^*)w=b$,

then from Theorem 1.1 (Cramer Rule) we obtain that:

$$z_{1} = \frac{\det\left((AA^{*})_{1}\right)}{\det(AA^{*})}, z_{2} = \frac{\det\left((AA^{*})_{2}\right)}{\det(AA^{*})}, \cdots, z_{n} = \frac{\det\left((AA^{*})_{n}\right)}{\det(AA^{*})},$$

where $(AA^*)_i$ is the matrix obtained by replacing the entries in the ith column of AA^* by the entries in the matrix

$$\begin{bmatrix} b_1 \\ b_2 \\ \vdots \\ b_n \end{bmatrix}.$$

Then, the solution $x=A^*(AA^*)^{-1}b$ of (1) can be written as follows

$$x = \begin{bmatrix} a_{1,1} & a_{2,1} & \cdots & a_{n,1} \\ a_{1,2} & a_{2,2} & \cdots & a_{n,2} \\ \vdots & \vdots & \ddots & \vdots \\ a_{1,m} & a_{2,m} & \cdots & a_{n,m} \end{bmatrix} \begin{bmatrix} \frac{\det\left(\left(AA^*\right)_1\right)}{\det\left(AA^*\right)} \\ \frac{\det\left(\left(AA^*\right)_2\right)}{\det\left(AA^*\right)} \\ \vdots \\ \frac{\det\left(\left(AA^*\right)_n\right)}{\det\left(AA^*\right)} \end{bmatrix} = \begin{bmatrix} \sum_{j=1}^{n} a_{j,1} \frac{\det\left(\left(AA^*\right)_j\right)}{\det\left(AA^*\right)} \\ \sum_{j=1}^{n} a_{j,2} \frac{\det\left(\left(AA^*\right)_j\right)}{\det\left(AA^*\right)} \\ \vdots \\ \sum_{j=1}^{n} a_{j,m} \frac{\det\left(\left(AA^*\right)_j\right)}{\det\left(AA^*\right)} \end{bmatrix}.$$

Now, we shall see that this solution has minimum norm. In fact, consider w in IR^m such that $Aw=b$ and

$$\|w\|^2 = \|x+(w-x)\|^2 = \|x\|^2 + 2\,\mathrm{Re}\langle x, w-x\rangle + \|w-x\|^2.$$

On the other hand,

$$\langle x, w-x\rangle = \left\langle A^*\left(AA^*\right)^{-1} b, w-x\right\rangle = \left\langle \left(AA^*\right)^{-1} b, Aw-Ax\right\rangle = \left\langle \left(AA^*\right)^{-1} b, b-b\right\rangle = 0.$$

Hence, $\|w\|^2 - \|x\|^2 = \|w-x\|^2 \geq 0$.

Therefore, $\|x\| \leq \|w\|$, and $\|x\| = \|w\|$ if $x = w$.

Proof of Theorem 1.5. Suppose the system is solvable for all $b \in IR^n$. Now, assume the existence of real numbers $c_i, i=1,2,\text{---}n$ such that

$$c_1 l_1 + c_2 l_2 + c_3 l_3 + \cdots + c_n l_n = 0.$$

Then, there exists $x \in IR^m$ such that

$$\begin{cases} a_{1,1}x_1 + a_{1,2}x_2 + \cdots + a_{1,m}x_m = c_1 \\ a_{2,1}x_1 + a_{2,2}x_2 + \cdots + a_{2,m}x_m = c_2 \\ \vdots \\ a_{n,1}x_1 + a_{n,2}x_2 + \cdots + a_{n,m}x_m = c_n \end{cases}.$$

In other words,

$$\langle l_i, x \rangle = c_i, \quad i = 1,2,\cdots,n.$$

Hence,

$$\langle c_i l_i, x \rangle = c_i^2, \; i = 1,2,\cdots,n.$$

So,

$$\langle c_1 l_1 + c_2 l_2 + c_3 l_3 + \cdots + c_n l_n, x \rangle = c_1^2 + c_2^2 + c_3^2 + \cdots + c_n^2 = 0.$$ There fore, $c_1 = c_2 = \cdots = c_n = 0$,which prove the independence of. $\{l_1, l_2, \cdots, l_n\}$.

Now, suppose that the set $\{l_1,l_2---l_n\}$ is linearly independent in IR^m. Using the Gram-Schmidt process we can find a set $\{v_1,v_2---v_n\}$ of orthogonal vectors in IR^m given by the formula:

$$\begin{aligned}
\upsilon_1 &= l_1 \\
\upsilon_2 &= l_2 - \frac{\langle l_2, \upsilon_1 \rangle}{\|\upsilon_1\|^2}\upsilon_1 \\
\upsilon_3 &= l_3 - \frac{\langle l_3, \upsilon_1 \rangle}{\|\upsilon_1\|^2}\upsilon_1 - \frac{\langle l_3, \upsilon_2 \rangle}{\|\upsilon_2\|^2}\upsilon_2 \; . \\
&\vdots \\
\upsilon_n &= l_n - \sum_{i=1}^{n-1} \frac{\langle l_n, \upsilon_i \rangle}{\|\upsilon_i\|^2}\upsilon_i
\end{aligned} \tag{16}$$

Then, system (1) will be equivalent to the following system:

$$\begin{cases}
\langle \upsilon_1, x \rangle = c_1 \\
\langle \upsilon_2, x \rangle = c_2 \\
\langle \upsilon_3, x \rangle = c_3 \\
\qquad \vdots \\
\langle \upsilon_n, x \rangle = c_n
\end{cases} , \tag{17}$$

where

$$\begin{aligned}
c_1 &= b_1 \\
c_2 &= b_2 - \frac{\langle l_2, \upsilon_1 \rangle}{\|\upsilon_1\|^2}c_1 \\
c_3 &= b_3 - \frac{\langle l_3, \upsilon_1 \rangle}{\|\upsilon_1\|^2}c_1 - \frac{\langle l_3, \upsilon_2 \rangle}{\|\upsilon_2\|^2}c_2 \; . \\
&\vdots
\end{aligned} \tag{18}$$

If we denote the vectors υ_i 's by

$$\upsilon_i = \begin{bmatrix} \upsilon_{i1} \\ \upsilon_{i2} \\ \upsilon_{i3} \\ \vdots \\ \upsilon_{im} \end{bmatrix}, \quad i = 1, 2, \cdots, n,$$

and the *n*×*m* matrix ?

by

$$\Upsilon = \begin{bmatrix} \upsilon_{1,1} & \upsilon_{1,2} & \cdots & \upsilon_{1,m} \\ \upsilon_{2,1} & \upsilon_{2,2} & \cdots & \upsilon_{2,m} \\ \vdots & \vdots & \ddots & \vdots \\ \upsilon_{n,1} & \upsilon_{n,2} & \cdots & \upsilon_{n,m} \end{bmatrix},$$

then, applying Theorem 1.3 we obtain that system (17) has solution for all $C \in IR^n$ if, and only if, $(\Upsilon\Upsilon^*) \neq 0$. But,

$$\Upsilon\Upsilon^* = \begin{bmatrix} \|\upsilon_1\|^2 & \langle \upsilon_1, \upsilon_2 \rangle & \cdots & \langle \upsilon_1, \upsilon_n \rangle \\ \langle \upsilon_2, \upsilon_1 \rangle & \|\upsilon_2\|^2 & \cdots & \langle \upsilon_2, \upsilon_n \rangle \\ \vdots & \vdots & \ddots & \vdots \\ \langle \upsilon_n, \upsilon_1 \rangle & \langle \upsilon_n, \upsilon_2 \rangle & \cdots & \|\upsilon_n\|^2 \end{bmatrix} = \begin{bmatrix} \|\upsilon_1\|^2 & 0 & 0 & \cdots & 0 \\ 0 & \|\upsilon_2\|^2 & 0 & \cdots & 0 \\ \vdots & \vdots & \vdots & \ddots & \vdots \\ 0 & 0 & 0 & \cdots & \|\upsilon_n\|^2 \end{bmatrix}.$$

So,

$$\det(\Upsilon\Upsilon^*) = \|\upsilon_1\|^2 \|\upsilon_2\|^2 \cdots \|\upsilon_n\|^2 \neq 0.$$

From here and using the formula (9) we complete the proof of this Theorem.

Examples and Particular Cases

In this section we shall consider some particular cases and examples to illustrate the results of this work.

Example 2.1. *Consider the following particular case of system* (1)

$$a_{1,1}x_1 + a_{1,2}x_2 + \cdots + a_{1,m}x_m = b. \tag{19}$$

In this case n=1 andA=[$a_{1,1}a_{1,2}$,---a_{1m}],Then, if we define the column vector

$$l_1 = \begin{bmatrix} a_{1,1} \\ a_{1,2} \\ \vdots \\ a_{1,m} \end{bmatrix},$$

$$AA^* = \begin{bmatrix} a_{1,1} & a_{1,2} & \cdots & a_{1,m} \end{bmatrix} \begin{bmatrix} a_{1,1} \\ a_{1,2} \\ \vdots \\ a_{1,m} \end{bmatrix} = \|l_1\|^2.$$

Then, $\left(AA^*\right)^{-1} b = b\|l_1\|^{-2}$ and

$$x = A^*\left(AA^*\right)^{-1} b = \begin{bmatrix} a_{1,1}b\|l_1\|^{-2} \\ a_{1,2}b\|l_1\|^{-2} \\ \vdots \\ a_{1,m}b\|l_1\|^{-2} \end{bmatrix}.$$

Therefore, a solution of the system (19) is given by:

$$x_i = \frac{a_{1i}b}{\|l_1\|^2} = \frac{a_{1i}b}{\sum_{j=1}^{m} a_{1j}^2}, \quad i = 1, 2, \cdots, m. \tag{20}$$

Example 2.2. Consider the following particular case of system (1)

$$a_{1,1}x_1 + a_{1,2}x_2 + \cdots + a_{1,m}x_m = b_1$$
$$a_{2,1}x_1 + a_{2,2}x_2 + \cdots + a_{2,m}x_m = b_2$$

(21)

In this case n=2 and

$$A = \begin{bmatrix} a_{1,1} & a_{1,2}x_2 & \cdots & a_{1,m} \\ a_{2,1} & a_{2,2} & \cdots & a_{2,m} \end{bmatrix},$$

Then, if we define the column vectors

$$l_1 = \begin{bmatrix} a_{1,1} \\ a_{1,2} \\ \vdots \\ a_{1,m} \end{bmatrix}, \quad l_2 = \begin{bmatrix} a_{2,1} \\ a_{2,2} \\ \vdots \\ a_{2,m} \end{bmatrix},$$

then

$$AA^* = \begin{bmatrix} a_{1,1} & a_{1,2}x_2 & \cdots & a_{1,m} \\ a_{2,1} & a_{2,2} & \cdots & a_{2,m} \end{bmatrix} \begin{bmatrix} a_{1,1} & a_{2,1} \\ a_{1,2} & a_{2,2} \\ \vdots & \vdots \\ a_{1,m} & a_{12,m} \end{bmatrix} = \begin{bmatrix} \|l_1\|^2 & \langle l_1, l_2 \rangle \\ \langle l_2, l_1 \rangle & \|l_2\|^2 \end{bmatrix}.$$

Hence, from the formula (10) we obtain that:

$$\begin{bmatrix} x_1 \\ x_2 \\ \vdots \\ x_m \end{bmatrix} = A^* \left(AA^* \right)^{-1} b = \begin{bmatrix} a_{1,1} & a_{2,1} \\ a_{1,2} & a_{2,2} \\ \vdots & \vdots \\ a_{1,m} & a_{12,m} \end{bmatrix} \begin{bmatrix} \dfrac{\det\left(\left(AA^* \right)_1 \right)}{\det\left(AA^* \right)} \\ \dfrac{\det\left(\left(AA^* \right)_2 \right)}{\det\left(AA^* \right)} \end{bmatrix}.$$

Therefore, a solution of the system (21) is given by:

$$x_1 = a_{11}\frac{b_1\|l_2\|^2 - b_2\langle l_1, l_2\rangle}{\|l_1\|^2\|l_2\|^2 - |\langle l_1, l_2\rangle|^2} + a_{21}\frac{b_2\|l_1\|^2 - b_1\langle l_2, l_1\rangle}{\|l_1\|^2\|l_2\|^2 - |\langle l_1, l_2\rangle|^2} \tag{22}$$

$$x_2 = a_{12}\frac{b_1\|l_2\|^2 - b_2\langle l_1, l_2\rangle}{\|l_1\|^2\|l_2\|^2 - |\langle l_1, l_2\rangle|^2} + a_{22}\frac{b_2\|l_1\|^2 - b_1\langle l_2, l_1\rangle}{\|l_1\|^2\|l_2\|^2 - |\langle l_1, l_2\rangle|^2} \tag{23}$$

$$\vdots \tag{24}$$

$$x_m = a_{1m}\frac{b_1\|l_2\|^2 - b_2\langle l_1, l_2\rangle}{\|l_1\|^2\|l_2\|^2 - |\langle l_1, l_2\rangle|^2} + a_{2m}\frac{b_2\|l_1\|^2 - b_1\langle l_2, l_1\rangle}{\|l_1\|^2\|l_2\|^2 - |\langle l_1, l_2\rangle|^2}. \tag{25}$$

Now, we shall apply the foregoing formula or (12) to find the solution of the following system

$$\begin{cases} x_1 + x_2 = 1 \\ -x_1 + x_2 + x_3 = -1 \end{cases}. \tag{26}$$

If we define the column vectors

$$l_1 = \begin{bmatrix} 1 \\ 1 \\ 0 \end{bmatrix}, \quad l_2 = \begin{bmatrix} -1 \\ 1 \\ 1 \end{bmatrix},$$

then $\det(AA^*) = \|l_1\|^2\|l_2\|^2 - |\langle l_1, l_2\rangle|^2 = \|l_1\|^2\|l_2\|^2 = 6$ and $x_1 = \frac{5}{6}$, $x_2 = \frac{1}{6}$, and $x_3 = \frac{-2}{6}$.

Example 2.3. Consider the following general case of system (1)

$$\begin{cases} a_{1,1}x_1 + a_{1,2}x_2 + \cdots + a_{1,m}x_m = b_1 \\ a_{2,1}x_1 + a_{2,2}x_2 + \cdots + a_{2,m}x_m = b_2 \\ \vdots \\ a_{n,1}x_1 + a_{n,2}x_2 + \cdots + a_{n,m}x_m = b_n \end{cases}. \tag{27}$$

Then, if $\{l_1, l_2 \cdots l_n\}$ is an orthogonal set in IR^m, we get

$$AA^* = \begin{bmatrix} \|l_1\|^2 & 0 & 0 & \cdots & 0 \\ 0 & \|l_2\|^2 & 0 & \cdots & 0 \\ \vdots & \vdots & \vdots & \ddots & \vdots \\ 0 & 0 & 0 & \cdots & \|l_n\|^2 \end{bmatrix},$$

and the solution of the system (1) is very simple and given by:

$$x_i = \sum_{j=1}^{n} a_{j,i} b_j \left\| l_j \right\|^{-2}, \quad i = 1, 2, \cdots, m. \tag{28}$$

Now, we shall apply the formula (28) or (12) to find solution of the following system:

$$\begin{cases} -x_1 - x_2 + x_3 + x_4 = 1 \\ -x_1 + x_2 - x_3 + x_4 = 1 \\ x_1 - x_2 - x_3 + x_4 = 1 \end{cases}. \tag{29}$$

If we define the column vectors

$$l_1 = \begin{bmatrix} -1 \\ -1 \\ 1 \\ 1 \end{bmatrix}, \quad l_2 = \begin{bmatrix} -1 \\ 1 \\ -1 \\ 1 \end{bmatrix}, \quad l_3 = \begin{bmatrix} 1 \\ -1 \\ -1 \\ 1 \end{bmatrix}.$$

Then, $\{l_1, l_2, L_3\}$ is an orthogonal set in IR^3 and the solution of this system is given by: $x_1 = \frac{-1}{4}$, $x_2 = \frac{-1}{4}$, $x_3 = \frac{-1}{4}$,and $x_4 = \frac{3}{4}$.

3. VARIATIONAL METHOD TO OBTAIN SOLUTIONS

Theorems 1.3, 1.4 and 1.5 give a formula for one solution of the system (1) which has minimum norma. But it is not the only way allowing to build solutions of this equation. Next, we shall present a variational method to obtain solutions of (1) as a minimum of the quadratic functional$J{:}IR^n \rightarrow IR$,

$$j(\xi) = \frac{1}{2}\left\|A^*\xi\right\|^2 - \langle b, \xi \rangle, \quad \forall \xi \in IR^n. \tag{30}$$

Proposition 3.1. For a given $b \in IR^n$ the Equation (1) has a solution $x \in IR^m$ if, and only if,

$$\langle x, A^*\xi \rangle - \langle b, \xi \rangle = 0, \quad \forall \xi \in IR^n. \tag{31}$$

It is easy to see that (31) is in fact an optimality condition for the critical points of the quadratic functional j define above.

Lemma 3.1. Suppose the quadratic functional j has a minimize $\xi_b \in IR^n$. Then,

$$x_b = A^*\xi_b \tag{32}$$

is a solution of (1).

Proof. First, we observe that j has the following form:

$$j(\xi)=\frac{1}{2}\langle AA^{*}\xi,\xi\rangle-\langle b,\xi\rangle, \quad \forall\xi\in IR^{n}.$$

Then, if ξ_b is a point where j achieves its minimum value, we obtain that:

$$\frac{\mathrm{d}}{\mathrm{d}\xi}\{j\}(\xi_b)=AA^{*}\xi_b-b=0.$$

So, $AA^{*}\xi_b=b$ and $x_b=A^{*}\xi_b$ is a solution of (1).

Remark 3.1. Under the condition of Theorem 1.3, the solution given by the formulas (32) and (9) coincide.

Theorem 3.1. The system (1) is solvable if, and only if, the quadratic functional j defined by (30) has a minimum for all $b\in IR^{n}$.

Proof. Suppose (8) is solvable. Then, the matrix A viewed as an operator from IR^m to IR^n is surjective. Hence, from Lemma 2.1. there exists ⍰>0 such that

$$\left\|A^{*}\xi\right\|^{2}\geq\gamma^{2}\left\|\xi\right\|^{2}, \ \xi\in IR^{n}.$$

Then,

$$j(\xi)\geq\frac{\gamma^{2}}{2}\left\|\xi\right\|^{2}-\left\|b\right\|\left\|\xi\right\|, \quad \xi\in IR^{n}.$$

Therefore,

$$\lim_{\|\xi\|\to\infty} j(\xi)=\infty.$$

Consequently, j is coercive and the existence of a minimum is ensured.

The other way of the proof follows as in proposition 3.1.

Now, we shall consider an example where Theorems 1.3, 1.4 and 1.5 can not be applied, but proposition 3.1 does.

Example 3.1. It considers the system with linearly independent rows

$$\begin{cases} x_1+x_2+x_3=1 \\ 2x_1+2x_2+2x_3=2 \end{cases}.$$

In this case n=2 and

$$A=\begin{bmatrix}1 & 1 & 1\\ 2 & 2 & 2\end{bmatrix} \text{ and } b=\begin{bmatrix}1\\ 2\end{bmatrix}.$$

Then

$$AA^{*}=\begin{bmatrix}1 & 1 & 1\\ 2 & 2 & 2\end{bmatrix}\begin{bmatrix}1 & 2\\ 1 & 2\\ 1 & 2\end{bmatrix}=\begin{bmatrix}3 & 6\\ 6 & 12\end{bmatrix}.$$

Therefore, the critical points of the quadratic functional j given by (30) satisfy the equation:

$$AA^{*}\xi = b,$$

i.e.,

$$\begin{aligned}3\xi_1+6\xi_2&=1\\ 6\xi_1+12\xi_2&=2\end{aligned}.$$

So, there are infinitely many critical points given by

$$\xi=\begin{bmatrix}\frac{1}{3}-2a\\ a\end{bmatrix}, \quad a\in IR.$$

Hence, a solution of the system is given by

$$x=A^{*}\xi=\begin{bmatrix}1 & 2\\ 1 & 2\\ 1 & 2\end{bmatrix}\begin{bmatrix}\frac{1}{3}-2a\\ a\end{bmatrix}=\begin{bmatrix}\frac{1}{3}\\ \frac{1}{3}\\ \frac{1}{3}\end{bmatrix}.$$

REFERENCES

1. Burgstahier, S. (1983) A Generalization of Cramer's Rulle. The Two-Year College Mathematics Journal, 14, 203-205. http://dx.doi.org/10.2307/3027088
2. 2.Iturriaga, E. and Leiva, H. (2007) A Necessary and Sufficient Conditions for the Controllability of Linear System in Hilbert Spaces and Applications. IMA Journal Mathematical and Information, 25, 269-280. http://dx.doi.org/ 10.1093 /imamci/dnm017
3. Curtain, R.F. and Pritchard, A.J. (1978) Infinite Dimensional Linear Systems. Lecture Notes in Control and Information Sciences, Vol. 8, Springer Verlag, Berlin

CHAPTER 3

L-Fuzzy Vector Subspaces and Its Fuzzy Dimension

Chun'e Huang[1]*, Yan Song[2], Xiruo Wang[2]

[1]*College of Biochemical Engineering, Beijing Union University, Beijing, China*

[2]*Mudanjiang Normal University, Mudanjiang, China*

ABSTRACT

In this paper, we introduce the definition of L-fuzzy vector subspace, define its dimension by an L-fuzzy natural number. For a finite-dimensional L-fuzzy vector subspace, we prove that the equality $\dim(\tilde{E}_1+\tilde{E}_2)+\dim(\tilde{E}_1\cap\tilde{E})=\dim\tilde{E}_1+\dim\tilde{E}_2$ holds without any restricted conditions. At the same time, we deduce that the formula $\dim(\widetilde{\mathrm{im} f})+\dim(\widetilde{\mathrm{ker} f})=\dim\tilde{E}$ holds.

Keywords: L-Fuzzy Sets, L-Fuzzy Vector Subspace, L-Fuzzy Dimension

1. INTRODUCTION

Firstly, fuzzy vector subspace was introduced by Katsaras and Liu [1]. Then its pro- perties and characters were investigated (see [2] [3] [4] [5] , etc). The dimension of a fuzzy vector space was defined as a n-tuple by Lowen [6]. Subsequently, it was defined as a non-negative real number or infinity by Lubczonok [5] , and proved that the for- mula

$$\dim\left(\tilde{E}_1+\tilde{E}_2\right)+\dim\left(\tilde{E}_1\cap\tilde{E}_2\right)=\dim\tilde{E}_1+\dim\tilde{E}_2 \tag{1}$$

is valid under certain conditions, where $\tilde{E}_1$ and $\tilde{E}_2$ are fuzzy vector spaces. Recently, basis and dimension of a fuzzy vector space were redefined as a fuzzy set and a fuzzy natural number by Shi and Huang [7] , respectively. Under the definitions, more pro- perties of (crisp) vector spaces were correct in fuzzy vector spaces.

In this paper, we generalize the results in [7] to L lattice, and prove that some for- mulas still hold in the lattice L. In particular, we present the definition of L-fuzzy vector subspace and its -fuzzy dimension. The L-fuzzy dimension of a finite dimensional fuzzy vector subspace is a fuzzy natural number. We prove that (1) holds without any re- stricted conditions and $\dim\left(\widetilde{\ker f}\right)+\dim\left(\widetilde{\operatorname{im} f}\right)=\dim\tilde{E}$ holds.

2. PRELIMINARIES

Given a set *X* and a completely distributive lattice L, we denote the power set of *X* and the set of all L-fuzzy sets on *X* (or L-sets for short) by 2^X and L^X, respec- tively . For any $A \subseteq X$, we denote the cardinality of *A* by $|A|$.

An element *a* in L is called a prime element if $a \geq b \wedge c$ implies $a \geq b$ or $a \geq c$. *A* in L is called co-prime if $a \leq b \vee c$ implies $a \leq b$ or $a \leq c$ [8]. The set of non- unit prime elements in L is denoted by *P*(*L*). The set of non-zero co-prime elements in L is denoted by *J*(*L*).

The binary relation < in L is defined as follows: for $a, b \in L$, $a<b$ if and only if for every subset $D \subseteq L$, the relation $b \leq \sup D$ always implies the existence of $d \in D$ with $a \leq d$ [9]. $\{a \in L : a < b\}$ is called the greatest minimal family of ***b*** in the sense of [10] , denoted by $\beta(b)$, and $\beta^*(b)=\beta(b)\cap J(L)$. Moreover, for $b \in L$, we define $\alpha(b)=\{a \in L : a <^{op} b\}$ and $\alpha^*(b)=\alpha(b)\cap P(L)$. In a completely distri- butive lattice *L*, there exist $a(b)$ and $\beta(b)$ for each $b \in L$, and $b=\vee\beta(b)=\wedge\alpha(b)$ (see [10]).

In [10] , Wang thought that $\beta(0)=\{0\}$ and $\alpha(1)=\{1\}$. In fact, it should be that $\beta(0)=\varnothing$ and $\alpha(1)=\varnothing$.

Throughout this paper, *L* denotes a completely distributive lattice, and *E* is a crisp vector space. We often do not distinguish a crisp subset *A* of *E* and its cha- racteristic function X_A.

If $A \in L^X$ and $a \in L$, we can define

$$A_{[a]} = \{x \in X : A(x) \geq a\}, \qquad A_{(a)} = \{x \in X : a \in \beta(A(x))\}.$$

$$A^{[a]} = \{x \in X : a \notin \alpha(A(x))\}, \quad A^{(a)} = \{x \in X : A(x) \nleq a\}.$$

Some properties of these cut sets can be found in [11] - [16].

In [17] Shi introduced the concept of L-fuzzy natural numbers(denoted by $\mathbb{N}(L)$), defined their operations and discussed the relation of a-cut sets. We simply recall as follows: for any $\lambda, \mu \in \mathbb{N}(L)$, $a \in L$

(1) $(\lambda+\mu)_{(a)} \subseteq \lambda_{(a)} + \mu_{(a)} \subseteq \lambda_{[a]} + \mu_{[a]} \subseteq (\lambda+\mu)_{[a]}$

(2) $(\lambda+\mu)^{(a)} \subseteq \lambda^{(a)} + \mu^{(a)} \subseteq \lambda^{[a]} + \mu^{[a]} \subseteq (\lambda+\mu)^{[a]}$;

(3) For any $\lambda, \mu \in \mathbb{N}(L)$ and $a \in P(L)$, it follows that $(\lambda+\mu)^{(a)} = \lambda^{(a)} + \mu^{(a)}$

3. *L*-FUZZY VECTOR SUBSPACES

Definition 3.1. L-fuzzy vector subspace is a pair $\tilde{E} = (E, \mu)$ where E is a vector space on field F, $\mu : E \to L$ is a map with the property that for any $x, y \in E, k, l \in F$, we have $\mu(kx + ly) \geq \mu(x) \wedge \mu(y)$.

In this definition, when $L = [0,1]$, L-fuzzy vector subspace is exactly the fuzzy vector subspace defined in [1]. We denote the family of L-fuzzy vector subspaces by **LFVS**.

Let $\tilde{E} = (E, \mu)$ be a member of **LFVS**, we denote

$$\tilde{E}_{[a]} = \mu_{[a]} = \{x \in E : \mu(x) \geq a\}, \qquad \tilde{E}_{(a)} = \mu_{(a)} = \{x \in E : a \in \beta(\mu(x))\}.$$

$$\tilde{E}^{[a]} = \mu^{[a]} = \{x \in E : a \notin \alpha(\mu(x))\}, \quad \tilde{E}^{(a)} = \mu^{(a)} = \{x \in E : \mu(x) \nleq a\}.$$

We can obtain some properties of **LFVS** analogous to fuzzy vector subspaces as follows.

Theorem 3.2. Let $\tilde{E} = (E, \mu)$ be a member of **LFVS**, then

(1) $\mu(0) = \sup_{x \in E} \mu(x)$

(2) For any $k \in F \setminus \{0\}$ and $x \in E, \mu(kx) = \mu(x)$

The prove is trivial and omitted.

Remark: Since *L* is a completely distributive lattice, the property that if $\mu(x) \neq \mu(y)$, then $\mu(x+y) = \mu(x) \wedge \mu(y)$ not holds for **LFVS**. This can be seen from the following example.

Example 3.3. Let *L* be a completely distributive lattice with four elements as fol- lows.

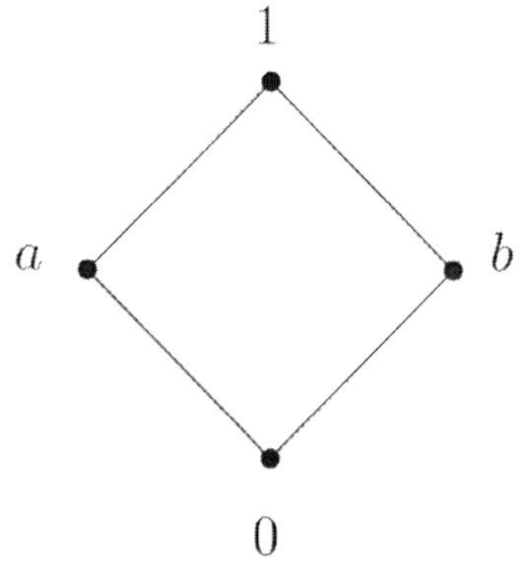

Let $\tilde{E} = (\mathbb{R}^2, \mu)$ be an *L*-fuzzy vector subspace on $\mathbb{R}^2$ where μ is defined by

$$\mu(x) = \begin{cases} 1, & x = (0,0). \\ a, & x \in \{(y,0) : y \in \mathbb{R} \setminus \{0\}\}. \\ b, & x \in \{(0,y) : y \in \mathbb{R} \setminus \{0\}\}. \\ 0, & \text{otherwise.} \end{cases}$$

We can easily check $\tilde{E}$ is an L-fuzzy vector subspace on $\mathbb{R}^2$. Suppose that *x*=(3, 2) and *y*= (0,-2) , then $\mu(x+y) = \mu(3,0) = a > \mu(x) \wedge \mu(y) = 0 \wedge b = 0.$ This example illustrates for *L*-fuzzy vector subspace $\mu(x) \neq \mu(y)$, $\mu(x+y) > \mu(x) \wedge \mu(y)$.

Theorem 3.4. Let *E* be a vector space, $\mu \in L^E$ and $\tilde{E} = (E, \mu)$. Then the follow- ing statements are equivalent:

(1) $\tilde{E}$ is an *L*-fuzzy vector subspace.

(2) (a) For all $x \in E$ and $k \in F, \mu(kx) \geq \mu(x)$

(b) For any $x, y \in E, \mu(x+y) \geq \mu(x) \wedge \mu(y)$

(3) For any $x_1, \cdots, x_r \in E$ and $k_1, \cdots, k_r \in F$, where r is a finite natural number, we have

$$\mu\left(\sum_{i=1}^{r} k_i x_i\right) \geq \bigwedge_{i=1}^{r} \mu(x_i)$$

The prove is trivial and omitted.

In the following paper, the vector spaces we discuss are finite-dimensional. For their L-fuzzy vector subspaces, the following observation will be useful.

Remark: Let $\tilde{E} = (E, \mu)$ be a member of **LFVS**. Suppose that $\mu(E) = \{\mu(x) : x \in E\}$. Since E is finite-dimensional vector space, denotes dim $E = n$, then $\mu(E)$ is a finite subset of L.

In the fact, let B be a basis of E, then $|B| = n$. Suppose that $\mu(E)$ is infinite, then for all $a \in L$, the total number of $\tilde{E}_{[a]}$ is infinite. Since $B \cap \tilde{E}_{[a]}$ is a basis of $\tilde{E}_{[a]}$, we have $\tilde{E}_{[a]} = \langle B \cap \tilde{E}_{[a]} \rangle$. Again since B is finite, the total number of $\tilde{E}_{[a]}$ is also finite. It contradicts with the hypothesis. Therefore $\mu(E)$ is a finite subset of L with at most 2^n+1 values; 2^n values which can be attained at the vectors of $E \setminus \{0\}$ and the maximum which is attained at 0.

Theorem 3.5. Let E be a vector space, $\mu \in L^E$ and $\tilde{E} = (E, \mu)$. Then the follow- ing statements equivalent:

(1) $\tilde{E}$ is an L-fuzzy vector subspace.

(2) For all $a \in L$, $\tilde{E}_{[a]}$ is a vector space.

(3) For all, $a \in J$, $\tilde{E}_{[a]}$ is a vector space.

(4) For all $a \in L$, $\tilde{E}^{[a]}$ is a vector space.

(5) For all $a \in P(L)$, $\tilde{E}^{[a]}$ is a vector space.

(6) For all $a \in P(L)$, $\tilde{E}^{[a]}$ is a vector space.

Proof. We prove $(1) \Leftrightarrow (4)$ and $(1) \Leftrightarrow (6)$, the others can be proved analogously.

$(1)\Rightarrow(4)$ We show that $\tilde{E}^{[a]}$ is a vector space as follows. Suppose that $x,y\in\tilde{E}^{[a]}$, then $a\notin\alpha(\mu(x))$ and $a\notin\alpha(\mu(y))$, i.e.. $a\notin\alpha(\mu(x))\cup\alpha(\mu(y))=\alpha(\mu(x)\wedge\mu(y))$

Since $\tilde{E}=(E,\mu)$ be an L-fuzzy vector subspace, then $\alpha(\mu(x)\wedge\mu(y))\supseteq\alpha(\mu((kx+ly))$, we have $a\notin\alpha(\mu(kx+ly))$, this means $kx+ly\in\tilde{E}^{[a]}$. Therefore $\tilde{E}^{[a]}$ is a vector space.

$(4)\Rightarrow(1)$ Suppose that for all $a\in L$, $\tilde{E}^{[a]}$ is a vector space. Let $x,y\in E$ and $k,l\in F$. Since $\tilde{E}^{[a]}$ is a vector space, then $kx+ly\in\tilde{E}^{[a]}$ if and only if $x\in\tilde{E}^{[a]}$ and $y\in\tilde{E}^{[a]}$. We have

$$\begin{aligned}
\mu(kx+ly) &= \bigwedge_{a\in L}\left(a\wedge\tilde{E}^{[a]}\right)(kx+ly)\\
&= \bigwedge_{a\in L}\left(a\vee\left(\tilde{E}^{[a]}(x)\wedge\tilde{E}^{[a]}(y)\right)\right)\\
&=\left(\bigwedge_{a\in L}\left(a\vee\tilde{E}^{[a]}(x)\right)\right)\wedge\left(\bigwedge_{a\in L}\left(a\vee\tilde{E}^{[a]}(y)\right)\right)\\
&=\mu(x)\wedge\mu(y).
\end{aligned}$$

Therefore $\tilde{E}$ is an L-fuzzy vector subspace.

$(1)\Rightarrow(6)$ Suppose that $x,y\in E^{(a)}$, then $\mu(x)\not\leq a$ and $\mu(y)\not\leq a$. Since $a\in P(L)$, then $\mu(x)\wedge\mu(y)\not\leq a$. Because $\tilde{E}=(E,\mu)$ is an L-fuzzy vector subspace, we can have $\mu(kx+ly)\not\leq a$, this implies $kx+ly\in E^{(a)}$. Thus $E^{(a)}$ is a vector space.

$(6)\Rightarrow(1)$ Let $x,y\in E$ and $k,l\in F$. Since $\tilde{E}^{(a)}$ is a vector space, then $kx+ly\in\tilde{E}^{(a)}$ if and only if $x\in\tilde{E}^{(a)}$ and $y\in\tilde{E}^{(a)}$. We have the following implications.

$$\begin{aligned}
\mu(kx+ly) &= \bigwedge_{a\in P(L)}\left(a\vee\tilde{E}^{(a)}\right)(kx+ly)\\
&= \bigwedge_{a\in P(L)}\left(a\vee\left(\tilde{E}^{(a)}(x)\wedge\tilde{E}^{(a)}(y)\right)\right)\\
&=\left(\bigwedge_{a\in P(L)}\left(a\vee\tilde{E}^{(a)}(x)\right)\right)\wedge\left(\bigwedge_{a\in P(L)}\left(a\vee\tilde{E}^{(a)}(y)\right)\right)\\
&=\mu(x)\wedge\mu(y).
\end{aligned}$$

Therefore $\tilde{E}$ is an L-fuzzy vector subspace.

Theorem 3.6. Let E be a vector space, $\mu : E \to L$ be a map, $\tilde{E} = (E, \mu)$, and for all $a, b \in L, \beta(a \wedge b) = \beta(a) \cap \beta(b)$ Then the following statements equivalent:

(1) $\tilde{E}$ is an L-fuzzy vector subspace.

(2) For all $a \in L$, $\tilde{E}_{(a)}$ is a vector space.

Proof. $(1) \Rightarrow (2)$ Suppose that $x, y \in \tilde{E}_{(a)}$, then $a \in \beta(\mu(x))$ and $a \in \beta(\mu(y))$, i.e. $a \in \beta(\mu((x)) \cap \beta(\mu(y))$ Since for all $a, b \in L, \beta(a \wedge b) = \beta(a) \cap \beta(b)$ and $\tilde{E}$ is an L-fuzzy vector subspace, we can know $a \in \beta(\mu(x) \wedge \mu(y)) \subseteq \beta(\mu(ax + by))$, this implies $ax + by \in \tilde{E}_{(a)}$. Therefore $\tilde{E}_{(a)}$ is a vector space.

$(2) \Rightarrow (1)$ Suppose that for all $a \in L$, $\tilde{E}_{(a)}$ is a vector space. Let $x, y \in E$ and $k, l \in F$. Since $\tilde{E}_{(a)}$ is a vector space, then $kx + ly \in \tilde{E}_{(a)}$ if and only if and only if $x \in \tilde{E}_{(a)}$ and $y \in \tilde{E}_{(a)}$. We have

$$\begin{aligned}
\mu(kx + ly) &= \vee_{a \in L}\left(a \wedge \tilde{E}_{(a)}\right)(kx + ly) \\
&= \vee_{a \in L}\left(a \wedge \left(\tilde{E}_{(a)}(x) \wedge \tilde{E}_{(a)}(y)\right)\right) \\
&= \left(\vee_{a \in L}\left(a \wedge \tilde{E}_{(a)}(x)\right)\right) \wedge \left(\vee_{a \in L}\left(a \wedge \tilde{E}_{(a)}(y)\right)\right) \\
&= \mu(x) \wedge \mu(y).
\end{aligned}$$

Therefore $\tilde{E}$ is an L-fuzzy vector subspace.

We can define the operations between two L-fuzzy vector subspaces analogous to fuzzy vector subspaces.

Definition 3.7. Let $\tilde{E}_1 = (E, \mu_1)$ and $\tilde{E}_2 = (E, \mu_2)$ be two L-fuzzy vector subspaces on ***E***. Define the intersection of $\tilde{E}_1$ and $\tilde{E}_2$ to be $\tilde{E}_1 \cap \tilde{E}_2 = (E, \mu_1 \wedge \mu_2)$. Define the sum of $\tilde{E}_1$ and $\tilde{E}_2$ to be $\tilde{E}_1 + \tilde{E}_2 = (E, \mu_1 + \mu_2)$ where $\mu_1 + \mu_2$ is defined by for all $x \in E$

$$\begin{aligned}
(\mu_1 + \mu_2)(x) &= \vee_{x = x_1 + x_2}\left(\mu_1(x_1) \wedge \mu_2(x_2)\right) \\
&= \vee_{x_1 \in E}\left(\mu_1(x_1) \wedge \mu_2(x - x_1)\right)
\end{aligned}$$

.

Definition 3.8. Let $\tilde{E}_1=(E_1,\mu_1)$ and $\tilde{E}_2=(E_2,\mu_2)$ be two members of **LFVS** and. $E=E_1\oplus E_2$ We define the direct sum of $\tilde{E}_1$ and $\tilde{E}_2$ to be $\tilde{E}_1\oplus\tilde{E}_2=(E,\mu_1\oplus\mu_2)$ where $\mu_1\oplus\mu_2$ is defined by for all $x\in E, x=x_1\oplus x_2, x_i\in E_i, i=1,2$

$$(\mu_1\oplus\mu_2)(x)=(\mu_1\oplus\mu_2)(x_1\oplus x_2)=\mu_1(x_1)\wedge\mu_2(x_2)$$

Theorem 3.9. Let $\tilde{E}_1=(E,\mu_1)$ and $\tilde{E}_2=(E,\mu_2)$ be two members of **LFVS** on ***E***. We have

(1) $\tilde{E}_1\cap\tilde{E}_2$ is a member of **LFVS** on ***E***.

(2) $\tilde{E}_1+\tilde{E}_2$ is a member of **LFVS** on ***E***.

The proof of the theorem is trivial and it is omitted.

Theorem 3.10. Let $\tilde{E}_1=(E,\mu_1)$ and $\tilde{E}_2=(E,\mu_2)$ be the members of **LFVS**. We have

(1) For all $a\in L$, $(\tilde{E}_1\cap\tilde{E}_2)_{[a]}=(\tilde{E}_1)_{[a]}\cap(\tilde{E}_2)_{[a]}$.

(2) For all $a\in L$, $(\tilde{E}_1\cap\tilde{E}_2)^{[a]}=(\tilde{E}_1)^{[a]}\cap(\tilde{E}_2)^{[a]}$.

(3) For any $a\in P(L)$, $(\tilde{E}_1\cap\tilde{E}_2)^{(a)}=(\tilde{E}_1)^{(a)}\cap(\tilde{E}_2)^{(a)}$.

(4) For any $a\in P(L)$, $(\tilde{E}_1+\tilde{E}_2)^{(a)}=(\tilde{E}_1)^{(a)}+(\tilde{E}_2)^{(a)}$.

Proof. The proofs of (1) and (2) are easy by the definition of $\tilde{E}_1\cap\tilde{E}_2$ and the properties of *L*-fuzzy sets.

(3) For any $a\in P(L)$, we have

$$\begin{aligned} x\in(\tilde{E}_1\cap\tilde{E}_2)^{(a)} &\Leftrightarrow \mu_1(x)\wedge\mu_2(x)\nleq a \\ &\Leftrightarrow \mu_1(x)\nleq a \text{ and } \mu_2(x)\nleq a \\ &\Leftrightarrow x\in(\tilde{E}_1)^{(a)}\cap(\tilde{E}_2)^{(a)} \end{aligned}$$

(4) By the definition of the sum of *L*-fuzzy vector subspaces, for any $a\in P(L)$ we have

$$x \in \left(\tilde{E}_1 + \tilde{E}_2\right)^{(a)} \Leftrightarrow \vee_{x = x_1 + x_2} \left(\mu_1(x_1) \wedge \mu_2(x_2)\right) \not\leq a$$

$$\Leftrightarrow \exists x_1, x_2 \in E \text{ and } x = x_1 + x_2, \text{ such that } \mu_1(x_1) \wedge \mu_2(x_2) \not\leq a$$

$$\Leftrightarrow \exists x_1, x_2 \in E, \mu_1(x_1) \not\leq a \text{ and } \mu_2(x_2) \not\leq a.$$

$$\Leftrightarrow x = x_1 + x_2 \in \tilde{E}_1^{(a)} + \tilde{E}_2^{(a)}.$$

Theorem 3.11. Let $\tilde{E}_1 = (E, \mu_1)$ and $\tilde{E}_2 = (E, \mu_2)$ be two members of **LFVS**. Suppose that for any $a, b \in L$, we have $\beta(a \wedge b) = \beta(a) \cap \beta(b)$. Then

(1) $\left(\tilde{E}_1 \cap \tilde{E}_2\right)_{(a)} = \left(\tilde{E}_1\right)_{(a)} \cap \left(\tilde{E}_2\right)_{(a)}$

(2) $\left(\tilde{E}_1 + \tilde{E}_2\right)_{(a)} = \left(\tilde{E}_1\right)_{(a)} + \left(\tilde{E}_2\right)_{(a)}$

The prove is trivial and omitted.

4. Fuzzy Dimension of L-Fuzzy Vector Subspaces

Definition 4.1. Let N(L) be the family of L-fuzzy natural number. The map **LFVS**→ N(L) is defined by

$$\dim \tilde{E}(n) = \vee_{a \in L} \left(a \wedge \dim \tilde{E}_{[a]}\right)(n)$$

is called the L-fuzzy dimensional function of the L-fuzzy vector subspace $\tilde{E}$, and $\dim \tilde{E}$ is called the L-fuzzy dimension of $\tilde{E}$, it is an L-fuzzy natural number. We usually use another form of $\dim \tilde{E}$ as follows.

$$\dim \tilde{E}(n) = \vee \left\{a \in L : \dim \tilde{E}_{[a]} \geq n\right\}$$

Theorem 4.2. For each $\tilde{E} \in \mathrm{LFVS}$ and $n \in \mathbb{N}$, we have

$$\dim \tilde{E}(n) = \vee_{a \in L} \left(a \wedge \dim \tilde{E}_{(a)}\right)(n) = \vee \left\{a \in L : \dim \tilde{E}_{(a)} \geq n\right\}$$

Proof. For any $n \in \mathbb{N}$, let $\lambda = \vee_{a \in L} \left(a \wedge \dim \tilde{E}_{(a)}\right)(n)$. Obviously $\lambda \leq \dim \tilde{E}(n)$. Next we show that $\lambda \geq \dim \tilde{E}(n)$ Suppose that $b \in L$ and $b \in \beta\left(\dim \tilde{E}(n)\right)$, then there

exists $a \in L$ and $\dim \tilde{E}_{[a]} \geq n$ such that $b \in \beta(a)$. In this case, $n \leq \dim \tilde{E}_{[a]} \leq \dim \tilde{E}_{(b)} \leq \dim \tilde{E}_{[b]}$ which implies $\lambda = \vee \left\{a \in L : \dim \tilde{E}_{(a)} \geq n\right\} \geq b$. Thus we have

$$\lambda \geq \vee \left\{b \middle| b \in \beta\left(\dim \tilde{E}(n)\right)\right\} = \dim \tilde{E}(n)$$

This completes the proof.

Theorem 4.3. Let the pair $\tilde{E}=(E,\mu)$ be a member of **LFVS**. Then for any $a \in L$

$$\left(\dim \tilde{E}\right)_{(a)} \le \dim \tilde{E}_{[a]} \le \left(\dim \tilde{E}\right)_{[a]}$$

If $\beta(a \wedge b)=\beta(a)\cap\beta(b)$ for all $a,b \in L$, then

$\left(\dim \tilde{E}\right)_{(a)} \le \dim \tilde{E}_{(a)} \le \dim \tilde{E}_{[a]} \le \left(\dim \tilde{E}\right)_{[a]}$

In particular, $\left(\dim \tilde{E}\right)_{(a)}=\dim \tilde{E}_{(a)}$ for any $a \in J(L)$.

Proof. In order to prove $\left(\dim \tilde{E}\right)_{(a)} \le \dim \tilde{E}_{(a)}$. Suppose that $n \le \left(\dim \tilde{E}\right)_{(a)}$, then $a \in \beta\left(\dim \tilde{E}(n)\right)$. Since β is a preserve-union map, there is $b \in L$ and $n \le \dim \tilde{E}_{[b]}$, such that $a \in \beta(b)$. Because $\tilde{E}_{[b]} \subseteq \tilde{E}_{(a)} \subseteq \tilde{E}_{[a]}$, thus $n \le \dim \tilde{E}_{(a)}$. Therefore $\left(\dim \tilde{E}\right)_{(a)} \le \dim \tilde{E}_{(a)}$.

$\dim \tilde{E}_{(a)} \le \dim \tilde{E}_{[a]}$ is obvious. Moreover, we can obtain that $\dim \tilde{E}_{[a]} \le \left(\dim \tilde{E}\right)_{[a]}$ from the definition of $\dim\left(\tilde{E}\right)$.

In order to prove for any $a \in J(L), \left(\dim \tilde{E}\right)_{[a]} = \dim \tilde{E}_{[a]}$, we only need to show $\left(\dim \tilde{E}\right)_{[a]} \subseteq \dim \tilde{E}_{[a]}$. Since the set $\mu(E)$ is finite, for any $a \in J(L)$ we have

$$\begin{aligned}
n \le \left(\dim \tilde{E}\right)_{[a]} &\Rightarrow \dim \tilde{E}(n) \ge a \\
&\Rightarrow \vee\left\{b \in L : \dim \tilde{E}_{[b]} \ge n\right\} \ge a \\
&\Rightarrow \exists a \le b, \text{ such that } n \le \dim \tilde{E}_{[b]} \\
&\Rightarrow n \le \dim \tilde{E}_{[a]}
\end{aligned}$$

Therefore $\left(\dim \tilde{E}\right)_{[a]} = \dim \tilde{E}_{[a]}$

Theorem 4.4. Let $\tilde{E}=(E,\mu)$ be a member of **LFVS**. Then

$$\left(\dim \tilde{E}\right)^{(a)} \le \dim \tilde{E}^{(a)} \le \dim \tilde{E}^{[a]} \le \left(\dim \tilde{E}\right)^{[a]}$$

In particular, $\left(\dim \tilde{E}\right)^{(a)} = \dim \tilde{E}^{(a)}$ for any $a \in P(L)$

Proof. $\left(\dim \tilde{E}\right)^{(a)} \le \dim \tilde{E}^{(a)}$ can be proved from the following implications.

$$\begin{aligned} n \le \left(\dim \tilde{E}\right)^{(a)} &\Leftrightarrow \dim \tilde{E}(n) \not\le a \\ &\Leftrightarrow \vee\left\{b \in L : \dim \tilde{E}_{[b]} \ge n\right\} \not\le a \\ &\Leftrightarrow \exists b \not\le a, \text{ such that } \quad n \le \dim \tilde{E}_{[b]} \\ &\Rightarrow \dim \tilde{E}^{(a)} = \dim\left(\bigcup_{b \not\le a} \tilde{E}_{[b]}\right) \ge n. \end{aligned}$$

Let $a \in P(L)$. In order to show $\dim \tilde{E}^{(a)} \le \left(\dim \tilde{E}\right)^{(a)}$, we need to show that $\dim\left(\bigcup_{b \not\le a} \tilde{E}_{[b]}\right) \le \bigvee_{b \not\le a} \dim \tilde{E}_{[b]}$ Suppose that $n \le \dim\left(\bigcup_{b \not\le a} \tilde{E}_{[b]}\right)$. Since the number of $\tilde{E}_{[a]}$ is finite, then when $b \not\le a$ the number of $\tilde{E}_{[b]}$ is finite, denotes $\tilde{E}_{[a_1]}, \tilde{E}_{[a_2]}, \cdots, \tilde{E}_{[a_r]}$, where $a_i \not\le a$ for any $i \in \{1,2,\cdots,r\}$ Thus $\bigcup_{b \not\le a} \tilde{E}_{[b]} = \bigcup_{i=1}^{r} \tilde{E}_{[a_i]}$ Since $a \in P(L)$, then we have $c = a_1 \wedge a_2 \wedge \cdots \wedge a_r \not\le a$. Further we have $\bigcup_{i=1}^{r} \tilde{E}_{[a_i]} \subseteq \tilde{E}_{[c]}$. Thus for any

$$n \le \dim\left(\bigcup_{b \not\le a} \tilde{E}_{[b]}\right) = \dim\left(\bigcup_{i=1}^{r} \tilde{E}_{[a_i]}\right) \le \dim \tilde{E}_{[c]} \le \bigvee_{b \not\le a} \dim \tilde{E}_{[b]} \le \bigvee_{b \not\le a} \left(\dim \tilde{E}\right)_{[b]} = \left(\dim \tilde{E}\right)^{(a)}$$

Therefore for any $a \in P(L)$, $\left(\dim \tilde{E}\right)^{(a)} = \dim \tilde{E}^{(a)}$

$\dim \tilde{E}^{(a)} \le \dim \tilde{E}^{[a]}$ is obvious. We show that $\dim \tilde{E}^{[a]} \le \left(\dim \tilde{E}\right)^{[a]}$ in the following implications.

$$\begin{aligned} \dim \tilde{E}^{[a]} &= \dim \bigcap_{\substack{a \in \alpha(b) \\ b \in P(L)}} \tilde{E}^{(b)} \le \bigwedge_{\substack{a \in \alpha(b) \\ b \in P(L)}} \dim \tilde{E}^{(b)} \\ &= \bigwedge_{\substack{a \in \alpha(b) \\ b \in P(L)}} \left(\dim \tilde{E}\right)^{(b)} = (\dim \tilde{E})^{[a]}. \end{aligned}$$

Theorem 4.5. Let $\tilde{E}_1 = (E, \mu_1)$ and $\tilde{E}_2 = (E, \mu_2)$ be two *L*-fuzzy vector subspaces. Then the following equality holds

$$\dim\left(\tilde{E}_1 + \tilde{E}_2\right) + \dim\left(\tilde{E}_1 \cap \tilde{E}_2\right) = \dim \tilde{E}_1 + \dim \tilde{E}_2.$$

Proof. We denote the sum of $\tilde{E}_1$ and $\tilde{E}_2$ by $\tilde{E}_1 + \tilde{E}_2 = (E, \mu)$. From Theorem 11, we know that $\tilde{E}_1 + \tilde{E}_2$ is a *L*-fuzzy vector subspace. By the properties of *L*-fuzzy natural numbers, Theorem 12 and the dimensional formulation of vector spaces, we know for any $a \in P(L)$,

$$\left(\dim\left(\tilde{E}_1+\tilde{E}_2\right)+\dim\left(\tilde{E}_1\cap\tilde{E}_2\right)\right)^{(a)}$$
$$=\left(\dim\left(\tilde{E}_1+\tilde{E}_2\right)\right)^{(a)}+\left(\dim\left(\tilde{E}_1\cap\tilde{E}_2\right)\right)^{(a)}$$
$$=\dim\left(\tilde{E}_1+\tilde{E}_2\right)^{(a)}+\dim\left(\tilde{E}_1\cap\tilde{E}_2\right)^{(a)}$$
$$=\dim\left(\tilde{E}_1^{(a)}+\tilde{E}_2^{(a)}\right)+\dim\left(\tilde{E}_1^{(a)}\cap\tilde{E}_2^{(a)}\right)$$
$$=\dim\tilde{E}_1^{(a)}+\dim\tilde{E}_2^{(a)}-\dim\left(\tilde{E}_1^{(a)}\cap\tilde{E}_2^{(a)}\right)+\left(\dim(\tilde{E}_1^{(a)}\cap\tilde{E}_2^{(a)}\right)$$
$$=\dim\tilde{E}_1^{(a)}+\dim\tilde{E}_2^{(a)}$$

Therefore $\dim\left(\tilde{E}_1+\tilde{E}_2\right)+\dim\left(\tilde{E}_1\cap\tilde{E}_2\right)=\dim\tilde{E}_1+\dim\tilde{E}_2$.

Definition 4.6. Suppose that $\tilde{E}=(E,\mu)$ is an L-fuzzy vector subspace. A map $f: E\to E$ is called an L-fuzzy linear transformation, if it satisfies the following conditions:

(1) f is a linear map on E.

(2) For all $x\in E$, $\mu(f(x))\geq\mu(x)$

Theorem 4.7. Suppose that $\tilde{E}=(E,\mu)$ is an L-fuzzy vector subspace, f is an L-fuzzy linear transformation on E, then $\widetilde{\ker f}=\left(\ker f,\mu|_{\ker f}\right)$ and $\widetilde{\operatorname{im} f}=\left(\operatorname{im} f,\mu|_{\operatorname{im} f}\right)$ are L-fuzzy vector subspaces.

The prove is trivial and omitted.

Theorem 4.8. Suppose that $\tilde{E}=(E,\mu)$ is an L-fuzzy vector subspace, $f: E\to E$ is an L-fuzzy linear transformation, then

$$\dim\left(\widetilde{\ker f}\right)+\dim\left(\widetilde{\operatorname{im} f}\right)=\dim\tilde{E}$$

Proof. Suppose that φ is a linear transformation on (crisp) vector spaces V, then the equality $\dim(\operatorname{im}\varphi)+\dim(\operatorname{kef}\varphi)=\dim V$ holds. Hence, for all $a\in P(L)$ we have

$$\left(\dim\left(\widetilde{\operatorname{im} f}\right)+\dim\left(\widetilde{\ker f}\right)\right)^{(a)}=\left(\dim(\widetilde{\operatorname{im} f})\right)^{(a)}+\left(\dim\left(\widetilde{\ker f}\right)\right)^{(a)}$$
$$=\dim\left(\widetilde{\operatorname{im} f}\right)^{(a)}+\dim\left(\widetilde{\ker f}\right)^{(a)}$$
$$=\dim\left(\tilde{E}^{(a)}\cap\operatorname{im} f\right)+\dim\left(\tilde{E}^{(a)}\cap\operatorname{kef} f\right)$$

Since $f|_{\tilde{E}^{(a)}}$ is a linear transformation on $\tilde{E}^{(a)}$, we have

$$\left(\dim\left(\widetilde{\mathrm{im}f}\right)+\dim\left(\widetilde{\ker f}\right)\right)^{(a)}=\dim\left(\mathrm{im}f|_{\tilde{E}^{(a)}}\right)+\dim\left(\mathrm{kef}\ f|_{\tilde{E}^{(a)}}\right)$$
$$=\dim\tilde{E}^{(a)}=\left(\dim\tilde{E}\right)^{(a)}.$$

Therefore $\dim\left(\widetilde{\ker f}\right)+\dim\left(\widetilde{\mathrm{im}f}\right)=\dim\tilde{E}.$

5. CONCLUSION

In this paper, *L*-fuzzy vector subspace is defined and showed that its dimension is an *L*-fuzzy natural number. Based on the definitions, some good properties of crisp vector spaces are hold in a finite-dimensional *L*-fuzzy vector subspace. In particular, the equality $\dim\left(\tilde{E}_1+\tilde{E}_2\right)+\dim\left(\tilde{E}_1\cap\tilde{E}_2\right)=\dim\tilde{E}_1+\dim\tilde{E}_2$ holds without any restricted conditions. At the same time, $\dim\left(\widetilde{\mathrm{im}f}\right)+\dim\left(\widetilde{\mathrm{kef}f}\right)=\dim\tilde{E}$ holds.

ACKNOWLEDGEMENTS

The authors would like to thank the reviewers for their valuable comments and sug- gestions.

FUND

The project is by the Science & Technology Program of Beijing Municipal Commission of Education (KM201611417007), the NNSF of China (11371002), the academic youth backbone project of Heilongjiang Education Department (1251G3036), the foundation of Heilongjiang Province (A201209).

REFERENCES

1. Katsaras, A.K. and Liu, D.B. (1977) Fuzzy Vector Spaces and Fuzzy Topological Vector Spaces. Journal of Mathematical Analysis and Applications, 58, 135-146. https://doi.org/10.1016/0022-247X(77)90233-5
2. Lubczonok, G. and Murali, V. (2002) On Flags and Fuzzy Subspaces of Vector Spaces. Fuzzy Sets and Systems, 125, 201-207.https://doi.org/ 10.1016 /S0165-0114(00)00129-9
3. Abdukhalikov, K.S., Tulenbaev, M.S. and Umirbarv, U.U. (1994) On Fuzzy Bases of Vector Spaces. Fuzzy Sets and Systems, 63, 201-206. https://doi.org/10.1016/0165-0114(94)90350-6
4. Abdukhalikov, K.S. (1996) The Dual of a Fuzzy Subspace. Fuzzy Sets and Systems, 82, 375-381. https://doi.org/10.1016/0165-0114(96)84977-3
5. Lubczonok, P. (1990) Fuzzy Vector Spaces. Fuzzy Sets and Systems, 38, 329-343. https://doi.org/10.1016/0165-0114(90)90206-L
6. Lowen, R. (1980) Convex Fuzzy Sets. Fuzzy Sets and Systems, 3, 291-310. https://doi.org/10.1016/0165-0114(80)90025-1
7. Shi, F.G. and Huang, C.E. (2010) Fuzzy Bases and the Fuzzy Dimension of Fuzzy Vector Spaces. Mathematical Communications, 15, 303-310.
8. Gierz, G., et al. (2003) Continuous Lattices and Domains. Encyclopedia of Mathematics and its Applications, Cambridge University Press, Cambridge. https://doi.org/10.1017/CBO9780511542725
9. Dwinger, P. (1982) Characterizations of the Complete Homomorphic Images of a Completely Distributive Complete Lattice I. Indagationes Mathematicae (Proceedings), 85, 403-414. https://doi.org/10.1016/1385-7258(82)90034-8
10. Wang, G.-J. (1992) Theory of Topological Molecular Lattices. Fuzzy Sets and Systems, 47, 351-376. https://doi.org/10.1016/0165-0114(92)90301-J
11. Huang, H.-L. and Shi, F.-G. (2008) L-Fuzzy Numbers and Their Properties. Information Sciences, 178, 1141-1151. https://doi.org/10.1016/j.ins.2007.10.001
12. Shi, F.-G. (2000) L-Fuzzy Relation and L-Fuzzy Subgroup. Journal of Fuzzy Mathematics, 8, 491-499.
13. Negoita, C.V. and Ralescu, D.A. (1975) Applications of Fuzzy Sets to Systems Analysis, Interdisciplinary Systems Research Series 11, Birkhaeuser, Basel. https://doi.org/10.1007/978-3-0348-5921-9
14. Shi, F.-G. (1995) Theory of Lβ-Nested Sets and Lα-Nested Sets and Its Applications. Fuzzy Systems and Mathematics, 4, 65-72. (In Chinese)
15. Shi, F.-G. (1996) L-Fuzzy Sets and Prime Element Nested Sets. Journal of Mathematical Research and Exposition, 16, 398-402. (In Chinese)
16. Shi, F.-G. (1996) Theory of Molecular Nested Sets and Its Applications. Journal of Yantai Teachers University, 1, 33-36. (In Chinese)
17. Shi, F.-G. (2009) A New Approach to the Fuzzification of Matroids. Fuzzy Sets and Systems, 160, 696-705. https://doi.org/10.1016/j.fss.2008.05.007

CHAPTER 4

Morphisms between Indecomposable Complexes in The Bounded Derived Category of a Gentle Algebra

Kristin KroghArnesen[a], RosannaLaking[b,*], DavidPauksztello[b,*]

[a] *Faculty of Teacher and Interpreter Education, Norwegian University of Science and Technology (NTNU), N-7491 Trondheim, Norway*

[b] *School of Mathematics, The University of Manchester, Oxford Road, Manchester, M13 9PL, United Kingdom*

ABSTRACT

In this article we provide a simple combinatorial description of morphisms between indecomposable complexes in the bounded derived category of a gentle algebra.

Keywords: Bounded derived category; Gentle algebra; Homotopy string and band; String combinatorics; Morphism

INTRODUCTION

Triangulated categories are of central importance in many branches of mathematics, providing a common framework for algebraists, geometers, topologists and theoretical physicists, amongst others. Perhaps the most famous illustration of their utility is Beilin-son's equivalences between the derived categories of coherent sheaves on projective spaces and representations of certain finite-dimensional algebras [7], which provided deep con-nections between algebra and geometry.

In algebra and geometry, the triangulated categories we have in mind are derived cat-egories and categories constructed from them, for example, cluster categories. However, despite their utility, there is a major drawback: the construction of derived categories is abstract and explicit computation is often difficult. Indeed, much intuition is often obtained from examples of homological dimension one (=hereditary), owing to par-ticularly nice homological properties which allow one to reduce computations to the (well-understood) abelian categories with which one starts.

Developing intuition in such an abstract setting requires a good collection of exam-ples, for which computation becomes straightforward and non-trivial phenomena can be observed concretely. In this article, we shall show that so-called *gentle algebras*provide a wide class of such examples. Moreover, given their central place in the current thrust of research in cluster-tilting theory, where they occur as surface algebras [3], concrete understanding of the derived categories of gentle algebras is both useful and timely.

*Main results.*A principal way in which one can understand the structure of a category is, firstly, to describe all its indecomposable objects and, secondly, the morphisms between them. This is demonstrated very successfully by Auslander–Reiten (AR) theory.

Let Λbe a gentle algebra and let $D\mathit{b}(\Lambda)$be its bounded derived category with shift functor Σ. The first step was accomplished by Bekkert and Merklen in [8]who, inspired by a classic paper [16], described the indecomposable objects of $D\mathit{b}(\Lambda)$by string combi-natorics: these include, in the terminology of [9], the so-called *homotopy string complexes*and (one-dimensional) *homotopy band complexes*, i.e. complexes which can be unfolded to look like an oriented copy of a Dynkin diagram of type A or type $\tilde{A}$, respectively, and whose differentials are paths in the quiver of Λ. Thus, a homotopy string or band is none other than a word whose letters consist of paths in the quiver of $\Lambda$and their inverses. We refer also to [15]for a similar approach in the case of nodal algebras.

In this article, we describe all morphisms between indecomposable complexes in $Db(\Lambda)$, thus completing the hands-on combinatorial framework that facilitates straightforward computation in these non-trivial categories. The main theorem is as follows.

Theorem A. *Let X and Y be homotopy string or one-dimensional band complexes in* $D^b(\Lambda)$. *Let* wX*and* wY*be the words corresponding to X and Y. There is a canonical basis of* $\mathrm{Hom}_{D^b(\Lambda)}(X, Y)$*given by the following three classes of maps:*

•***graph maps:*** *corresponding to the maximal overlaps in wX and wY satisfying certain compatibility conditions at the endpoints;*

•***quasi-graph maps:*** *corresponding to maximal overlaps in wX and w*$\Sigma^{-1}Y$ *satisfying certain non-degeneracy conditions at the endpoints; these give rise to homotopy classes of maps X→Y;*

•***singleton maps:*** *certain special maps which can be detected easily from the word combinatorics of wX and wY.*

The remaining indecomposable complexes in $D^b(\Lambda)$are constructed from homotopy band complexes. Homotopy band complexes sit at the mouths of homogeneous tubes: atube is indexed by a homotopy band and a non-zero scalar. The object of length *n* in a given tube is specified by the additional data of an *n*-dimensional vector space. We call these 'higher-dimensional' band complexes; we refer to Section5for more precise details. TheoremAdeals with one-dimensional band complexes and string complexes, which are always 'one-dimensional'. What about maps involving higher dimensional band complexes? Our second main theorem tells us that we don't have to worry about them:

Theorem B. *Suppose Xand Yare string or one-dimensional band complexes in* $D^b(\Lambda)$. *For Xa homotopy band, let Xn be its 'n-dimensional' version, otherwise take n =1and X1=X; similarly for Y. Then, generically,* dimHom*(Xm, Yn) =mn* ·dimHom*(X, Y).*

In TheoremBcare must be taken when $X \cong Y$ or $X \cong \Sigma^{-1}Y$; see Section 5 for precise details. The module category analogues of these results are classical; see [17,18,21].

In summary: TheoremsAand Breduce some difficult homological algebra to ele-mentary word combinatorics. Combined with the description of indecomposables, this opens up a wide and natural class of examples of triangulated categories to explicit computation.

*Applications and context.*Before continuing, we make some remarks on the potential applications of these results. Gentle algebras present us with particularly good candidates to begin a systematic 'hands on' study of derived categories for a number of reasons:

•Certain general aspects of the structure of $D^b(\Lambda)$are already known: for instance, the Avella Alaminos–Geiß invariant describes certain fractionally Calabi–Yau triangulated subcategories whose AR components have a boundary [5].

•The AR theory of $D^b(\Lambda)$can be computed: using the Happel functor $F: D^b(\Lambda) \to \mathrm{mod}(\hat{\Lambda})$[19], where $\hat{\Lambda}$ is the repetitive algebra of Λ, Bobiński [9]gave an algorithm which computes the AR triangles in $D^b(\Lambda)$. Indeed, as an application of our results, we recover Bobiński's algorithm without recourse to the Happel functor, and the often unpleasant computations that ensue. Examples of the kinds of results we have in mind are the structural results on the AR quiver in [1,2], which use the combinatorial invariants of [6]to parametrise the AR components completely. Can one use these to extend the results of Avella Alaminos and Geiß to get better derived invariants?

•Vossieck [22]introduced the family of *derived-discrete algebras*, for which we now understand various non-trivial homological properties [10–14]. These algebras are gentle, and thus they can be used as a template for further study of derived categories of gentle algebras. Here as an application of our results, we recover the universal Hom-dimension bound of [13].

We expect our results to be useful in the classification of tilting and, more gener-ally, silting objects for gentle algebras, particularly for surface algebras; such objects are very closely related to cluster-tilting objects. We therefore expect that the string combinatorics here will be adapted to cluster combinatorics.

1. PRELIMINARIES AND NOTATION

Let $\Gamma = (\Gamma_0, \Gamma_1)$ be a finite connected quiver. Note that we read paths in $\Gamma$from right to left. Recall from [4]that a bound path algebra $\Lambda \cong \mathbf{k}\Gamma/I$ is called *gentle* if:

(1)for each vertex x of Γ, there are at most two arrows starting at x and at most two arrows ending at x;\

(2)for any arrow a in Γ there is at most one arrow b in Γ such that $ab \notin I$ and at most one arrow c in Γ such that $ca \notin I$;

(3)for any arrow a in Γthere is at most one arrow b in Γsuch that $ab \in I$ and at most one arrow c in Γsuch that $ca \in I$;

(4)the ideal I is generated by paths of length 2.

Let $P(x)$be the indecomposable projective left Λ-module corresponding to $x \in \Gamma 0$. We recall the following useful property of gentle algebras; see, for instance [8, Section 3].

Proposition 1.1. *There is a bijection*

$$\{paths\ p\colon x \rightsquigarrow y\ in\ \Gamma\} \overset{1-1}{\longleftrightarrow} \{basis\ elements\ of\ \mathsf{Hom}_\Lambda(P(y), P(x))\};$$

$$p \longmapsto (u \mapsto up).$$

Convention 1.2. From now on, by abuse of notation, we shall identify a path $p\colon x \rightsquigarrow y$ with its corresponding basis element in $\mathrm{Hom}\Lambda(P(y), P(x))$.

Throughout this article, we shall fix a gentle algebra $\Lambda = \mathbf{k}\Gamma/I$ over an algebraically closed field **k**. Algebraic closure of **k** is not strictly necessary, but it significantly simplifies the presentation of the combinatorics.

All modules in this paper will be left modules. We shall be interested in three cate-gories:

•C $:= C^{-,b}$ (proj(Λ)): the category of right bounded complexes of finitely generated projective Λ-modules whose cohomology is bounded;

•K $:= K^{b}$(proj(Λ)): the homotopy category of bounded complexes of finitely generated projective Λ-modules – the so-called *perfect complexes*;

•D $:= D^{b}(\Lambda) = D^{b}$(mod(Λ)): the bounded derived category of finitely generated Λ-modules.

Throughout the paper, we shall identify the bounded derived category Dwith the trian-gle equivalent category $K^{-,b}$(proj(Λ)), consisting of right bounded complexes of finitely generated projective Λ-modules whose cohomology is bounded. This identification allows us to use the combinatorics of homotopy strings and bands, which will be described in the next section, throughout the paper. We direct the reader to consult [19]for background on derived and homotopy categories.

2. INDECOMPOSABLE OBJECTS IN D

In this section we give an overview of Bekkert and Merklen's description of the in-decomposable objects in D. Their crucial observation is that it is enough to consider complexes where the differential is given by matrices whose entries are either zero or a path (cf. Convention 1.2). The indecomposable objects are obtained by unravelling the differential into 'homotopy strings and bands' corresponding to perfect complexes, and 'infinite homotopy strings' for the unbounded complexes. The reader is encouraged to have the following example in mind when reading this section:

Running Example. Let $\Lambda = \mathbf{k}\Gamma/I$ be given by the quiver

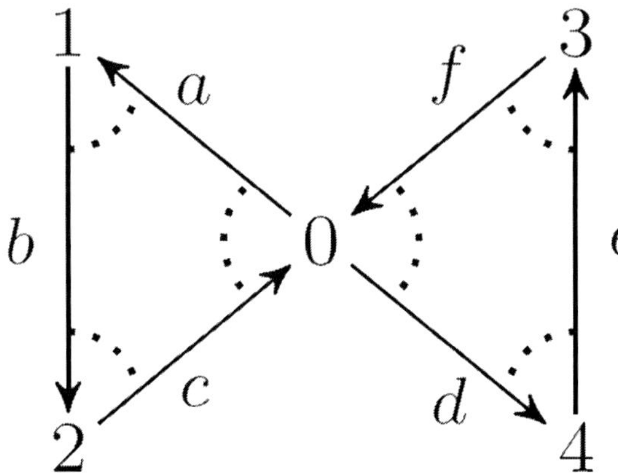

The following complex, with the leftmost non-zero term in cohomological degree 0, is an indecomposable object of K:

$$0 \longrightarrow P(0) \xrightarrow{\left(\begin{smallmatrix} c & f \end{smallmatrix}\right)} P(2) \oplus P(3) \xrightarrow{\left(\begin{smallmatrix} & \\ 0 & e \end{smallmatrix}\right)} P(1) \oplus P(4) \xrightarrow{\left(\begin{smallmatrix} \\ 0 \end{smallmatrix}\right)} P(3) \longrightarrow 0$$

Observe that the criterion $d^2 = 0$is obtained by either passing through a relation, or by having 0in the differential. We notice that this complex 'unfolds' as

$$\begin{array}{ccccccccccc} 2 & & 1 & & 0 & & 1 & & 2 & & 3 \\ P(4) & \xleftarrow{e} & P(3) & \xleftarrow{f} & P(0) & \xrightarrow{c} & P(2) & \xrightarrow{b} & P(1) & \xrightarrow{af} & P(3) \end{array}$$

where the cohomological degrees are written above each module. Moreover, the modules appearing are uniquely determined by the endpoints of the maps, so all information in this complex is communicated by the diagram

$$\begin{array}{ccccccccccc} 2 & & 1 & & 0 & & 1 & & 2 & & 3 \\ \bullet & \xleftarrow{e} & \bullet & \xleftarrow{f} & \bullet & \xrightarrow{c} & \bullet & \xrightarrow{b} & \bullet & \xrightarrow{af} & \bullet \end{array}$$

This is what we will later define as a 'homotopy string'. Another way of encoding this object is as a 'string with degrees' $(e, 2, 1)(f, 1, 0)(c, 0, 1)(b, 1, 2)(af, 2, 3)$.

Remark 2.1. By composing maps from left to right we ensure that composition of maps (i.e. matrix multiplication) fits with composition of paths. Therefore, all matrices are transposed. For instance; $\left(\begin{smallmatrix} c & f \end{smallmatrix}\right)\left(\begin{smallmatrix} b & 0 \\ 0 & e \end{smallmatrix}\right) = \left(\begin{smallmatrix} cb & fe \end{smallmatrix}\right) = 0$.

2.1. Homotopy strings

A *homotopy letter*is a triple $(p, i, j)$where p is a path in $\Gamma$with no subpath in I, and i, j are integers such that $|i - j| \leq 1$. The homotopy letter $(p, i, j)$is called *direct*

if $i < j$ and *inverse*if $i > j$. If $i = j$ then pmust be a stationary path and is called a *trivial*homotopy letter. We set $(p, i, j)^{-1} := (p, j, i)$. The starting and ending vertices of a homotopy letter are defined as

$$s(p, i, j) = \begin{cases} s(p) & \text{if } (p, i, j) \text{ is direct;} \\ t(p) & \text{otherwise;} \end{cases} \quad \text{and} \quad t(p, i, j) = \begin{cases} t(p) & \text{if } (p, i, j) \text{ is direct;} \\ s(p) & \text{otherwise.} \end{cases}$$

The composition $(p, i, j)(p', I', j')$is defined if $j = i'$ and $s(p, i, j) = t(p', I', j')$. If $(p, i, j)$is trivial in this situation, we write $(p, i, j)(p', i', j') := (p', i', j)$; similarly if $(p', i', j)$is trivial then $(p, i, j)(p', i', j) := (p, i, j)$.

Convention 2.2. We shall often write pas shorthand for (p, i, j). However, the degrees i, j should always be considered to be implicitly present.

A *homotopy string*is a sequence of pairwise composable homotopy letters

$$w = \prod_{r=n}^{1} (w_r, i_r, j_r) = (w_n, i_n, j_n)(w_{n-1}, i_{n-1}, j_{n-1}) \cdots (w_2, i_2, j_2)(w_1, i_1, j_1)$$

such that the following holds:

(1)whenever $(w_r, i, i+1)(w_{r-1}, i+1, i+2)$occurs, $w_r w_{r-1}$ has a subpath in I;

(2)whenever $(w_r, i, i-1)(w_{r-1}, i-1, i-2)$occurs, $w_{r-1} w_r$ has a subpath in I;

(3)whenever $(w_r, i, i+1)(w_{r-1}, i+1, i)$occurs, w_r and w_{r-1}do not start with the same arrow;

(4)whenever $(w_r, i, i-1)(w_{r-1}, i-1, i)$occurs, w_r and w_{r-1} do not end with the same arrow.

Write $s(w) = s(w_1, i_1, j_1)$ and $t(w) = t(w_n, i_n, j_n)$ if w is non-trivial, and for a trivial homotopy string $w = (1_x, i, i)$we write $s(w) = x = t(w)$. Its inverse is given by

$$((w_n, i_n, j_n) \cdots (w_1, i_1, j_1))^{-1} := (w_1, j_1, i_1) \cdots (w_n, j_n, i_n)$$

.

Another way of expressing a homotopy string $w = \prod_{r=n}^{1} (w_r, i_r, j_r)$ is by a diagram

$$\underset{i_n}{\bullet} \xrightarrow{\quad w_n \quad} \underset{j_n}{\bullet} \text{———} \cdots \text{———} \underset{j_2}{\bullet} \xrightarrow{\quad w_1 \quad} \underset{j_1}{\bullet},$$

wherethe line labelled wris an arrow pointing to the right if wris direct, and an arrow pointing to the left if wris inverse. Note that each vertex $\xrightarrow{w_r} \bullet \xrightarrow{w_{r-1}}$ corresponds to a unique indecomposable projective Λ-module, namely *P*(*s*(*wr, ir, jr*)) =*P*(*t*($w_{r-1}, i_{r-1}, j_{r-1}$)).

Let $w = \prod_{r=n}^{1}(w_r, i_r, j_r)$ be a homotopy string. The *string complex Pw* corresponding to wis constructed as follows. We define indexing sets

$$\mathcal{I}_i := \begin{cases} \{r \mid (w_r, i, i \pm 1) \text{ is in } w\} \sqcup \{0\} & \text{if } i = j_1; \\ \{r \mid (w_r, i, i \pm 1) \text{ is in } w\} & \text{otherwise} \end{cases}.$$

Sitting in degree i, the corresponding complex has the object P^i_w given by

$$\bigoplus_{r \in \mathcal{I}_i} P(\varphi_w(r)), \quad \text{where } \varphi_w(r) = t(w_r, i_r, j_r) \text{ for } r > 0 \text{ and } \varphi_w(0) = s(w_1, i_1, j_1).$$

The differentials are defined componentwise: Any direct homotopy letter *wr* yields the component $P(\varphi_w(r)) \xrightarrow{w_r} P(\varphi_w(r-1))$, and any inverse homotopy letter *wr* yields the component $P(\varphi_w(r-1)) \xrightarrow{w_r} P(\varphi_w(r))$. Note that $P_w \cong P_{w^{-1}}$.

2.2. Homotopy bands

A non-trivial homotopy string $w = \prod_{r=n}^{1}(w_r, i_r, j_r)$ is a *homotopy band*if *s*(*w*) =*t*(*w*), i_n=*j*1, one of {w_n, w_1} is direct and the other inverse, and wis not a proper power of another homotopy string.

We now describe how to construct *one-dimensional band complexes*. The higher dimen-sional band complexes will be studied in Section5, and the definition is given there. Fix a homotopy band wand an element λ ∈ **k***. Again we define indexing sets

$$\mathcal{I}_i := \{r \mid (w_r, i, i \pm 1) \text{ is in } w\},$$

and $B^i_{w,\lambda,1}$ is defined as for string complexes, that is,

$B^i_{w,\lambda,1} = \bigoplus_{r \in \mathcal{I}_i} P(\varphi_w(r))$ where $\varphi_w(r)$ is as before.

For band complexes with $n \geq 2$, the components of the differential are determined as for string complexes, except for that corresponding to w1, which acquires the scalar λ: $P(\varphi_w(1)) \xrightarrow{\lambda w_1} P(\varphi_w(n))$ for w1direct, and $P(\varphi_w(n)) \xrightarrow{\lambda w_1} P(\varphi_w(1))$ for w_1 inverse. When n =2, the only non-zero component of the differential of $B_{w,\lambda,1}$ is $\lambda w_1 + w_2$. Note that $B_{w,\lambda,1} \cong B_{w^{-1},\frac{1}{\lambda},1}$.

The diagram notation of a homotopy band is an infinite repeating diagram, with the scalar λattached to w1as in the complex:

$$\cdots \xrightarrow{\lambda w_1} \overset{i_n}{\bullet} \xrightarrow{w_n} \overset{j_n}{\bullet} \longrightarrow \cdots \longrightarrow \overset{j_2}{\bullet} \xrightarrow{\lambda w_1} \overset{j_1 = i_n}{\bullet} \xrightarrow{w_n} \cdots.$$

Example 2.3. Consider the Running Example. The homotopy band

$z = (d, 3, 2)(e, 2, 1)(f, 1, 0)(c, 0, 1)(b, 1, 2)(a, 2, 3)$,

together with a fixed scalar λ, gives rise to the complex Bz,λ,1, where again the leftmost component lies in degree 0,

$$0 \longrightarrow P(0) \xrightarrow{(c\ f)} P(2) \oplus P(3) \xrightarrow{\begin{pmatrix} b & 0 \\ 0 & e \end{pmatrix}} P(1) \oplus P(4) \xrightarrow{\begin{pmatrix} \lambda a \\ d \end{pmatrix}} P(0) \longrightarrow 0$$

.

We write this as the diagram

$$\cdots \xrightarrow{\lambda a} \overset{3}{\bullet} \xleftarrow{d} \overset{2}{\bullet} \xleftarrow{e} \overset{1}{\bullet} \xleftarrow{f} \overset{0}{\bullet} \xrightarrow{c} \overset{1}{\bullet} \xrightarrow{b} \overset{2}{\bullet} \xrightarrow{\lambda a} \overset{3}{\bullet} \xleftarrow{d} \cdots.$$

Convention 2.4. From now on, unless explicitly needed for emphasis, we shall omit the degrees from unfolded diagrams. However, all homotopy letters in this article carry de-grees, therefore, the reader should be aware that they are implicitly always present.

2.3. Infinite homotopy strings

These indecomposable complexes only occur when Λhas infinite global dimension: when the global dimension of Λis finite all the complexes in D are isomorphic to perfect complexes. If Λ =**k**Γ/I has infinite global dimension, then Γcontains oriented cycles with 'full relations'. Let C(Λ)denote the collection of arrows a ∈ Γ1such that there exists a repetition-free cyclic path $a_n \cdots a_2 a_1$in Γsuch that $ai+1a_i \in I$ for $1 \le i \le n$ and $a_1 an \in I$, where a_1=a. We need the following definition from [9].

Definition 2.5. A *direct (resp. inverse) antipath*is a homotopy string where all homotopy letters are direct (resp. inverse) and arrows in the quiver.

Definition 2.6. Let $w = \prod_{k=n}^{1}(w_k, i_k, j_k)$ be a homotopy string. We say *w* is

(1)*left resolvable*if (*wn, in, jn*) is direct (i.e. *jn* = *in* + 1), $i_n \le i_k, j_k$ for all $1 \le k < n$, and there exists $a \in C(\Lambda)$ such that $(a, i_n - 1, i_n)w$ is is a homotopy string. We shall say that wis *left resolvable by* a;

(2)*primitive left resolvable*if there is no direct antipath $\prod_{k=n}^{t}(w_k, i_k, j_k)$ such that $\prod_{k=t-1}^{1}(w_k, i_k, j_k)$ is left resolvable.

One can write down the obvious dual definitions of *(primitive) right resolvable.* Aho-motopy string will be called*(primitive) two-sided resolvable*if it is both (primitive) left resolvable and (primitive) right resolvable.

If *w* is left (resp. right) resolvable by *a*, then gentleness of Λensures that this is the unique arrow in $\Gamma_1$by which wis left (resp. right) resolvable. If *w* is left resolvable then $w^{-1}$is right resolvable, and vice versa. We shall call a left or right resolvable homotopy string that is not two-sided resolvable *one-sided resolvable.* There is then the obvious notion of *primitive one-sided resolvable.*

Suppose $w = \prod_{k=n}^{1}(w_k, i_k, j_k)$ is left resolvable by $a_1 \in C(\Lambda)$, which sits in the repetition-free cyclic path $a_m \ldots a_1$ in Γ. We form the *left infinite homotopy string* ∞w by concatenating infinitely many appropriately shifted copies of the cycle on the left of *w*, i.e. the unfolded diagram of ∞*w* has the form

$$^{\infty}w: \quad \cdots \xrightarrow{a_1} \bullet \xrightarrow{a_m} \bullet \cdots \bullet \xrightarrow{a_2} \bullet \xrightarrow{a_1} \bullet \xrightarrow{w_n} \bullet \xrightarrow{w_{n-1}} \bullet \cdots \bullet \xrightarrow{w_1} \bullet.$$

Analogously, we form the *right infinite homotopy string* w^∞ and the *two-sided infinite homotopy strings* $^\infty w^\infty$ from a right resolvable homotopy string or two-sided resolvable homotopy string, respectively.

If *w* is one-sided left resolvable we obtain the corresponding infinite string complex from the left infinite homotopy string $^\infty w$, or equivalently the right infinite homotopy string $(w^{-1})\infty$. Dually for *w* right resolvable. For *w* two-sided resolvable we represent it by $^\infty w^\infty$ and $^\infty(w^{-1})^\infty$.

Briefly we justify the term resolvable: the corresponding infinite string complex is the projective resolution of a complex closely related to the resolvable perfect complex (one puts the kernel of the left/rightmost differential in the minimal degree). This is made transparent in the following example.

EXAMPLE 2.7

Consider the algebra Λ given by the following bound quiver.

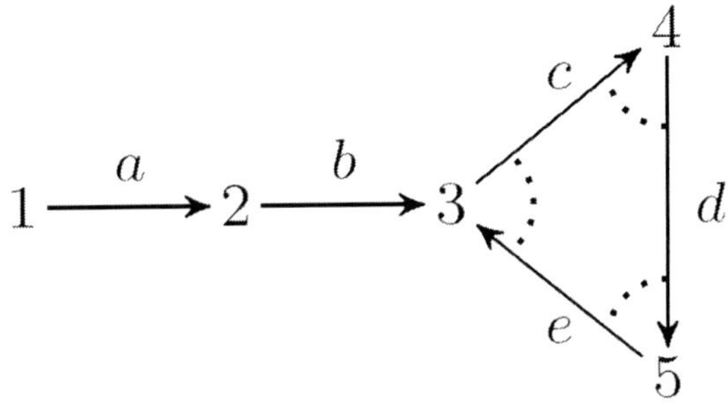

There are five primitive one-sided resolvable strings (up to shift and equivalence) and no two-sided resolvable strings. The five primitive one-sided resolvable strings are:

$$\bullet \xrightarrow{c} \bullet\, ,\ \bullet \xrightarrow{d} \bullet\, ,\ \bullet \xrightarrow{e} \bullet\, ,\ \bullet \xrightarrow{cb} \bullet \ \text{and}\ \bullet \xrightarrow{cba} \bullet$$

.

Consider the homotopy string $w = cb$. Then $^{\infty}w$ is the string

$$\cdots\cdots\cdots\cdots \bullet \xrightarrow{d} \bullet \xrightarrow{c} \bullet \xrightarrow{e} \bullet \xrightarrow{d} \bullet \xrightarrow{cb} \bullet,$$

which one can see is the projective resolution of the following complex

$$\ker(cb) \hookrightarrow P(4) \xrightarrow{cb} P(2).$$

Note that the homotopy string $P\infty_w$ is also the projective resolution of the
2
module with composition series 3.

2.4. The indecomposable objects of D

We first set up some notation. A *cyclic rotation* of the homotopy band w is a homotopy band of the form

$$(w_k, i_k, j_k)(w_{k-1}, i_{k-1}, j_{k-1}) \cdots (w_1, i_1, j_1)(w_n, i_n, j_n) \cdots (w_{k+1}, i_{k+1}, j_{k+1})$$

.

Consider the equivalence relation $\sim^{-1}$ generated by identifying a homotopy string with its inverse, and the equivalence relation $\sim^{r}$ generated by identifying a homotopy band with its cyclic rotations and their inverses. The following will denote complete sets of representatives of the specified objects under the given equivalence relations:

St := allhomotopystringsunder $\sim^{-1}$;

Ba := allhomotopybandsunder $\sim^{r}$;

St_1 := allrightandleftinfinitehomotopystringsunder $\sim^{-1}$;

St_2 := alltwo-sidedinfinitehomotopystringsunder $\sim^{-1}$,

where the left (resp. right, resp. two-sided) infinite homotopy strings are precisely those constructed from the primitive left (resp. right, resp. two-sided) resolvable homotopy strings in the manner described in the previous subsection.

REMARK 2.8

A homotopy band can always be considered as a homotopy string, thus we must take the disjoint union: a band gives rise both to a string complex and a family of band complexes. Moreover, every primitive two-sided resolvable homotopy string *w* yields three infinite homotopy strings: ${}_{\infty}w, w^{\infty}, {}_{\infty}w_{\infty}$.

THEOREM 2.9 *([8, Theorem 3]). There are bijections*

$$\mathrm{ind}(\mathsf{K}) \overset{1-1}{\longleftrightarrow} \mathrm{St} \sqcup (\mathrm{Ba} \times \mathbf{k}^* \times \mathbb{N}) \quad \textit{and} \quad \mathrm{ind}(\mathsf{D}\backslash\mathsf{K}) \overset{1-1}{\longleftrightarrow} \mathrm{St}_1 \sqcup \mathrm{St}_2.$$

We mention here that a homotopy band can always be considered as a homotopy string, thus we must take the disjoint union: a band gives rise both to a string complex and a family of band complexes. Moreover, every primitive two-sided resolvable homotopy string *w* yields three infinite homotopy strings: ${}_{\infty}w, w^{\infty}, {}_{\infty}w_{\infty}$.

3. MORPHISMS BETWEEN INDECOMPOSABLE OBJECTS OF D

In this section, we shall describe a canonical basis for the set of homomorphisms between (finite or infinite) string and/or one-dimensional band complexes; higher dimensional bands are dealt with in Section 5. The proof that the maps described here do indeed form a basis is contained in Section 4.

We first set up some notation and define three canonical classes of maps. Fix

$w \in \mathsf{St} \sqcup \mathsf{Ba} \sqcup \mathsf{St}_1 \sqcup \mathsf{St}_2$ and $\lambda_w \in \mathbf{k}^*$ Define

$$Q_w := \begin{cases} P_w & \text{if } w \in \mathsf{St} \sqcup \mathsf{St}_1 \sqcup \mathsf{St}_2; \\ B_{w,\lambda_w,1} & \text{if } w \in \mathsf{Ba}. \end{cases}$$

We have abused notation here by allowing the scalar λ_w to disappear when dealing with Q_w in the case that w is a homotopy band; it should be treated as implicitly present.

For $f \in \mathrm{Hom}_{\mathcal{C}}(Qv,Qw)$ and a degree t, the map f^t: $f^t : Q_v^t \to Q_w^t$ can be written as a matrix between the finitely-many indecomposable summands of Q_v^t and Q_w^t. Each entry of this matrix is a linear combination of paths (see Proposition 1.1). We refer to a single term in this sum as a *component* of f^t. Moreover, a *component of f* is taken to be a component of ft for some degree t. Throughout this section we fix two homotopy strings or bands v and w and the corresponding complexes Qv and Qw. We consider 'maps' between the unfolded diagrams of Q_v and Q_w which, at each projective, will look like:

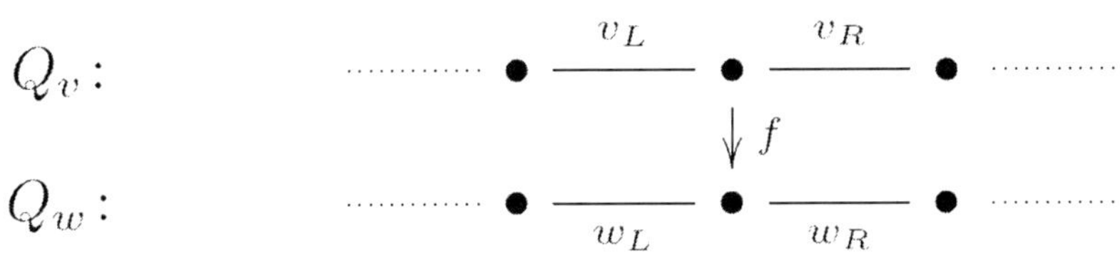

If f is at the leftmost end of the unfolded string complex Qv we say that v_L is *zero*; likewise for v_R, w_L, and w_R. Again, f is a linear combination of paths and we will use the term *component* to refer to a single summand of f.

In the next subsections we shall define maps occurring in a canonical basis of $\mathrm{Hom}_{\mathcal{C}}(Q_v,Q_w)$.

3.1. Single and double maps

Suppose we are in the following situation:

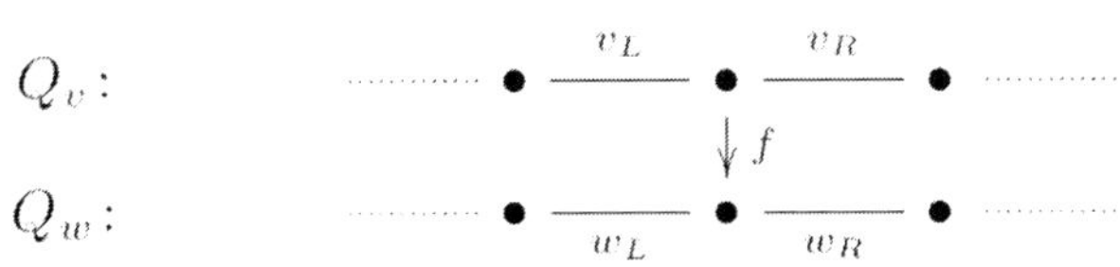

(1)

where f is some non-stationary path in the quiver.

DEFINITION 3.1

A map in $\mathrm{Hom}_C(Qv,Qw)$ will be called a *single map* if it has only one non-zero component whose unfolded diagram is as above and satisfies the following conditions:

(**L1**) If v_L is direct, then $v_L f = 0$.

(**L2**) If w_L is inverse, then $fw_L = 0$.

(**R1**) If v_R is inverse, then $v_R f = 0$.

(**R2**) If w_R is direct, then $fw_R = 0$.

Write $S_{v,w}$ for the set of single maps $Qv \to Qw$.

Example 3.2. Again consider the Running Example. Consider the homotopy strings

$v = (f, 0, 1)(e, 1, 2)(dc, 2, 3)(b, 3, 4)(a, 4, 5)(d, 5, 4)$, and

$w = (c, 0, -1)(f, -1, 0)(e, 0, 1)(dc, 1, 2)(b, 2, 3)(a, 3, 4)$.

Then we have the following:

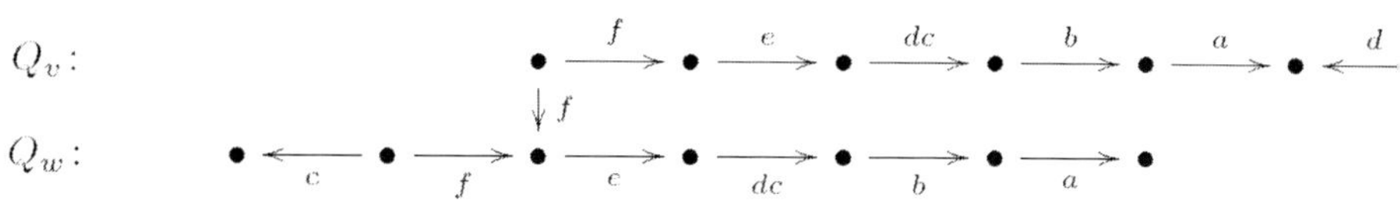

which gives rise to a single map $Qv \to Qw$:

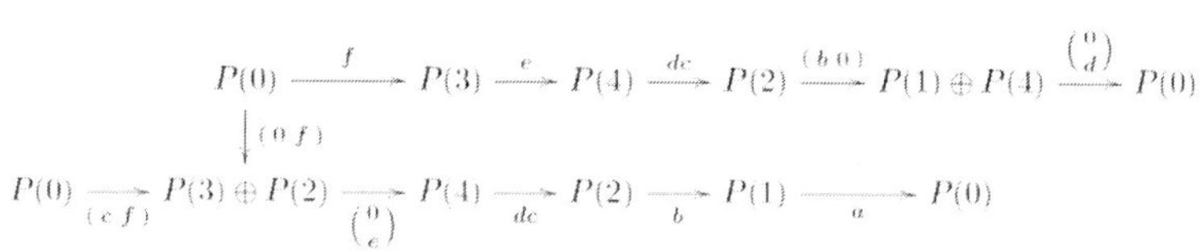

Suppose we have the following situation:

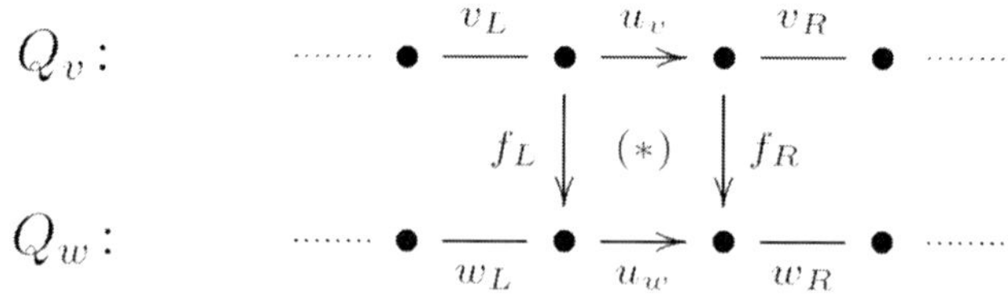

(2)

such that (∗)commutes for non-stationary paths fL, fR.

Definition 3.3. If (**L1**)and (**L2**)hold for fL and (**R1**)and (**R2**)hold for fR, then the diagram above induces a map $Qv \to Qw$ with two non-zero components in consecutive degrees given by *fL* and *fR*. We call such a map a *double map*; write *Dv,w*for the set of double maps *Qv* → *Qw*.

Remark 3.4. To obtain all double maps *Qv*→*Qw* in terms of diagrams as above, it is necessary to consider overlaps between unfolded diagrams of *Qv* and any *Qw'* such that *Qw* ≅ *Qw'*; see Section2.4for the relevant equivalence relations on homotopy strings and bands.

Example 3.5. Returning to the Running Example, consider the homotopy strings:

v = (*dc*,−2,−1)(*b*,−1, 0)(*a*, 0, 1) and *w* = (*af*, 3, 2)(*b*, 2, 1)(*c*, 1, 0)(*f*, 0, 1)(*e*, 1, 2).

The unfolded diagram

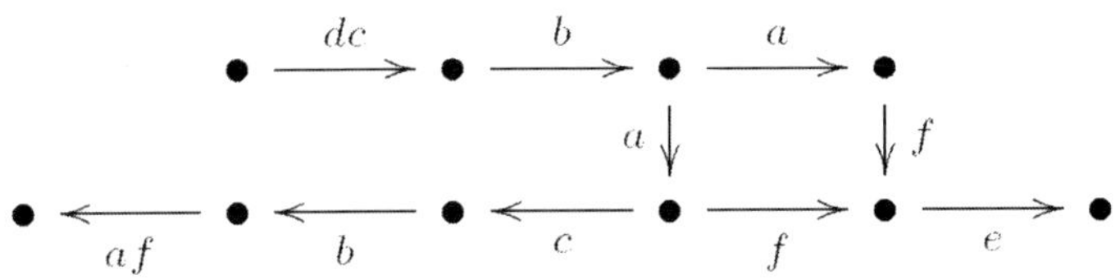

gives rise to the double map:

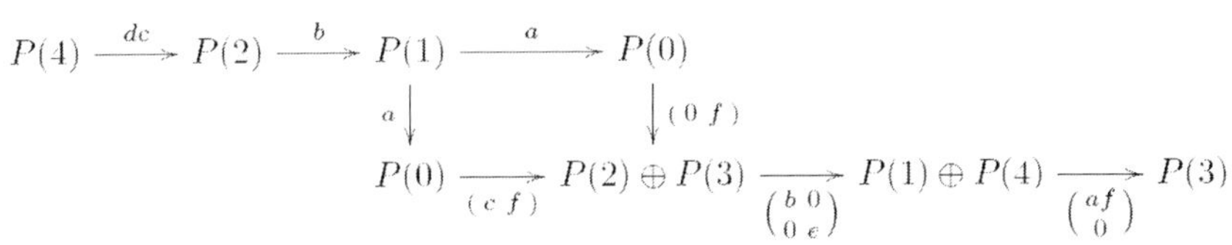

Notation 3.6. Let *f* ∈ *Sv,w* and suppose the unique non-zero component of fcorresponds to the path $p\colon x \rightsquigarrow y$ in Γ. Then we write *f* = (*p*). Similarly, for *f* ∈ $D_{v,w}$whose two non-zero components correspond to the paths $p\colon x \rightsquigarrow y$ and $q\colon x' \rightsquigarrow y'$, we write *f* = (*p*, *q*). The notation should be suggestive of an infinite vector, in which all entries are zero apart from those which are written.

We next highlight two important classes of single and double maps.

Definition 3.7. Recall the setup in (1). A map *f* ∈ $S_{v,w}$ will be called a *singleton single map*when its non-zero component is given by a path pand we are in the situation of the following unfolded diagrams (up to inverting one of the homotopy strings):

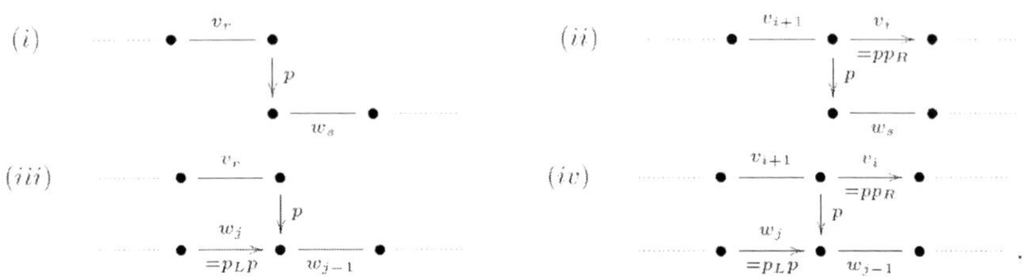

where $r \in \{1,n\}$, $s \in \{1,m\}$, either v_r is inverse (or zero) or $v_r p = 0$, either w_s is inverse (or zero) or pw_s =0, and p is not a homotopy subletter of v_r or w_s. We denote the set of all singleton single maps $Q_v \to Q_w$ by $\mathcal{S}^1_{v,w}$.

DEFINITION 3.8

Definition 3.8. Recall the setup in (2). A map $f = (fL, fR) \in D_{v,w}$will be called a *singleton double map*if it satisfies the following condition (or its dual).

(**D**)There exists a non-stationary path f' such that $u_v = fLf'$ and $u_w = f'fR$.

We denote the set of all singleton double maps $Qv \to Qw$ by $\mathcal{D}^1_{v,w}$.

Condition (**D**)means that to find singleton double maps, it is sufficient to look for homotopy letters v_i of v and w_j of wsuch that v_i and w_j sit in the same degrees, with the same orientation, and a 'proper right substring' of v_i is a 'proper left substring' ofw_j.

3.2. Graph maps and quasi-graph maps

Suppose the unfolded diagrams of Q_v and Q_w overlap as follows:

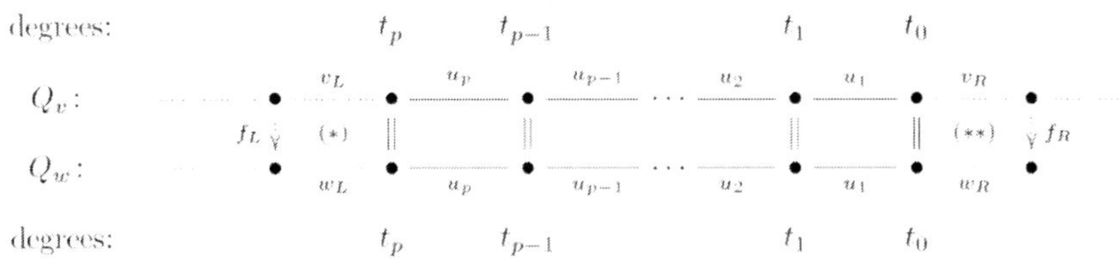

such that $v_L \neq w_L$and $v_R \neq w_R$. The double lines represent isomorphisms and all of the squares with solid lines commute as paths in the quiver. Consider the following *left endpoint conditions.*

(**LG1**) The arrows v_L and w_L are either both direct or both inverse and there exists some (scalar multiple of a) non-stationary path f_L such that the square $(*)$ commutes.

(**LG2**) The arrows v_L and w_L are neither both direct nor both inverse. In this case, if v_L is non-zero then it is inverse and if wLis non-zero then it is direct.

If both strings continue infinitely to the left, we have an additional left endpoint condi-tion:

(**LG∞**) The diagram continues infinitely to the left with commuting squares in which the vertical maps are isomorphisms.

There are dual *right endpoint conditions,* (**RG1**), (**RG2**) and (**RG∞**), respectively.

Definition 3.9. If one of (**LG1**), (**LG2**)or (**LG∞**)hold and one of (**RG1**), (**RG2**)or (**RG∞**) hold, the diagram induces a map $Q_v \to Q_w$, whose non-zero components are exactly those described by the diagram. Such maps are called *graph maps*; write Gv,wfor the set of graph maps $Q_v \to Q_w$.

Example 3.10. Consider the algebra in the Running Example. Let

$v = (dc, 0, 1)(b, 1, 2)(a, 2, 3)$ and $w = (e, 2, 1)(f, 1, 0)(c, 0, 1)(b, 1, 2)(af, 2, 3)$.

The unfolded diagram on the left gives rise to the graph map on the right.

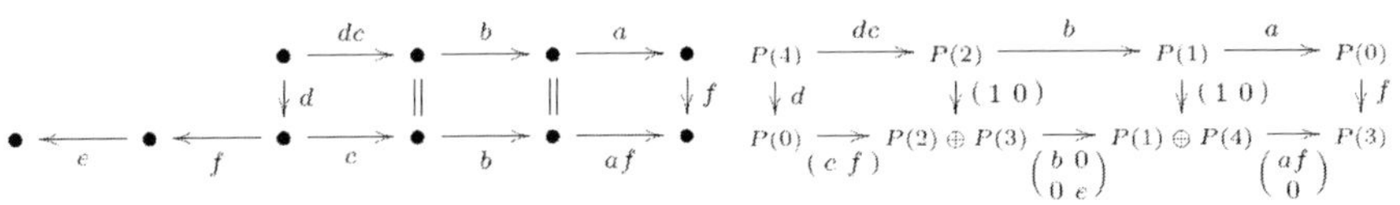

Definition 3.11. If none of the conditions (**LG1**), (**LG2**), (**LG∞**), (**RG1**), (**RG2**) or (**RG∞**) hold, then the diagram no longer induces a map, however we shall say that there is a *quasi-graph map*from Q_v to Q_w.

Definition 3.12. Consider the diagram representing a quasi-graph map $Q_v \to \Sigma^{-1}Q_w$:

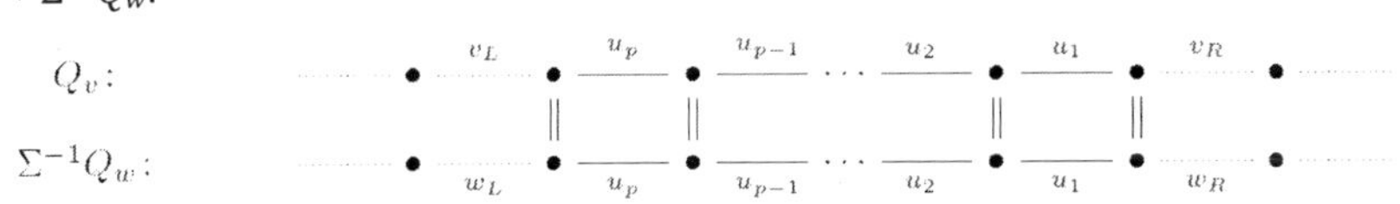

Then there exist psingle maps $Q_v \to Q_w$ given by the paths $u_p, u_{p-1}, ..., u_1$ in the ap-propriate degrees. There are also two (single or double) maps with non-zero components given by v_L, w_L, v_R or wRin the appropriate degree. For example, if $v_L$is direct and w_L is inverse than there is a double map (v_L, w_L). We call these maps the associated *quasi-graph map representatives*. Let $Q_{v,w}$ be a fixed set of quasi-graph map representatives $Q_v \to Q_w$, one for each quasi-graph map from Q_v to $\Sigma^{-1}Q_w$.

Example 3.13. Let us return to Example3.2. The following diagram describes a quasi-graph map $Q_v \to \Sigma^{-1}Q_w$:

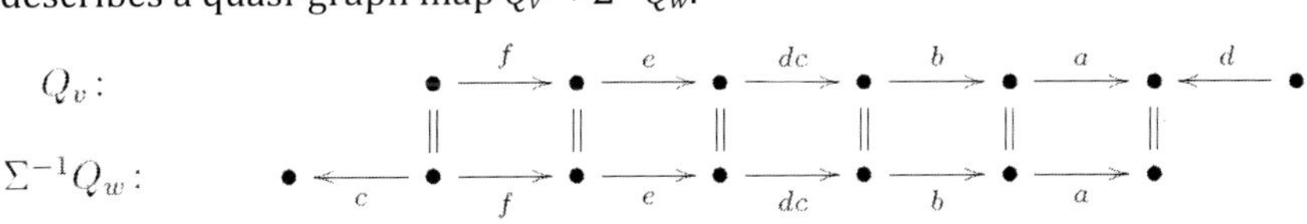

and the associated quasi-graph map representatives $Q_v \to Q_w$ are

$\{(-c), (f), (-e), (dc), (-b), (a), (-d)\}$.

Remarks 3.14. We highlight the following.

(1)For the convenience of the reader, we spell out the construction of (quasi-)graph maps for infinite homotopy strings: if a left (respectively right) endpoint condition holds and the substrings to the right (respectively left) are equal and infinite then we can define a (quasi-)graph map with infinitely many components which are isomorphisms. Cf. [18]for infinite graph maps between modules.

(2)To standardise what we mean by a (quasi-)graph map we take the following conven-tions. If vand ware both homotopy strings, then each isomorphism is an identity. If one of vand wis a homotopy band, then the leftmost isomorphism in the di-agram is an identity and the remaining isomorphisms will be determined by this. Note that any other choice of isomorphisms will result in a scalar multiple of such a standardised map.

(3)It is necessary to replace wby an equivalent homotopy string/band to obtain all of the possible (quasi)-graph maps $Q_v \to Q_w$; cf. Remark3.4.

(4)There is one pathological example arising from the identity map $B_{w,\lambda,1} \to B_{w,\lambda,1}$. This is a graph map since it 'travels' the whole way around the band. In particular, there are no real endpoint conditions. As such it also defines a quasi-graph map $B_{w,\lambda,1} \to \Sigma{-1}B_{w,\lambda,1}$. This situation will be treated in more detail in Section5.

3.3. The main theorem

We have now assembled all the maps that occur in a canonical basis of HomD (Q_v, Q_w) and can state our main result succinctly as the following.

Theorem 3.15. *The set* $\mathcal{B}^{\mathsf{D}}_{v,w} := \mathcal{S}^{1}_{v,w} \cup \mathcal{D}^{1}_{v,w} \cup \mathcal{G}_{v,w} \cup \mathcal{Q}_{v,w}$ *is a* **k**-*linear basis for* HomD(Q_v, Qw).

The next section concerns the proof of this result.

Example 3.16. Consider the following homotopy strings over the Running Example:

$v = (a,-1, 0)(c, 0, 1)(b, 1, 2)$

$w = (e, 2, 3)(d, 3, 4)(a, 4, 3)(b, 3, 2)(cd, 2, 1)(e, 1, 0)(f, 0,-1)(c,-1, 0)(b, 0, 1)(a, 1, 2)$

The set $\mathcal{B}^{\mathsf{D}}_{v,w}$ has two elements. The first is a singleton single map (af) in degree 2from the right end-point of v to the left end-point of w. The second is given by the following quasi-graph map $Q_v \to \Sigma^{-1}Q_w$, where the associated quasi-graph map representatives $Q_v \to Q_w$ are indicated by dashed arrows (single maps) and dotted arrows (double map):

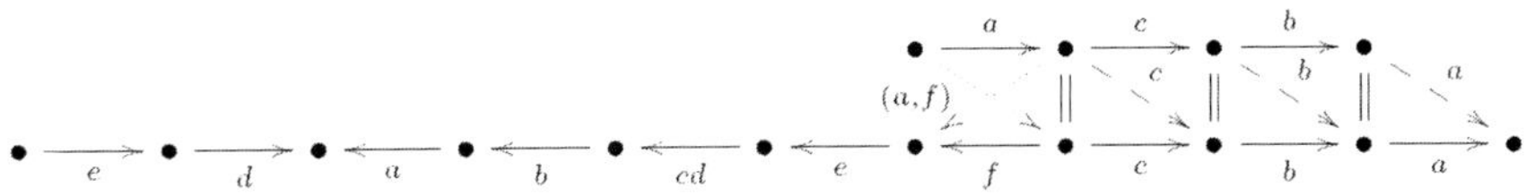

4. PROOF OF THE MAIN THEOREM

Let $v, w \in \mathrm{St} \sqcup \mathrm{Ba}$ and Q_v, Q_w be as in the previous section. We shall split up the proof of Theorem3.15into two parts. The first part establishes a canonical basis for $\mathrm{Hom}_{\mathrm{C}}(Q_v, Q_w)$. In the second part, we identify which elements of this basis are homotopic or null-homotopic.

4.1. A basis at the level of complexes

In this section, we establish the following:

Proposition 4.1. *The set* $\mathcal{B}^{\mathrm{C}}_{v,w} := \mathcal{G}_{v,w} \cup \mathcal{S}_{v,w} \cup \mathcal{D}_{v,w}$ *is a* **k**-*linear basis for* $\mathrm{HomC}(Q_v, Q_w)$.

The proof of Proposition4.1is inspired by [18, Section 1.4]. We first need two technical lemmas from which we will deduce that $\mathcal{B}^{\mathrm{C}}_{v,w}$ is a linearly independent set.

Lemma 4.2. *Suppose we have the following situation:*

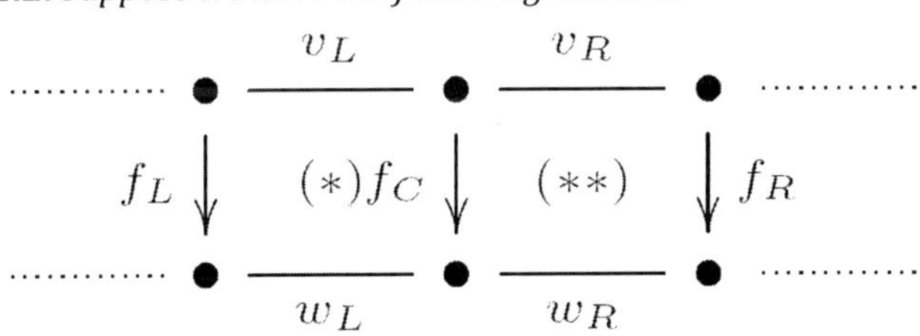

Proof. We analyse the case when all four homotopy letters vL, vR, wLand wRare direct; the remaining cases are analogous.

Suppose $(*)$ has a non-zero commutativity relation. Then $v_L f_C = f L w L \neq 0$. It follows that the paths f_C and w_L start with in the same arrow. By the definitions of homotopy strings and bands, $w_L w_R = 0$and, by condition (4) of gentleness, we must also have that $f_C w_R$=0. That is, $f_C w_R = v_R f_R = 0$and $(**)$ has a zero commutativity relation.

Dually, whenever (∗∗) has a non-zero commutativity relation, (∗) has a zero commu-tativity relation. □

Lemma 4.3. *The unfolded diagram giving rise to a graph, single or double map is com-pletely determined by any non-zero component.*

Proof. We give a proof for the case where we have a non-zero component of a graph map. The arguments for single and double maps are analogous.

We make use of the diagram from Lemma4.2. Suppose that f_C is a non-zero compo-nent of a graph map. We show that, if it exists, f_L is completely determined. Note that if v_L and w_L have different orientations then the arrow f_L does not make sense since the degrees will not match. So suppose v_L and w_L are either both direct or both inverse.

Suppose f_C is an isomorphism and v_L and w_L are both direct. Then the square (∗) must commute so fLis the unique path such that $f_L w_L = v_L f_C$ (if the path is stationary then fLis an isomorphism). Similarly, when v_L and w_L are both inverse then f_L is the unique path such that $v_L f_L = f_C w_L$.

Suppose f_C is given by a path, then according to Lemma4.2we must be at the left or right endpoint of the diagram. If v_L is direct and $v_L f_C = 0$, then we must be at the left end of the diagram and $f_L = 0$. If $v_L f_C \neq 0$, then we areat the right end of the diagram and f_L is an isomorphism. Similarly, when w_L is inverse and $f_C w_L = 0$, then $f_L = 0$. If $f_C w_L \neq 0$, then fLis an isomorphism.

We can apply dual arguments to conclude that the diagramis also completely deter-mined to the right (i.e. f_C determines f_R). □

Corollary 4.4. *The set* $\mathcal{B}^{\mathsf{C}}_{v,w} = \mathcal{G}_{v,w} \cup \mathcal{S}_{v,w} \cup \mathcal{D}_{v,w}$ *is linearly independent.*

Proof. Let $0 \neq b \in \mathcal{B}^{\mathsf{C}}_{v,w}$. We show that bcannot be written as a linear combination of other elements of $\mathcal{B}^{\mathsf{C}}_{v,w}$. Suppose $b = \sum_{i=1}^{n} k_i b_i$ for pairwise different $b_1, \ldots, b_n \in \mathcal{B}^{\mathsf{C}}_{v,w}$ and $k_i \in \mathbf{k}$. Let f_c be a non-zero component of b. Then some b_i, $1 \leq i \leq n$, must also have this non-zero component because the algebra is gentle and so only has zero relations. By Lemma4.3, $b = b_i$ and so k_i =1and $\sum_{j=1}^{i-1} k_j b_i + \sum_{j=i+1}^{n} k_j b_j = 0$ as required. □

PROOF OF PROPOSITION 4.1

It suffices to show that $\mathcal{B}^{\mathsf{C}}_{v,w}$ spans $\mathrm{Hom}_{\mathsf{C}}(Q_v, Q_w)$. Let $0 \neq h \in \mathrm{Hom}_{\mathsf{C}}(Q_v, Q_w)$. Then there is some degree t such that ht: $Q^t_v \to Q^t_w$ has a non-zero component $h^t_{ab}\colon P(\varphi_v(a)) \to P(\varphi_w(b))$. By the shape of homotopy strings and bands, $P(\varphi v(a))$ and $P(\varphi w(b))$ are each connected to at most two non-zero components of the differential. We must therefore consider the unfolded diagrams (as in Lemma 4.2 but with h^t_{ab} in place of f_C). Without loss of generality, assume that h^t_{ab} is an isomorphism or a scalar multiple of a path. By Lemma 4.3, there is a unique (scalar multiple of an) element of $\mathcal{B}^{\mathsf{C}}_{v,w}$ with this component and, by Lemma 4.2, this must be a summand of h. If this is not the whole of h, then we choose another non-zero component of h and continue until we have found a complete decomposition of h. Thus $h \in \mathsf{Span}\mathcal{B}^{\mathsf{C}}_{v,w}$. □

Proposition4.1gives us canonical bases for the Hom spaces between indecomposable complexes of Dconsidered as objects of C. We next turn our attention to homotopy classes of these maps. The following section highlights the strategy of our approach.

4.2. The strategy for constructing homotopy classes

We first recall the general definition; we direct the reader to a standard textbook on homological algebra, for example [20,23]for more information regarding homotopies.

Definition 4.5. Let (P•, d•) and (Q•, ∂•) be complexes in C. Then maps f, g:P•→ Q•are said to be *homotopic*, written $f \cong g$, if there are maps $h^i : P^i \rightarrow Q^{i-1}$ such that $f^i - g^i = d^i h^{i+1} + h^i \partial^{i-1}$ (where, as before, the maps are composed from left to right, see Remark2.1). The family of maps $\{h^i\}$ is called a *homotopy*from fto g. If g=0then fis called *null-homotopic.*

Consider a map $f \in \mathcal{B}^{\mathsf{C}}_{v,w}$ and let pbe a component of f. We can write down an unfolded diagram representation of this component as follows.

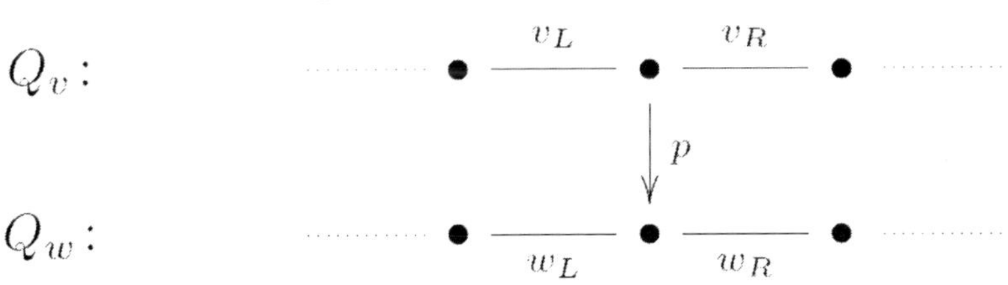

Thus, the corresponding component of a homotopic map must be given by

$$q = p + \left(\alpha v_L a + \beta v_R b + \gamma c w_L + \delta d w_R\right),$$

for some scalars α, β, γ, $\delta \in \mathbf{k}$ and paths a, b, c, d in the quiver. If the composition of paths does not make sense we take the corresponding scalar to be zero. For instance, if v_L is direct then α=0. Therefore, in order to construct homotopies between maps in $\mathcal{B}^{\mathsf{C}}_{v,w}$ it is enough to look at ways to construct the path p by 'completing' differential components.

Definition 4.6. We denote by $H(f)$ the set of maps f' such that $f \simeq f'$ and $f' = \lambda g$ for same $g \in \mathcal{B}^{\mathsf{C}}_{v,w}$ and $\lambda \in \mathbf{k}^*$.

Remark 4.7. The set $H(f)$is not the same as the homotopy class of f, however, if $g \simeq f$ then the decomposition of g into a linear combination of elements of $\mathcal{B}^{\mathsf{C}}_{v,w}$ ill consist of elements of $H(f) \cup H(0)$ only. Hence, it suffices to determine the sets $H(f)$. If f is non-zero and $H(f)$is a singleton set then we will say that f belongs to a *singleton homotopy class.*

4.3. Basic maps f such that H(f)is not singleton

Here we start with $f \in S_{v,w} \cup D_{v,w}$; we shall see in Section 4.4 that we do not need to consider $f \in G_{v,w}$. The reader may find it helpful to recall Definitions 3.9 and 3.11 and the corresponding endpoint conditions.

Proposition 4.8. *Suppose* $f \in S_{v,w} \cup D_{v,w}$ *such that* $H(f) \cong \{f\}$. *Then* f *is not null-homotopic if and only if the elements of* $H(f)$ *are in one-to-one correspondence with the representatives of a quasi-graph map* $Q_v \to \Sigma^{-1}Q_w$. *Moreover,*

(1) $H(f)$*has a double map on the left if* $v_L \cong 0$ *is direct and* $w_L \cong 0$ *is inverse.*

(2) $H(f)$*has a double map on the right if* $v_R \cong 0$ *is inverse and* $w_R \cong 0$ *is direct.*

(3) Otherwise, $H(f)$*ends with a single map on the left or right.*

Proof. For simplicity, we consider only the case $f \in S_{v,w}$. The case $f \in D_{v,w}$ is similar. The setup is the following, where by abuse of notation we have denoted the map and its unique non-zero component by f:

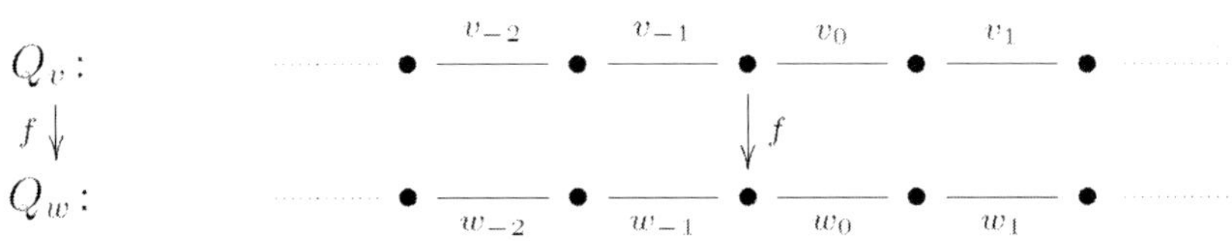

As explained in Section 4.2, the components corresponding to fin any homotopic map can be constructed only from four possible paths v'_{-1}, v'_0, w'_{-1} or w'_0 illustrated in the following diagrams:

Observe that f can be immediately seen to be null-homotopic in the following cases:

(N1) If any of v'_{-1}, v'_0, w'_{-1} and w'_0 (exist and) are non-stationary paths making one of the triangles commute.

(N2) If there are no arrows out of $*_w$ and either $f = v_{-1}$ and there is no other arrow in v into $*_1$, or, $f = v_0$ and there is no other arrow in vinto $*_2$.

(N3) If there are no arrows into $*^v$ and either $f = w_{-1}$ and there is no other arrow in w out of $*^1$, or $f = w_0$and there is no other arrow in wout of $*^2$.

The conditions **(N1)**–**(N3)**correspond to the endpoint conditions (**LG1**), (**LG2**), (**RG1**)and (**RG2**)of Definition3.9for graph maps.

Suppose we are not in any of the cases (**N1**)–(**N3**). Since $H(f) \cong \{f\}$, at least one, but possibly both, of the following must hold:

- f is *built from the source differential*, i.e. $f = v_{-1}$ or $f = v_0$ (but not both); or
- f is *built from the target differential*, i.e. $f = w_{-1}$ or $f = w_0$(but not both).

Suppose fcan be built from the source differential; the argument when fcan be built from the target differential is dual. Without loss of generality, we assume $f = v_0$; if $f = v_{-1}$, invert the homotopy string vand relabel v_{-1}as v_0. The differential v0will be called a *used differential*, because it has already been used to construct one of the single or double maps, namely f in this instance. There are two cases.

Case: There are no arrows out of $*_w$.We must be in the situation that w_{-1} is direct, w_0 is inverse, and v_1 exists and is inverse – for otherwise we would be in case (**N2**)above. The following diagram describes the situation.

In particular, $f = (v_0)$ is homotopic to the single map $-h = -(v_1)$. We can now see part of the quasi-graph map constructed: namely, the equality written diagonally. We now wish to continue by building hfrom a differential. The differential v_1 has already been used, so in order to continue, we must see whether $h = w_0$ or $h = w_{-1}$, and then use the dual argument for maps built from the target differential.

Case: There is an arrow out of $*_w$.Without loss of generality, assume $w_0 \cong 0$is direct. Note that only one of w_0 or w_{-1} may be an arrow out of $*_w$ – otherwise f cannot be a well-defined single map. Then there exists

- a single map $g = w_0 : P(s(v_0)) \to P(s(w_0))$if v_1 is direct or zero; or
- a double map $(h, g) = (v_1, w_0)$, whenever v_1 is inverse.

The used differentials are v_0 and w_0 in the first case; in the second case v1is additionally a used differential. The situation is illustrated below:

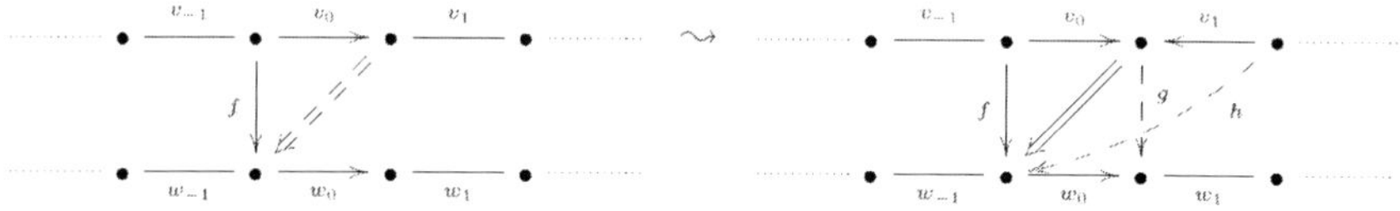

This gives $-g \in H(f)$ or $-(h, g) \in H(f)$. If the map we obtain at this step is a single map then we carry on, using the dual argument if necessary, to obtain further ele-ments of $H(f)$. The algorithm terminates when we reach one of the following threecases.

- We reach a single map g for which one of the conditions (**N1**), (**N2**)or (**N3**)is satisfied. In this case f is null-homotopic.

- We reach a single map g which is not equal to any of the unused differentials with respect to the already constructed elements of $H(f)$. This places us in case (3) of the proposition.

- We reach a double map; here there are insufficiently many unused differentials to continue to use to construct a homotopy. This places us in case (1) or (2) of the proposition.

In the case where we have two infinite strings and the algorithm does not terminate, then we have produced a 'quasi-graph map' satisfying either (LG∞) or (RG∞). Indeed, we have actually produced a graph map $Q_v \to \Sigma^{-1}Q_w$, whence the algorithm defines a null-homotopy class.

Now, if f can also be built from the target differential, we must now return to f and carry out the dual algorithm. □

REMARK 4.9 Null-homotopic single and double maps correspond to 'quasi-graph maps' which satisfy one of the graph map endpoint conditions, including in the case that both strings are infinite, the infinite 'endpoint' conditions.

EXAMPLE 4.10 Recall the quasi-graph map exhibited in Examples 3.2 and 3.13. Below is an explicit homotopy from (c) to (dc); the top line shows this at the level of complexes, the bottom line at the level of unfolded diagrams. In particular, the bottom line indicates how to use a quasi-graph map to construct families of homotopic maps.

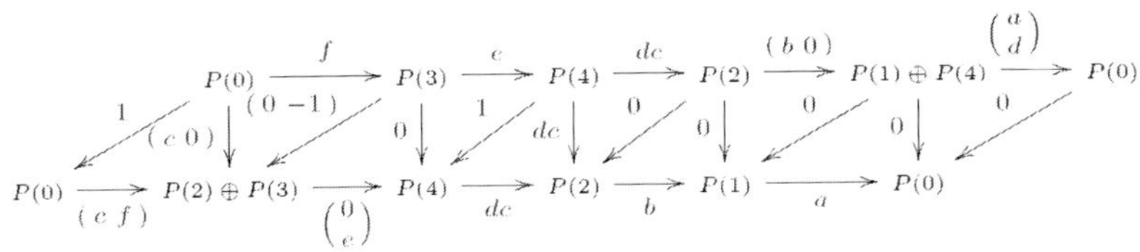

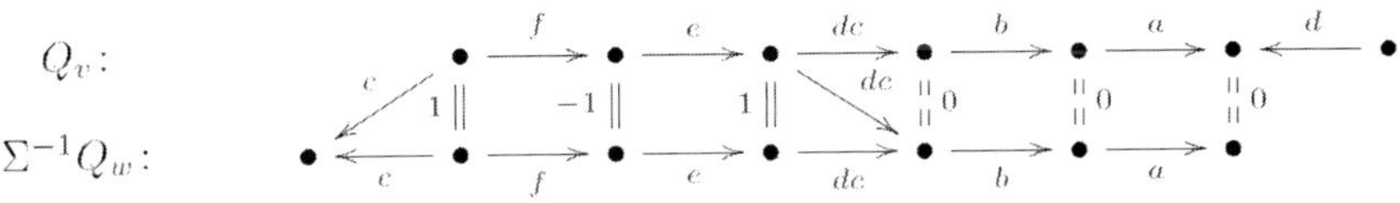

Example 4.11. Consider the following homotopy strings over the running example:

$v = (dc, 0, 1)(b, 1, 2)(a, 2, 3)$ and $w = (f, 0,-1)(c,-1, 0)(b, 0, 1)(af, 1, 2)$.

The diagrams below illustrate a homotopy, constructed as in the proof of Proposition 4.8, which makes the single map (b) null-homotopic. On the left hand side this is done at the level of unfolded diagrams, and on the right hand side, explicitly in terms of complexes.

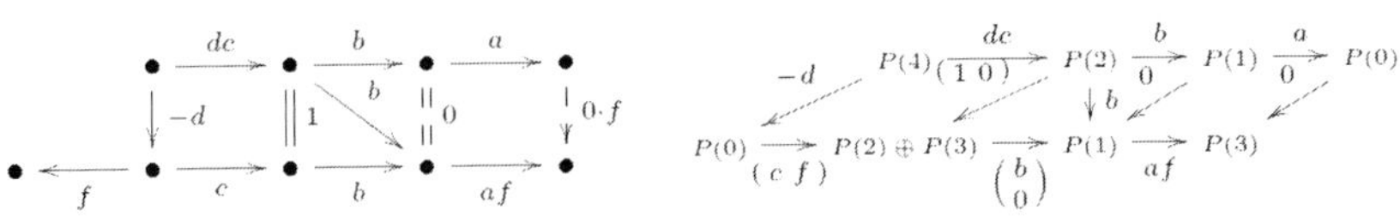

Observe that this is a 'quasi-graph' map satisfying **(N1)**. Indeed, this is essentially the graph map of Example 3.10 shifted by one degree.

4.4. Basic maps f such that $H(f)=\{f\}$

We start by observing that graph maps belong to singleton homotopy classes, and thus are never null-homotopic.

LEMMA 4.12

Suppose $f \in G_{v,w}$ *Then* $H(f) = \{f\}$.

PROOF. Suppose $f \simeq g{:}Q_v \to Q_w$. Since f is a graph map, there is an unfolded diagram:

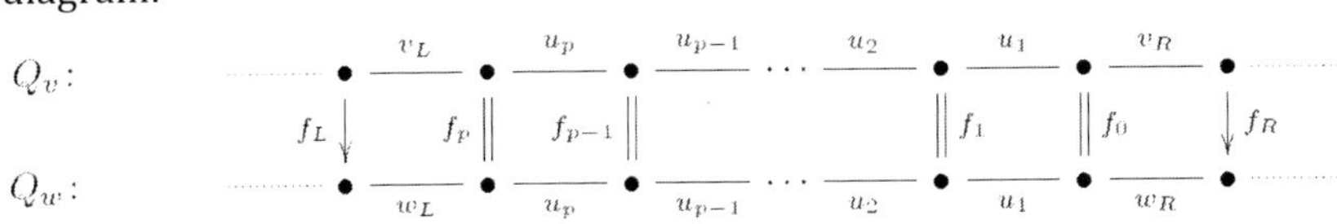

Without loss of generality, assume that $f_i = 1$ for $0 \leq i \leq p$. Consider the component f_i. Denote the corresponding component of the map g by g_i, which may be zero. Existence of a homotopy between f and g means that the difference between f_i and g_i is a linear combination,

$$f_i - g_i = 1 - g_i = \alpha u_i a + \beta u_{i+1} b + \gamma c u_i + \delta d u_{i+1},$$

where the $\alpha, \beta, \gamma, \delta$ are scalars and the a, b, c, d are paths in the quiver corresponding to the homotopy maps. If the composition does not make sense, we take the corresponding scalar to be zero. Since components of the differential are never zero, the compositions $u_i a$, u_i+1^b, cu_i and du_{i+1} are either zero or non-stationary paths in the quiver. It follows that $\alpha=\beta=\gamma=\delta=0$ and $g_i=1$. Lemma 4.3 gives $f=g$, whence $H(f)=\{f\}$. □

Next we consider singleton homotopy classes of single and double maps. It is clear that a single map $f \in S_{v,w}$ is in a singleton homotopy class exactly when its unfolded diagram corresponds to one of (*i*)–(*iv*) in Definition 3.7. Therefore, we need only examine when a double map occurs in a singleton homotopy class.

Recall the setup from Section 3.1(2) on page 13. We say that $f=(fL,fR)$ has *no common substring* if $u_v=f_L f'$ and $u_w=f'f_R$, and has *common substring* s if $f_L=u_v s$ and $f_R=su_w$. Note that these are the only ways in which the commutative square (*) in diagram (2) can decompose. Furthermore, s may be a stationary path, in which case f has *trivial common substring.*

The following lemma shows that any double map in a singleton homotopy class satisfies condition (D) of Definition 3.8. This then completes the proof of Theorem 3.15.

LEMMA 4.13. *Let $f=(f_L,f_R)$ be a double map.*

(1) *If $H(f)\neq\{f\}$, then f has a common substring.*

(2) *If f has a non-trivial common substring, then f is null-homotopic.*

PROOF. Suppose $H(f)$ is not a singleton set. Then one of the following must hold: $v_L v'_L=f_L$, $u_v u'_v=f_L$ or $w'_L w_L=$fL for some v'_L, u'_v, w'_L paths in the quiver. If $u_v u'_v=f_L$, then since $f_L u_w=u_v f_R$ it follows that u_v and v_L start with the same arrow, a contradiction. If $w'_L w_L=f_L$, then since $f_L u_w\neq 0$, we must have that $w_L u_w\neq 0$, a contradiction. Thus $u_v u'_v=f_L$ and u'_v is a non-zero component in the homotopy. But then $u'_v u_w=f_R$ and so f has a common substring u_v.

Let s be a non-trivial common substring for f, then we can take the required family of maps to be zero everywhere except for the component $P(s(u_v))\to P(t(u_w))$ which is taken to be s. □

5. HIGHER-DIMENSIONAL BAND COMPLEXES

Each pair $(w,\lambda)\in Ba\times k^*$ determines a homogeneous tube in K⊂D:

$$B_{w,\lambda,1} \rightleftarrows B_{w,\lambda,2} \rightleftarrows B_{w,\lambda,3} \rightleftarrows B_{w,\lambda,4} \rightleftarrows \cdots,$$

where we refer the reader to Section 5.1 for a precise definition of the higher dimensional band $B_{w,\lambda,r}$ for $r>1$. We shall show that the dimensions of the Hom spaces involving a higher dimensional band complex can be determined using the dimension of the Hom space of the corresponding one-dimensional band occurring at the mouth of the tube.

We start by making these definitions precise and describing unfolded diagrams for higher dimensional tubes. For simplicity, in this section we shall assume that **K** is an algebraically closed field.

5.1. Definition, example and unfolded diagrams

Let $(w,\lambda,r)\in Ba\times k^*\times N$ and recall that $B_{w,\lambda,1} := (B^i_{w,\lambda,1}, D^i)$, with $D^i = (d^i_{jk})$, denotes the one-dimensional band complex. The *r-dimensional band complex* is defined as follows:

$$B_{w,\lambda,r} := \left((B^i_{w,\lambda,1})^r, D^i_{w,\lambda,r}\right),$$

Where

$$D^i_{w,\lambda,r} = \begin{pmatrix} D^i & A^i & & 0 \\ 0 & D^i & & 0 \\ & & & \\ 0 & 0 & \ddots & A^i \\ 0 & 0 & & D^i \end{pmatrix}, \text{ and } A^i = (a^i_{kl}) \text{ is given by } a^i_{kl} = \begin{cases} w_1 & \text{if } d^i_{kl} = \lambda w_1; \\ 0 & \text{otherwise.} \end{cases}$$

The *unfolded diagram of $B_{w,\lambda,r}$* consists of r aligned copies of the unfolded diagram of $B_{w,\lambda,1}$ arranged from top to bottom of the page called *layers*, which are connected by *downwards* arrows corresponding to the non-zero entries of A^i, called *links*.

EXAMPLE 5.1. Let z be the homotopy band from the Running Example. The two-dimensional band complex $B_{z,\lambda,2}$ is

$$(P(0))^2 \xrightarrow{\begin{pmatrix} c & f & 0 & 0 \\ 0 & 0 & c & f \end{pmatrix}} (P(2)\oplus P(3))^2 \xrightarrow{\begin{pmatrix} b & 0 & 0 & 0 \\ 0 & c & 0 & 0 \\ 0 & 0 & b & 0 \\ 0 & 0 & 0 & c \end{pmatrix}} (P(1)\oplus P(4))^2 \xrightarrow{\begin{pmatrix} \lambda a & a \\ d & 0 \\ 0 & \lambda a \\ 0 & d \end{pmatrix}} (P(0))^2.$$

The corresponding unfolded diagram is:

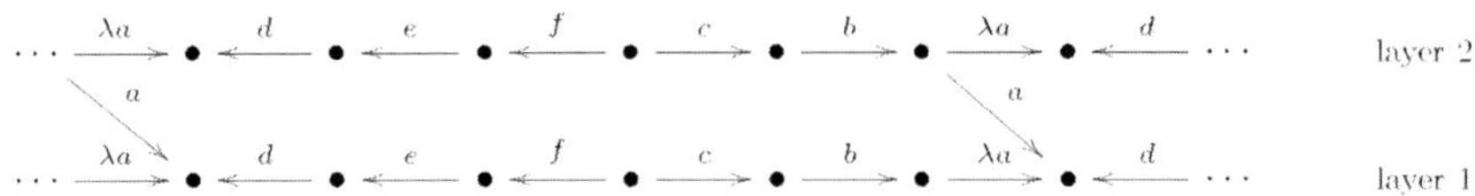

5.2. Passing through the link

DEFINITION 5.2. Let $1 \leq m \leq r$ and $1 \leq n \leq s$. We say a map $f \in \mathrm{Hom}_{\mathrm{D}}(B_{v,\lambda,r}, B_{w,\mu,s})$ is *lifted from a map* $f' \in \mathrm{Hom}_{\mathrm{D}}(B_{v,\lambda,1}, B_{w,\mu,1})$ *to the pair*(m,n) if the components of f' from layer m of $B_{v,\lambda,r}$ to layer n of $B_{w,\mu,s}$ are exactly the same as the components of f' and f is the minimal such map in terms of number of non-zero components.

For $f' \in \mathrm{Hom}_{\mathrm{D}}(B_{v,\lambda,1}, B_{w,\mu,1})$, we shall count the number of (homotopy classes of) maps in $\mathrm{Hom}_{\mathrm{D}}(B_{v,\lambda,r}, B_{w,\mu,s})$ which are lifted from f'. The idea is to put a copy of f' between the pair (m,n) of layers and see if a map arises; such a map will be called a *candidate map*. If a component of f' composes non-trivially with a link arrow, we say that the (candidate) map *passes through the link*. This will cause lifted maps to have non-zero components between more layers than just the pair (m,n).

LEMMA 5.3. *If* $f \in \mathrm{Hom}_{\mathrm{D}}(B_{v,\lambda,r}, B_{w,\mu,s})$ *then there is a map* $f' \in \mathrm{Hom}_{\mathrm{D}}(B_{v,\lambda,1}, B_{w,\mu,1})$ *such that* f *is lifted from* f' *to a pair* (m,n).

PROOF. As with one-dimensional maps, all maps in $\mathrm{Hom}_{\mathrm{D}}(B_{v,\lambda,r}, B_{w,\mu,s})$ are completely determined by any of their non-zero components. Ignoring the link arrows between layers, we simply have r copies of the band in $B_{v,\lambda,r}$ and s copies in $B_{w,\mu,s}$. It follows that if there is a map $f: B_{v,\lambda,r} \to B_{w,\mu,s}$ with a non-zero component from layer m' of $B_{v,\lambda,r}$ to layer n' of $B_{w,\mu,s}$, then there is a map $f' \in B_{v,\lambda,1} \to B_{w,\mu,1}$ with the same non-zero component and f is lifted from f'. □

The following lemma is straightforward.

LEMMA 5.4. *Suppose* $f \in \mathrm{HomD}(\mathrm{B}v,\lambda,r,\mathrm{B}w,\mu,s)$ *is lifted from* $f' \in \mathrm{HomD}(\mathrm{B}v,\lambda,1,\mathrm{B}w,\mu,1)$ *to the pair of layers*(m,n). *Then:*

(1) if f' *is a single map, then* f *does not pass through any link;*

(2) if $m=r$, *then* f *does not pass through a link in* $B_{v,\lambda,r}$*;*

(3) if $n=1$, *then* f *does not pass through a link in* $B_{w,\mu,s}$.

Recall the notation in Notation 3.6. To take care of homotopies for higher-dimensional homotopy bands, we need to modify Definition 4.6 slightly. Recall that the link in $B_{v,\lambda,r}$ is given by the homotopy letter v_1.

DEFINITION 5.5. Let $f: B_{v,\lambda,r} \to B_{w,\mu,s}$ be a map lifted from a map f' to the pair (m,n) as in Definition 5.2. We denote by $\widetilde{\mathcal{H}}^{(m,n)}(f)$ the set of **K**-linear combinations $\rho_1 g_1 + \rho_2(\widetilde{v_1}) + \rho_3(\widetilde{w_1})$ of maps, such that $\rho_1 g_1 + \rho_2(\widetilde{v_1}) + \rho_3(\widetilde{w_1}) \simeq f$ with $\rho_1 g_1 \in \mathcal{H}(f')$ and $(\widetilde{v_1})$, $(\widetilde{w_1})$ are the single maps (v_1), (w_1) lifted to pairs $(m+1,n)$ and $(m,n-1)$, respectively. When m and n are understood, we simply write $\widetilde{\mathcal{H}}(f)$ for $\widetilde{\mathcal{H}}^{(m,n)}(f)$.

Note that $\rho_2 \neq 0$ if and only if the homotopy map passes through the link in $B_{v,\lambda,r}$; similarly for ρ_3. Thus, the homotopy class of f can be determined by $\widetilde{\mathcal{H}}(f)$ and $\mathcal{H}(f')$.

5.3. A worked example

Before discussing the general behaviour of maps involving higher-dimensional homotopy bands, it is useful to examine an example in detail. This example will exhibit all possibilities regarding lifting of maps and homotopy classes and clarify the strategy in the proofs of the general results.

Throughout this worked example, Λ will be given by the following bound quiver.

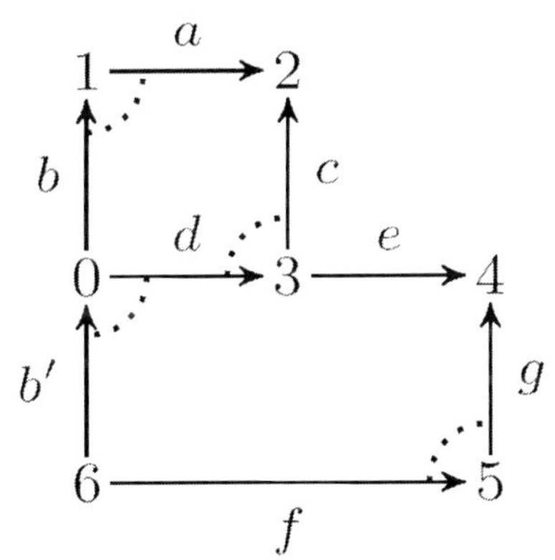

We consider the homotopy bands $v=(e,0,1)(c,1,0)\ (a,0,1)(bb',1,2)\ (f,2,1)(g,1,0)$ and $w=(d,i,i-1)(c,i-1,i-2)(a,i-2,i-1)(b,i-1,i)$, where the degree i will be specified by the diagrams occurring in each example in the context of the particular map or homotopy class we are interested in.

Our first example indicates the typical situation of lifting a singleton homotopy class.

EXAMPLE 5.6. The candidate map is a graph map $h' \in \mathrm{Hom}(B_{w,\lambda,1}, B_{v,\mu,1})$ lifted to the pair (1,1) of layers in $\mathrm{Hom}(B_{w,\lambda,2}, B_{v,\mu,1})$. The components of the lifted map h which are forced by passing through the link are drawn as broken lines (they are all identities).

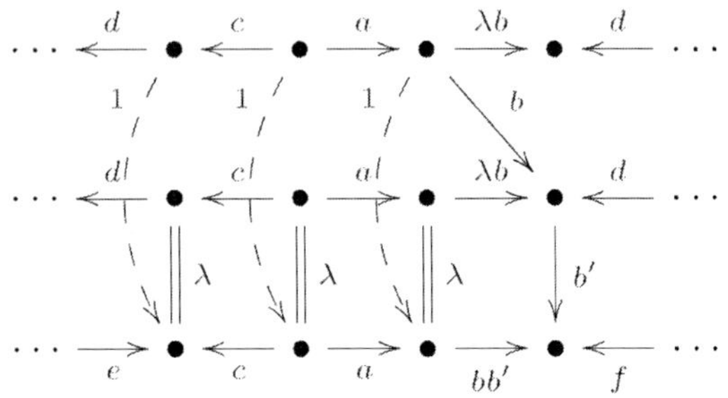

Note that h' also lifts to a map which includes a copy of h' from layer 2 of $Bw,\lambda,2$ to the unique layer of $B_{v,\mu,1}$. Thus, the graph map h' lifts to two maps in $\mathrm{Hom}_D(B_{w,\lambda,2},B_{v,\mu,1})$.

Our next example examines the case of lifting a non-singleton homotopy class. Recall the notation for homotopy equivalent basis maps in Definition 5.5.

EXAMPLE 5.7. We consider the homotopy set $\mathcal{H}((b)) = \{(b), -(\frac{1}{\mu}a), (\frac{1}{\mu}c), -(\frac{1}{\mu}e, \frac{1}{\mu}d)\}$ in $\mathrm{Hom}_C(B_{v,\lambda,1},B_{w,\mu,1})$, contributing one map in $\mathrm{Hom}_D(B_{v,\lambda,1},B_{w,\mu,1})$, which is indicated in the following diagram (drawn without signs or scalars).

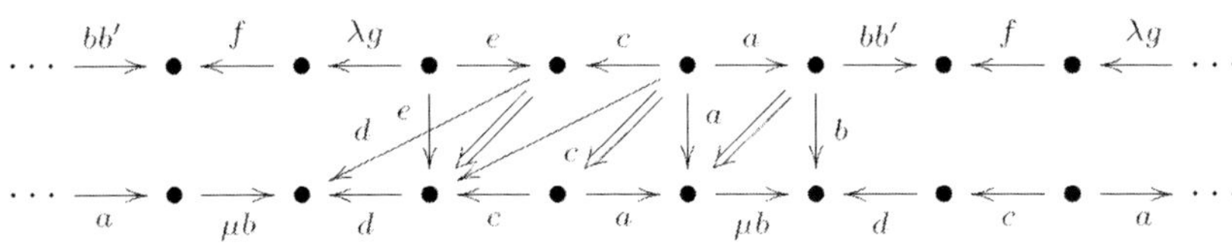

We shall now lift $H((b))$ to $\mathrm{Hom}_D(B_{v,\lambda,1},B_{w,\mu,2})$. From Lemma 5.4 (1) it is clear that each map in $H((b))$ gives rise to two maps in $\mathrm{Hom}_C(B_{v,\lambda,1},B_{w,\mu,2})$ without adding any extra components. For convenience, decorate maps to the first layer of $B_{w,\mu,2}$ with a tilde, i.e. $\tilde{b}$ and so on, and maps to the second layer with a hat, i.e. $\hat{b}$ and so on.

By Lemma 5.4, we get $\widetilde{\mathcal{H}}((\tilde{b})) = \{(\tilde{b}), (-\frac{1}{\mu}\tilde{a}), (\frac{1}{\mu}\tilde{c}), -\frac{1}{\mu}(\tilde{e}, \tilde{d})\}$. We use the following diagram to determine $\widetilde{\mathcal{H}}((\hat{b}))$.

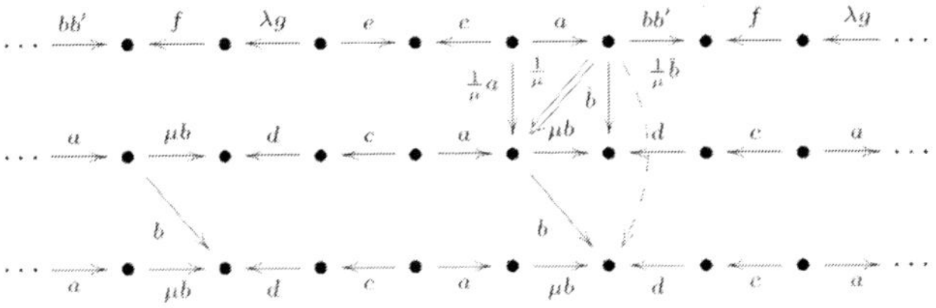

The homotopy passes through the link, so we have $(\check{b}) \simeq (-\frac{1}{\mu}\hat{a}) - (\frac{1}{\mu}\tilde{b})$. It is easy to see that $(\hat{a}) \simeq (-\hat{c}) \simeq (\hat{e}, \hat{d}) \simeq (-\mu b)$. Therefore, we determine $\widetilde{\mathcal{H}}((\hat{b}))$ and $\widetilde{\mathcal{H}}((\hat{a}))$ as follows:

$$\widetilde{\mathcal{H}}((\hat{b})) = \{(\hat{b}), -(\frac{1}{\mu}\hat{a}) - (\frac{1}{\mu}\tilde{b}), (\frac{1}{\mu}\hat{c}) - (\frac{1}{\mu}\tilde{b}), -\frac{1}{\mu}(\hat{e}, \hat{d}) - (\frac{1}{\mu}\tilde{b})\}$$

$$\widetilde{\mathcal{H}}((\hat{a})) = \{(\hat{a}), -(\hat{c}), (\hat{e}, \hat{d}), -(\mu\hat{b}) - (\tilde{b})\}.$$

Now, we have three different homotopy classes, but each of them is a **K**-linear combination of the other two. This shows that such a homotopy class lifts to two homotopy classes of maps in $\mathrm{Hom}_{\mathrm{C}}(B_{v,\lambda,1}, B_{w,\mu,2})$, thus giving rise to two maps in $\mathrm{Hom}_{\mathrm{D}}(B_{v,\lambda,1}, B_{w,\mu,2})$.

In the next example we look at what happens when one tries to lift an isomorphism. The following example shows that one cannot lift an identity morphism on a one-dimensional homotopy band to every pair of layers (m,n).

EXAMPLE 5.8. Consider the homotopy band w above and let $\lambda \in k^*$. Taking a copy of the identity from layer 1 of $B_{w,\lambda,2}$ to $B_{w,\lambda,1}$, we are forced to take the dashed components as before. However, once we reach the end of the band we must add dashed arrows to the right-hand side of the link as well:

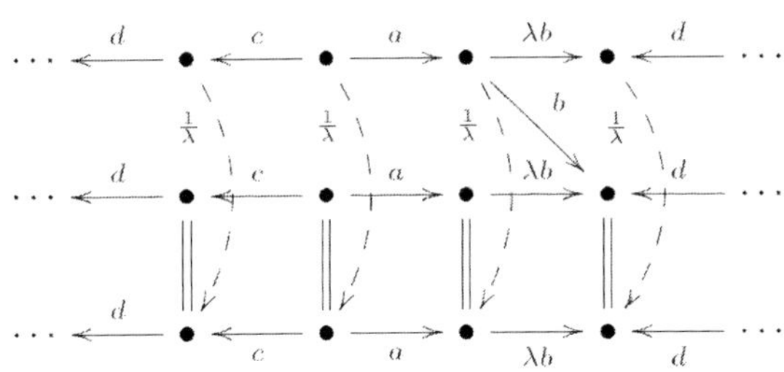

but then the square involving the link does not commute: we have $2b+\lambda b=b+\lambda b$. Therefore, we cannot lift a copy of the identity to layer 1 in $B_{w,\lambda,1}$ and get a well-defined map of homotopy band complexes.

In the final example of this section, we consider the homotopy class arising from the identity map on a one-dimensional band complex.

EXAMPLE 5.9. Consider the complex $B_{w,\lambda,1}$. The identity map on $B_{w,\lambda,1}$ gives rise to a homotopy class in $\mathrm{Hom}_D(B_{w,\lambda,1},\Sigma B_{w,\lambda,1})$ which is non-zero:

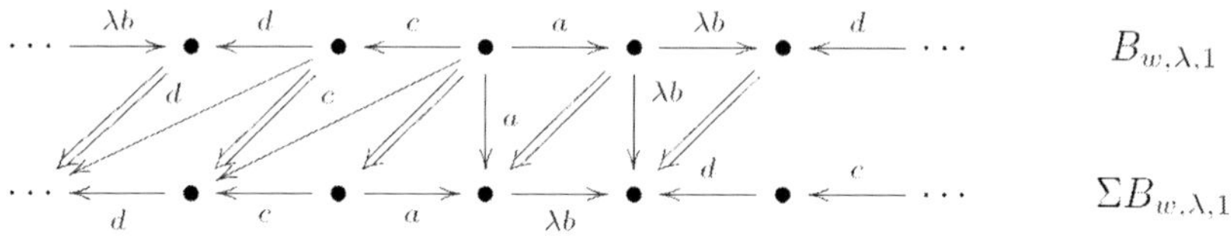

We have $H((b))=\{(b),-\lambda^{-1}(a),\lambda^{-1}(c),-\lambda^{-1}(d),(b)\}$, where we have written the (b) twice to emphasise the following key point: this gives a non-trivial way in which to obtain the tautologous homotopy equivalence $b-b\simeq 0$. Note that there can be other (homotopy classes of) maps in $\mathrm{Hom}(B_{w,\lambda,1},\Sigma B_{w,\lambda,1})$; these behave as in Examples 5.6 and 5.7.

5.4. Maps which are not self-extensions

In this section we shall make a number of statements regarding dimensions of Hom spaces. It is useful to first set up some notation. Let $\mathrm{RadD}(P^\bullet,Q^\bullet)$ denote the space of non-isomorphisms $P^\bullet\to Q^\bullet$. Following standard notation in algebraic geometry, we write

$\hom_D(P^\bullet,Q^\bullet)=\dim \mathrm{HomD}(P^\bullet,Q^\bullet)$ and $\mathrm{rad}_D(P^\bullet,Q^\bullet)=\dim \mathrm{Rad}_D(P^\bullet,Q^\bullet)$.

We now state the main result of this section.

THEOREM 5.10. *Let $r,s\in\mathbb{N}, w,v\in\mathrm{Ba}, u\in\mathrm{St}$ and $\lambda,\mu\in k^*$ be such that $B_{v,\lambda,1}\not\cong\Sigma B_{w,\mu,1}$. Let $\delta_{B_{v,\lambda,1},B_{w,\mu,1}}$ be the Kronecker delta. Then*

(1) $\hom_D(B_{v,\lambda,r},B_{w,\mu,s})=\min\{r,s\}\cdot\delta_{B_{v,\lambda,1},B_{w,\mu,1}}+rs\cdot\mathrm{rad}(B_{v,\lambda,1},B_{w,\mu,1})$;

(2) $\hom_D(B_{v,\lambda,r},P_u)=r\cdot\hom_D(B_{v,\lambda,1},P_u)$ and $\hom_D(P_u,B_{v,\lambda,r})=r\cdot\hom_D(P_u,B_{v,\lambda,1})$.

We now prove the first assertion of Theorem 5.10 in a sequence of lemmas; the second assertion is proved similarly. From now on assume that $v,w\in\mathrm{Ba}$, $\lambda,\mu\in k^*$ and $B_{v,\lambda,1}\not\cong\Sigma B_{w,\mu,1}$.

LEMMA 5.11. *Let $0\neq f'\in\mathrm{Hom}_D(B_{v,\lambda,1},B_{w,\mu,1})$. If $H(f')=\{f'\}$ and f' is not an isomorphism, then f' can be lifted to any pair of layers in $B_{v,\lambda,r}$ and $B_{w,\mu,s}$.*

PROOF. This follows directly from Lemma 5.4, noting that the resulting maps may acquire extra components if they pass through a link. □

The following lemmas deal with the generic case of Theorem 5.10; the example to bear in mind is Example 5.7.

LEMMA 5.12. *Let v and w be homotopy bands, and $\lambda,\mu\in k^*$ such that $B_{v,\lambda,1}\neq\Sigma B_{w,\mu,1}$ and $B_{v,\lambda,1}\neq B_{w,\mu,1}$. Let $f(m,n):B_{v,\lambda,1}\to B_{w,\mu,1}$ be a lift of $f\in S_{v,w}\cup D_{v,w}$, which is not null-homotopic, to the layers (m,n). Consider the lift $g^{(m,n)}$ of any* *map $g \in H(f)$. Then $g^{(m,n)}$ is homotopic to linear combination of representatives from only* $\widetilde{\mathcal{H}}^{(i,j)}(f)$ *for $i \geq m$ and $j \leq n$.*

PROOF. Choose the quasi-graph map in $Q_{v,w}$ of which f is a representative. Starting from the left of the quasi-graph map, we write down the homotopy set

$$\mathcal{H}(f_1) = \{f_1, \ldots, \lambda f_p, \ldots, \lambda\mu f_q, \ldots, \lambda\mu f_k\}$$

Without loss of generality, we may assume $f=f_1$; taking a different choice simply requires multiplication by the appropriate scalar. Note that our definition above places the link in the source band at f_p and in the target band at f_q with $p<q$. Other choices of p and q can be considered analogously. We consider the following lifts of $H(f1)$.

$$\widetilde{\mathcal{H}}^{(i,j)}(f_1) = \{\tilde{f}_1^{(i,j)}, \ldots, \tilde{f}_p^{(i,j)}, \ldots, \tilde{f}_q^{(i,j)}, \ldots, \tilde{f}_k^{(i,j)}\}$$

Where

$$\tilde{f}_t^{(i,j)} := \begin{cases} f_t^{(i,j)} & \text{for } 1 \leq t < p; \\ \lambda f_t^{(i,j)} + f_p^{(i+1,j)} & \text{for } p \leq t < q; \\ \lambda\mu f_t^{(i,j)} + f_p^{(i+1,j)} + f_q^{(i,j-1)} & \text{for } q \leq t \leq k, \end{cases}$$

where we interpret $f_p^{(i+1,j)}=0$ for $i=r$ and $f_q^{(i,j-1)} = 0$ for $j=1$.

The problem is to write the maps $f_t^{(i,j)}$ as linear combinations of maps from these sets. For $f_t^{(i,j)}$ with $1 \leq t < p$, there is nothing to show. For $f_t^{(i,j)}$ with $p \leq t < q$, we can write this map as a linear combination of maps from the sets $\widetilde{\mathcal{H}}^{(i,j)}(f_1), \ldots, \widetilde{\mathcal{H}}^{(r,j)}(f_1)$. For $q \leq t \leq k$, we get $f_t^{(i,j)}$ as a linear combination of representatives from the sets as above and $\widetilde{\mathcal{H}}^{(i,j-1)}(f_1), \ldots, \widetilde{\mathcal{H}}^{(i,1)}(f_1)$, as required. □

LEMMA 5.13. *Let v and w be homotopy bands, and $\lambda, \mu \in k^*$ such that $B_{v,\lambda,1} \neq \Sigma B_{w,\mu,1}$ and $B_{v,\lambda,1} \neq B_{w,\mu,1}$. Then* $\hom_D(B_{v,\lambda,r}, B_{w,\mu,s}) = rs \cdot \hom_D(B_{v,\lambda,1}, B_{w,\mu,1})$.

PROOF. By Lemmas 5.3 and 5.11, we need only check that we have no more than rs linearly independent non-singleton homotopy classes. Suppose $f \in S_{v,w} \cup D_{v,w}$ is such that $H(f) \neq \{f\}$. If the quasi-graph map determining the representatives of $H(f)$ does not pass through the link then it is clear that f contributes rs homotopy classes of maps to $\mathrm{Hom}_D(B_{v,\lambda,r}, B_{w,\mu,s})$. So we may assume that the quasi-graph map passes through the link in one of the two homotopy bands. Here we assume we are in the generic situation where the quasi-graph map passes through the link in both bands.

Consider the homotopy sets and notation as in the statement and proof of Lemma 5.12 above. Lemma 5.12 shows that any lift of a single or double map homotopic to *f* is a linear combination of representatives from $\widetilde{\mathcal{H}}^{(i,j)}(f_1)$ for $1 \leq i \leq r$ and $1 \leq j \leq s$. We just need to establish linear independence.

We claim that every lift of a map in *H*(*f*) is homotopic to a linear combination of the fq(i,j). First note that any representative in the $\widetilde{\mathcal{H}}^{(r,j)}(f_1)$ is homotopic to a linear combination of the $f_q^{(r,j)}$ for 1≤*j*≤*s* since $f_p^{(r+1,j)} = 0$ for each *j*. Now consider any representative in the $\widetilde{\mathcal{H}}^{(r-1,j)}(f_1)$. Each of these is homotopic to a linear combination of the maps $f_q^{(r-1,j)}$, $f_q^{(r,j)}$ and $f_p^{(r,j)}$ for $1 \leq j \leq s$. However, we have just shown that $f_p^{(r,j)}$ is homotopic to a linear combination of the maps $f_p^{(r,j)}$. Continuing in this way gives the claim.

Thus, to get linear independence, it is sufficient to show that if $\sum_{i,j} \alpha_{i,j} f_q^{(i,j)} \simeq 0$ then $\alpha_{i,j}=0$ f or 1≤*i*≤*r* and 1≤*j*≤*s*. This can be seen by carrying out the algorithm given in the proof of Proposition 4.8□ : the required zeros are obtained by the fact that the quasi-graph maps do not satisfy the required null-homotopic (= graph map) endpoint conditions.

Next we deal with special case $B_{v,\lambda,1}=B_{w,\mu,1}$.

LEMMA 5.14. *Let r and s be positive integers, v a homotopy band and λ∈k*. Then*

$$\mathrm{hom}_{\mathrm{D}}(B_{v,\lambda,r}, B_{v,\lambda,s}) = \min\{r,s\} + rs \cdot \mathrm{rad}(B_{v,\lambda,1}, B_{v,\lambda,1}).$$

PROOF. By the same arguments as in the proof of Lemma 5.13, any non-isomorphism from $B_{v,\lambda,1}$ to $B_{v,\lambda,1}$ can be lifted to any pair of layers (*m*,*n*). So we must have that $\mathrm{hom}_{\mathrm{D}}(B_{v,\lambda,r},B_{v,\lambda,s})$ is at least $rs \cdot \mathrm{rad}(B_{v,\lambda,1},B_{v,\lambda,1})$.

It remains to consider how many maps can be lifted from the identity map. Clearly the identity will pass through the link but, in contrast to non-isomorphism case, the non-zero components acquired in the layers above and below *n* and *m* will occur on both sides of the link (as in Example 5.8). Thus, if *f* is lifted from the identity to (*m*,*n*) for 1<*m*<*r* and 1<*n*<*s*, then *f* is also the identity lifted to (*m*+1,*n*+1) and to (*m*−1,*n*−1).

The identity map does not lift to (m,s) for $m \neq r$ nor to $(1,n)$ for $n \neq 1$ because it is not possible to satisfy the required commutativity relations, cf. Example 5.8. It follows from this that there are $\min\{r,s\}$ maps lifted from the identity, each projecting from the top-most layers of $B_{v,\lambda,r}$ onto the bottom-most layers of $B_{v,\lambda,s}$ □ .

5.5. Self-extensions of bands

As in Lemma 5.14, we must be careful when considering the homotopy class in $\mathrm{Hom}_C(B_{v,\lambda,1},\Sigma B_{v,\lambda,1})$ corresponding to the identity in $\mathrm{Hom}_C(B_{v,\lambda,1},B_{v,\lambda,1})$. Here, the example to keep in mind is Example 5.9.

LEMMA 5.15. *Let $1 \leq i \leq r$ and $1 \leq j \leq s$. Denote the lift of (v_1) to layers (i,j) by $(v_1)^{(i,j)}$. Then, $(v_1)^{(i,1)} \simeq 0$ for $i>1$ and $(v_1)^{(i,j-1)} \simeq -(v_1)^{(i+1,j)}$ whenever $j>1$, where we take $(v_1)^{(r+1,j)}=0$.*

PROOF. The single map $(v1){:}B_{v,\lambda,1} \to \Sigma B_{v,\lambda,1}$ is a representative of the quasi-graph map corresponding to the identity $B_{v,\lambda,1} \to B_{v,\lambda,1}$. Moreover, this quasi-graph map determines a non-trivial homotopy from (v_1) to (v_1). Lifting this non-trivial homotopy to the pair of layers (i,j) gives $(v_1)^{(i,j)} \simeq (v_1)^{(i,j)} - \lambda - 1(v_1)^{(i,j-1)} - \lambda - 1(v_1)^{(i+1,j)}$, which outputs the homotopy equivalences as claimed. □

PROPOSITION 5.16. *Let $r,s \in \mathbb{N}$, $v \in$ Ba and $\lambda \in k^*$. Then*

$\mathrm{hom}_D(B_{v,\lambda,r},\Sigma B_{v,\lambda,s}) = \min\{r,s\} + rs\cdot(\mathrm{hom}_D(B_{v,\lambda,1},\Sigma B_{v,\lambda,1})-1)$

Proof. Observe that every basis element of $\mathrm{HomD}(B_{v,\lambda,1},\Sigma B_{v,\lambda,1})$ passes through the link at most once except for the quasi-graph map corresponding to the shifted identity. As before, each of these basis elements can be lifted to (m,n) for any m, n and so $\mathrm{hom}_D(B_{v,\lambda,r},\Sigma B_{v,\lambda,s})$ is at least $rs\cdot(\mathrm{hom}_D(B_{v,\lambda,1},\Sigma B_{v,\lambda,1})-1)$. It remains to determine how many linearly independent homotopy classes are lifted from this remaining homotopy class.

Repeated application of Lemma 5.15 yields the following homotopy equivalences:

$$(v_1)^{(1,1)} \simeq -(v_1)^{(2,2)} \simeq (v_1)^{(3,3)} \simeq \cdots \simeq \pm(v_1)^{(s,s)};$$
$$(v_1)^{(1,2)} \simeq -(v_1)^{(2,3)} \simeq (v_1)^{(3,4)} \simeq \cdots \simeq \pm(v_1)^{(s-1,s)};$$
$$\vdots$$
$$(v_1)^{(1,s)}.$$

Therefore, if $r \geq s$, then none of these homotopy equivalences includes a zero map, and therefore none of the homotopy classes above are null. In particular, there are s homotopy classes. If $r<s$, the first $s-r$ homotopy classes are actually null-homotopic, which leaves $s-(s-r)=r$ non-null homotopy classes.

To see that all other lifts of (v_1) are null-homotopic, we again apply Lemma 5.15. In this case, it yields that $0 \simeq (v_1)^{(i,1)} \simeq -(v_1)^{(i+1,2)} \simeq (v_1)^{(i+2,3)} \simeq \cdots \simeq \pm(v_1)^{(r,s+2-i)}$ for $2 \leq i \leq r$ when $r \geq s$, and $0 \simeq (v_1)^{(i,1)} \simeq -(v_1)^{(i+1,2)} \simeq (v_1)^{(i+2,3)} \simeq \cdots \simeq \pm(v_1)^{(r,r+1-i)}$ for $2 \leq i \leq r$ when $r<s$.

All that remains to show is that lifts of all other single maps occurring as representatives of the quasi-graph map corresponding to $\mathrm{id}_{B_{v,\lambda,1}}$ are linear combinations of representatives of the homotopy classes described above. This follows directly from Lemma 5.12□ .

Many tubes occurring in representation theory are hereditary. It is natural to ask whether the same is true for the homogeneous tubes that arise from homotopy bands. This can be seen not to be the case using the following straightforward example.

EXAMPLE 5.17

Consider the algebra and homotopy band from the Running Example:

$z=(d,3,2)(e,2,1)(f,1,0)(c,0,1)(b,1,2)(a,2,3)$.

The unfolded diagram below shows a graph map of degree 3, thus $\mathrm{Ext}^3(B_{z,\lambda,1},B_{z,\lambda,1}) \neq 0$.

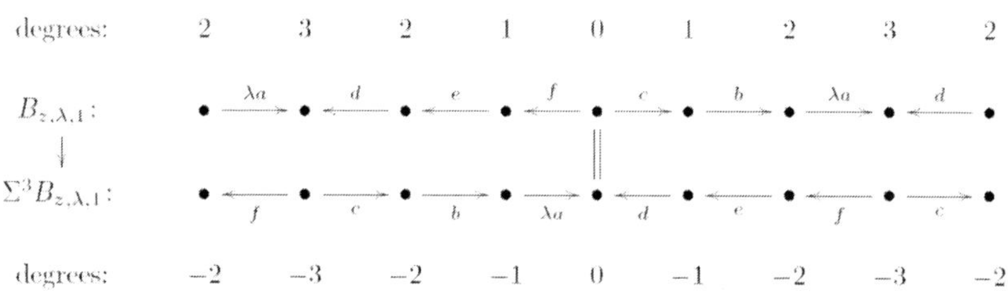

6. APPLICATION: IRREDUCIBLE MORPHISMS BETWEEN STRING COMPLEXES

In this section, we recover Bobiński's description [9] of the irreducible maps in K and extend it to D. In [9], Bobiński employed the Happel functor $F\colon \mathrm{D} \hookrightarrow \underline{\mathrm{mod}}(\hat{\Lambda})$ to determine the AR structure of K↪D. We present an algorithm intrinsic to $\mathrm{K}^{-,b}(\mathrm{proj}(\Lambda))$, which allows one to compute all irreducible maps in D. (Note that only the full subcategory of perfect complexes K↪D admits AR triangles.) By [16], it is known that for any $P \in \text{ind}(K)$, there are at most two irreducible maps starting and ending at P.

Recall from [2,9] that there exist *string functions* $S,T:St \to \{-1,1\}$ satisfying certain properties (see [2,9] for precise details). We remark only on the consequences.

- We must consider formal inverses of trivial homotopy strings, say $(1_x,i,i)^{-1}=(\overline{1_x},i,i)$, as distinct homotopy strings; of course they give rise to the same complex.
- For each homotopy string w, there are unique trivial homotopy strings 1_x and 1_y such that the compositions $w1_x$ and $1_y w$ are defined.

We set up some notation and terminology.

NOTATION 6.1. Let $f=(f_r,\ldots,f_0) \in \text{Hom}_D(P_w,P_v)$ be as illustrated below.

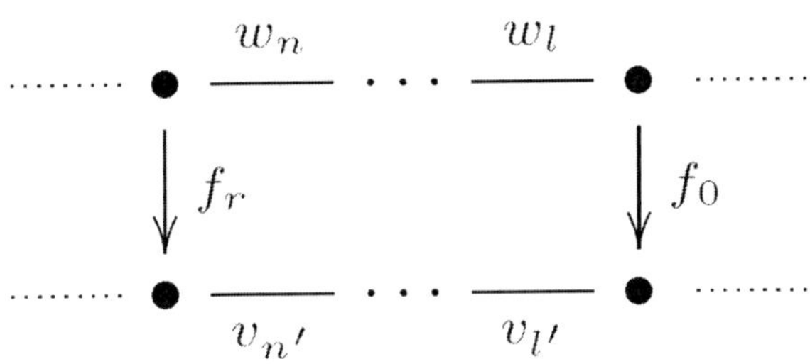

The *support of* f is the homotopy substring $\text{supp}(f) := \prod_{k=n}^{l}(w_k,i_k,j_k)$ of w. For $l \leq i \leq n$ we will also say that f is supported at $P(\varphi w(i))$. If f has only one non-zero component, say $f_0:P(\varphi w(k)) \to P(\varphi v(l))$, we say that f is supported at $P(\varphi w(k))$.

Recall the definition of antipath from Definition 2.5. A direct (resp. inverse) antipath θ is called *maximal* if there is no $a \in \Gamma 1$ such that $\theta(a,i,i+1)$ is a direct antipath (resp. $(a,i,i-1)\theta$ is an inverse antipath).

REMARK 6.2. (Maximal) antipaths do not have to be finite; however, we cannot compose a homotopy string with an infinite antipath if this takes us outside D. In particular, if $\theta=\theta_1\theta_2\cdots$ is an infinite direct antipath and w is a finite homotopy string such that $w\theta 1$ is defined, then the composition $w\theta$ is not in D. Similarly, if $\theta=\cdots\theta_2\theta_1$ is an infinite inverse antipath and $\theta_1 w$ is defined, then the composition θw is not in D. Note that infinite antipaths correspond to oriented cycles in the quiver with 'full relations'.

6.1. Algorithm for determining irreducible maps

The algorithm is stated here for K. We extend it to infinite strings in Section 6.2. Let $w = \prod_{k=n}^{1}(w_k, i_k, j_k) \in \mathrm{St}$. The strategy of the algorithm is as follows: consider the identity map on P_w:

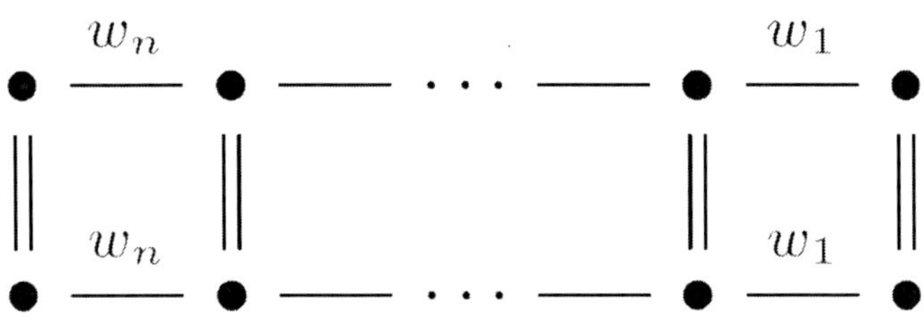

We alter the map 'from the left' in a minimal way to get a new map. This gives a new string w^+ and a new map $f^+ : P_w \to P_{w^+}$, which may each be zero. However, when they are nonzero, the resulting map will turn out to be irreducible. This deals with one of the two possible irreducible maps. To get the other, we alter the map in a minimal way 'from the right'. This gives a new string w_+ and a new map $f_+ : P_w \to P_{w+}$, which again may each be zero. At least one of f^+ or f_+ will be nonzero. We explicitly give the algorithm which alters the map in a minimal way 'from the left'. The algorithm doing this 'from the right' is dual.

ALGORITHM 6.3. The algorithm proceeds as follows, where we carry out each step in sequence unless instructed otherwise. The map f' is defined in each step in Fig. 1.

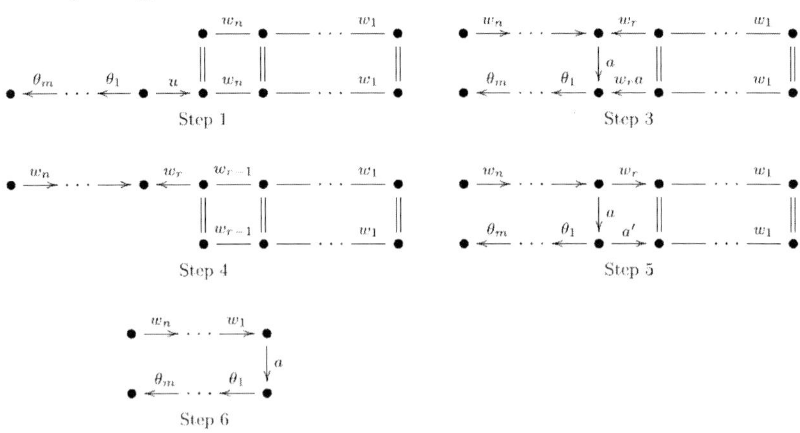

Figure 1. Diagrams defining the homotopy strings w' and maps f produced in each step of Algorithm 6.3, with the maximal inverse antipath θ of Step 8 also added where appropriate.

Step 1: If there is a direct homotopy letter u such that uw is a homotopy string and u is a maximal path, we set $w'=uw$, and go to Step 8.

Step 2: Remove the longest direct antipath which is a left substring of w. Write $\psi w:=wn\cdots wr+1$ for this antipath.

Step 3: If (w_r,i_r,j_r) is inverse and there exists $a\in\Gamma_1$ such that $w_ra\neq 0$, then set $w' = (w_r a, i_r, j_r)\prod_{k=r-1}^{1}(w_k, i_k, j_k)$, and go to Step 8.

Step 4: If (w_r,i_r,j_r) is inverse but there is no $a\in\Gamma 1$ such that $w_ra\neq 0$, then set $w' = \prod_{k=r-1}^{1}(w_k, i_k, j_k)$, and go to Step 9.

Step 5: If (wr,ir,jr) is direct, then decompose $w_r=aa'$ with $a\in\Gamma_1$ and set $(a', i_r, j_r)\prod_{k=r-1}^{1}(w_k, i_k, j_k)$, and go to Step 8.

Step 6: If w is a direct antipath and there exists $a\in\Gamma_1$ with $t(a)=\varphi w(0)$ with $w_1a=0$, then set w' to be the trivial homotopy string such that ww' is defined, and go to Step 8.

Step 7: Set $f'=f+$ to be the zero map $P_w\rightarrow 0$ and terminate the algorithm.

Step 8: If there is a maximal inverse antipath $\theta = \prod_{k=m}^{1}(\theta_k, i'_k, j'_k)$ such that $\theta w'$ is defined as composition of homotopy strings, we set $w^+=\theta w'$ and let f^+ have the same components as f' and terminate the algorithm.

Step 9: Set $f^+=f'$ and $w^+=w'$.

NOTATION 6.4. We shall refer to the dual algorithm that alters a homotopy string in a minimal way 'from the right' as Algorithm 6.3′; a prime will also be affixed to denote the dual of each of the corresponding steps, i.e. the dual of Step i is Step i'.

Note that Step 7 only occurs for homotopy strings that are either trivial or antipaths.

REMARK 6.5. We highlight the following:

(1) The maps f^+ and f_+, when non-zero, are either graph maps or single maps.

(2) If both f^+ and f_+ are non-zero, then the two maps are different at the level of chain maps. If w or w' are trivial, then this is taken care of by the functions S and T.

(3) Whenever there is an $a\in\Gamma_1$ with $s(a)=x$, then Step 1 or its 'right' dual will occur when considering the trivial homotopy string $(1_x,i,i)$. If there are two arrows a and b with $s(a)=x=s(b)$ then both Step 1 and its dual will occur.

(4) A case analysis shows that if a maximal antipath θ is added in Step 8, then this antipath is finite.

THEOREM 6.6. *Let w be a homotopy string, and let f^+ and f_+ be the outputs produced byAlgorithm 6.3and its dual, respectively. If f^+(respectively f_+) is non-zero, then it is irreducible.*

In the remainder of this section, we verify Theorem 6.6 in the case that $f^+ \neq 0$. The case $f_+ \neq 0$ is dual.

6.2. Irreducible maps involving infinite homotopy strings

Let $w \in St_1 \cup St_2$. We extend the above algorithm by possibly extending graph maps to infinite graph maps, and by adding the following step between Step 8 and Step 9:

Step 8.5: If there is a maximal, infinite inverse antipath $\theta = \cdots(\theta_1, i'_1, j'_1)$ such that $(\theta_1, i'_1, j'_1)\ w'$ is defined as composition of homotopy strings, we set f^+ to be the zero map $P_w \to 0$ and terminate the algorithm.

As commented above, the condition of this step is impossible if w is a finite homotopy string. Moreover, this step ensures that we never go outside D.

LEMMA 6.7. *If $w^+ \neq 0$ for some homotopy string w, then w^+ is in*St*if and only if w is in* St.

PROOF. The 'if' direction is already established. Consider $v \in St_1$ such that $w := {}^{\infty}v$ is left infinite. When we run the algorithm, we perform Step 2 and remove the left infinite part of w, which corresponds to a cyclic path ρ. We will then be able to add an arrow a in one of Step 3, Step 5 or Step 6 (take a to be the arrow preceding w_{r-1} in ρ). Next Step 8.5 will occur, since we can always add infinitely many copies of ρ, but this time as inverse homotopy letters.

If $v \in St_1$ and $w := v^{\infty}$, we will never remove the infinite part from w as it consists of inverse homotopy letters. Hence, if the output w^+ is non-zero, it is also in St_1. □

COROLLARY 6.8. *If w is a two-sided infinite homotopy string, then there are no irreducible maps in* D *with* P_w *as source.*

In the remainder of the section, we consider only irreducible maps in K, but it is clear that the irreducible maps involving one-sided homotopy strings will behave in the same way.

6.3. Non-irreducible maps

Before verifying Theorem 6.6, we eliminate some maps which can be easily seen to be non-irreducible.

PROPOSITION 6.9. *Let* $v,w \in \mathrm{St}$ *and* $g \in \mathrm{Hom}_K(P_v,P_w)$. *Then* g *is **not** irreducible if:*

(1) *g is a double map;*

(2) *$g \in G_{v,w}$ is such that both endpoints are non-isomorphisms;*

(3) *$g=(g_s,\ldots,g_0)$ is such that $g_s=ab$ with a, b non-stationary paths in Γ;*

(4) *$g=(g_s,\ldots,g_0)$ is such that $g_0=ab$ with a, b non-stationary paths in Γ.*

PROOF. (1) Write $g=(c,d)$ where c and d are paths in the quiver. Then g is, up to inverting v and w, given by the diagram to the left, and the factorisation of g is given by the diagram to the right.

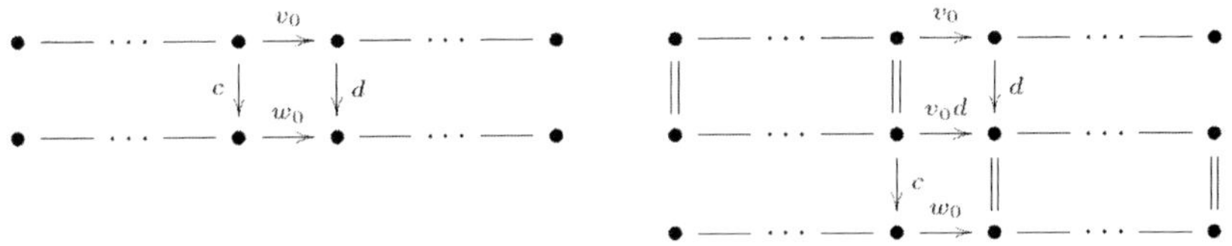

Cases (2), (3) and (4) are similar and can be verified by drawing the corresponding diagrams, noting in (3) and (4) one needs to check only single and graph maps. □

As a consequence, no single map in the homotopy class of a double map is irreducible. We now examine single maps more closely, giving a necessary condition for irreducibility, which is then used to eliminate further possibilities for irreducible single maps.

PROPOSITION 6.10. *If $g:P_v \to P_w$ is an irreducible single map, then g is of the form*

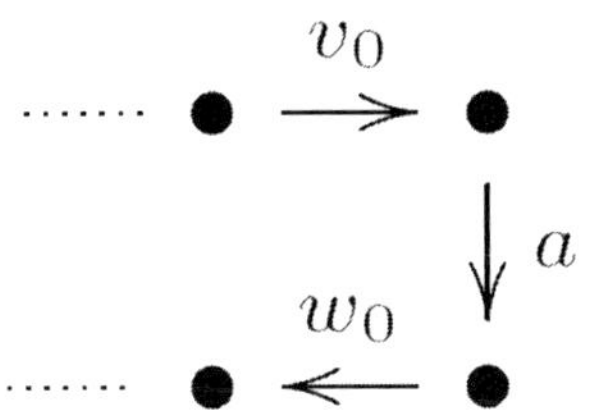

(3)

up to inverting w or v, where $a \in \Gamma_1$, and v_0 and/or w_0 may be zero. Moreover, g is in a singleton homotopy class; see Definition 3.7.

PROOF. Suppose $g \in S_{v,w}$ with non-zero support, by abuse of notation, denoted by $g:P \to P'$ with $P,P' \in \mathrm{ind}(\mathrm{proj}(\Lambda))$. This situation is indicated below.

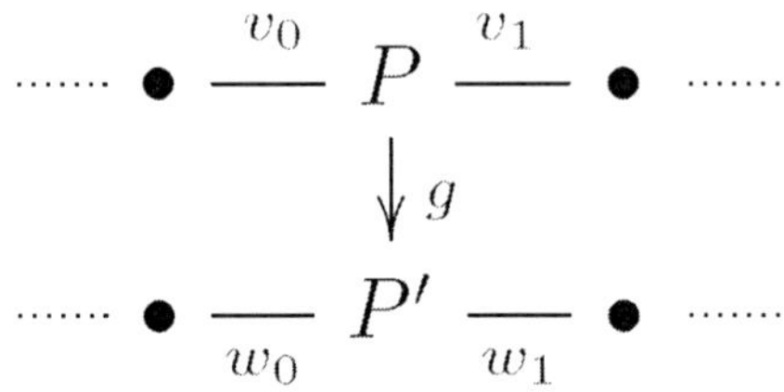

One can easily check that in the following cases, g is not irreducible:

- v_0 and v_1 are both non-zero, or precisely one is non-zero and it has P as source.
- w_0 and w_1 are both non-zero, or precisely one is non-zero and it has P' as target.

Indeed, if both conditions hold g factors as two graph maps and one single map (with g as its non-zero support), and if only one holds then g factors as one graph map and one single map. □

Corollary 6.11. *Suppose $g \in S_{v,w}$ has unfolded diagram taking the form(3)inProposition 6.10. Then g is* **NOT** *irreducible if:*

(1) *v (or w) is not a uniformly oriented homotopy string;*

(2) *any homotopy letter of v and w is a path of length longer than* 1.

PROOF. The following diagrams indicate the factorisations.

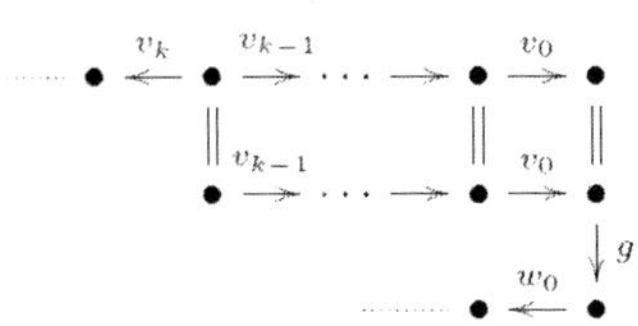

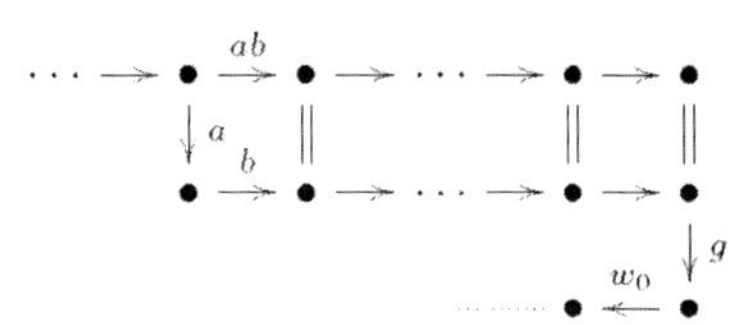

□

6.4. *Proof of* Theorem 6.6

We start by setting up the notation for the section. Throughout $w \in \mathrm{St}$, w' and map $f' : P_w \to P_{w'}$ will be the homotopy string and map output at Steps 1–7, and w^+ and $f^+ : P_w \to P_{w+}$ the homotopy string and map output at Steps 7–9. Note that f' differs from f^+ if and only if $w^+ = \theta w'$, where $\theta = (\theta_m, j_m, i_m) \cdots (\theta_1, j_1, i_1)$ is an inverse antipath. Write $f^+ = (f_k, \ldots, f_0)$ where, using the notation from Algorithm 6.3, $k \in \{0, r-1, r, n\}$.

LEMMA 6.12

We have the following:

(1) *f^+ is a well-defined map.*

(2) *f' factors through f^+.*

(3) *For each $v\in$St such that vw' is defined and such that the components of f' also determine a map $g:P_v\rightarrow P_{vw'}$, the map g factors through f^+.*

PROOF. This is easily verified. Observe that if f_k is the leftmost component of f^+, and if an antipath was added in Step 8 such that $f_k\theta_1$ occurs in the diagram, then $f_k\theta_1=0$. □

The following lemma shows that if there is a graph map with source P_w satisfying certain criteria, then at least one of the maps f^+ and f_+ is a graph map.

LEMMA 6.13. *Let w be a non-trivial homotopy string. If there exists $g\in G_{w,v}$ starting after the left endpoint of w and stopping with an isomorphism at the right endpoint of w, then f^+ is a graph map and g factors through f^+.*

PROOF. It follows from the hypotheses of the lemma that w is not an antipath, so neither Steps 6 nor 7 occur. A straightforward case analysis shows that g factors through f^+ even if v is not a homotopy substring of w^+. □

LEMMA 6.14. *Suppose* $f^+\in G_{w,w+}$ *and $g:P_w\rightarrow P_u$ is a map such that* supp(g) *is a homotopy substring of* supp(f^+). *If g is not supported on the source of f_k, then g factors through f^+.*

PROOF. If $g\in G_{w,u}$, then since $f_{k-1},\ldots,f_0$ are isomorphisms, supp(g) is also a homotopy substring of w^+. It is straightforward to check that the restriction $g':P_{w+}\rightarrow P_u$ of g is also a graph map, and $g=g'f^+$. Similarly for $g\in S_{w,u}$. □

There is a dual statement of Lemma 6.14 for f_+.

LEMMA 6.15. *Let $w\in$St be such that $f^+\in G_{w,w+}$. If $g:P_w\rightarrow P_u$ is such that* supp(f^+) *and* supp(g) *have no overlapping part, then $f_+\neq 0$ and g factors through f_+.*

PROOF. Write $g=(g_s,\ldots,g_0)$. Recall the notation from Algorithm 6.3. Only the outputs of Steps 1, 3, 4 and 5 of Algorithm 6.3 are graph maps. We consider these possible termination points in turn.

- At Step 1: Algorithm 6.3 cannot output f^+ at Step 1 under these conditions, since supp(f^+)$=w$.
- At Steps 3 or 5: in these cases supp(g) is a substring of ψ_w. Thus w contains direct letters and the duals of Steps 6 and 7 cannot occur. In particular, the dual of Algorithm 6.3 outputs a graph map f_+. Now apply Lemma 6.14.

- At Step 4: the rightmost possible support of g_0 is $P(\varphi_w(r))$. If this is not attained, then supp(g) is a substring of ψ_w and one argues as in Step 3 and 5 above.

Suppose the support of g_0 is $P(\varphi_w(r))$. If $g \in S_{w,u}$ then $w_n = w_r$ and g factors through f_+: Steps 1′, 3′ and 4′ yield a common substring of supp(f_+) and supp(f^+), whence we apply Lemma 6.14; and, if Step 6′ occurs then we must have $g_0 = aa'$ where a is the non-zero component of f_+, and this gives the desired conclusion. In particular Step 7′ never occurs and $f_+ \neq 0$.

If $g \in G_{w,u}$ we have isomorphisms in all components to the left of g_0, and the conclusion follows by Lemma 6.13.

Thus, $f_+ \neq 0$ and g factors through f_+ , as desired. □

We are now ready to prove Theorem 6.6.

PROOF OF THEOREM 6.6. Let $w \in$St and $f^+ = (f_k, \ldots, f_0)$ and w_+ be the outputs of Algorithm 6.3. The proof is a case analysis. We start with some generalities. Let $g: P_w \to P_v$ be another candidate for an irreducible map and note the following.

- Lemmas 6.14 and 6.15 say that any map g not supported on the source of fk factors through f^+ or f_+, and so is not irreducible.
- Proposition 6.9(1) says that g must be a graph map or a single map.
- Proposition 6.9(3)(4) and Corollary 6.11(2) say that all components of g which are not isomorphisms are arrows.

We therefore write $g = (g_s, \ldots, g_0, \ldots, g_{-t})$, where g_0 is the component supported on the source of f_k; if g is a single map then $g = g_0$. If $g \in G_{w,v}$, we fix the orientation of v such that the common substring of v and w has the same orientation.

We treat the three possible cases for each step of Algorithm 6.3 which outputs f_+.

Case 1: The map g is a graph map and g_0 is an arrow.

Case 2: The map g is a graph map and g_0 is an isomorphism.

Case 3: The map g is a single map, in which case g_0 must be an arrow.

It may be useful for the reader to keep Fig. 1 on page 36 in mind.

THE MAP f^+ IS OUTPUT AT STEP 1. In this case, g0 is the leftmost component of g.

Case 1: Clearly, g factors through f^+ since $ug_0 = 0$.

Case 2: If $t<n$ then we simply apply Lemma 6.13 to get g factoring through f_+. If $t=n$ and g_{-n} is an isomorphism, then g factors through f^+ by Lemma 6.12. If $t=n$ and g_{-n} is an arrow, then it is clear that g factors through f_+.

Case 3: For $g=g_0$ to be irreducible we need w_n to be inverse or trivial by Proposition 6.10. If w_n is inverse, g cannot factor through f^+ since $ug\neq 0$. If f_+ is a graph map then g factors through f_+ by Lemma 6.14. If f_+ is a single map, then it is the output of Step 6′, in which case g_0 is the non-zero component of f_+ and the factorisation follows. If w_n is trivial then w is trivial and P_w is a stalk complex. It can then be easily checked that g factors through f^+ or f_+.

THE MAP f^+ IS OUTPUT AT STEP 3. Here w is non-trivial and g_0 need not be the leftmost non-zero component of g.

Case 1: Here g_0 is either the leftmost or rightmost non-zero component of g. In the former case $w_r g_0\neq 0$ and $g_0=a$ (in the notation of Algorithm 6.3). By Proposition 6.9(2), we may assume that g_{-t} is an isomorphism. Now g factors through f^+ by Lemma 6.12 if $t<r$. If $t=r$, f_+ is a graph map (argue as in Lemma 6.13), and the conclusion follows by Lemma 6.14.

Case 2: In this case g does not factor through f^+. The components $g_{-1},\dots,g_{-t+1}$ are isomorphisms; g−t may be an arrow. The $g_s,\dots,g_1$ are all isomorphisms if ψ_w is non-trivial, and zero otherwise. If f_+ is a graph map then g factors through f_+ by Lemma 6.13. If f_+ is not a graph map, then $\prod_{k=r}^{1}(w_k, i_k, j_k)$ is an inverse antipath, which implies g_{-t} is an isomorphism. It follows that $w=u$ since adding anything to the endpoints forces either f^+ to be output in Step 1, or a graph map f_+ in Step 1′, whence g is an isomorphism.

Case 3: Apply the same argument as when g_0 was the rightmost non-zero component of g in Case 1.

THE MAP f^+ IS OUTPUT AT STEP 4. Here Cases 1 and 3 are straightforward. For Case 1 apply Lemma 6.14. For Case 3 it is immediate that g factors through f_+.

Case 2: If g0 is the leftmost non-zero component then by previously given arguments, g factors through f^+. So assume not: we must have that g_1 is an isomorphism, for otherwise f^+ would have been output at Step 3. Applying the argument as in Case 2 above, $g_s,\dots,g_1$ are each isomorphisms. If $t<r-1$ then g factors through f_+ by Lemma 6.14. If $t=r-1$, g is either an isomorphism or else Algorithm 6.3′ outputs f_+ at Step 1′, whence g clearly factors through f_+.

THE MAP f^+ IS OUTPUT AT STEP 5. Here f_+ cannot be output at Step 6′ of Algorithm 6.3′ since w_r is direct, so f_+ is a graph map. For Case 2, argue as in Case 2 above. Case 3 is again a straightforward verification.

Case 1: If g_0 is the leftmost non-zero component, then $g_0=a$ (in the notation of Algorithm 6.3) factors through f^+. If g_0 is the rightmost non-zero component then use Lemma 6.13.

THE MAP f^+ IS OUTPUT AT STEP 6. Note that the direct antipath w may be trivial.

Case 1 and 3: If $g_0 \neq a$ (in the notation of Algorithm 6.3) then in the graph map case g factors through f^+ by Lemma 6.13. In the single map case, w is a trivial homotopy string and g factors through f_+, which is output either at Step 1′ or Step 6′ of Algorithm 6.3′. Similarly when $g_0=a$.

Case 2: Here, the isomorphism g_0 is dragged down the entire homotopy string so that v has the form $v=\cdots cwd\cdots$ for some homotopy letters c and d. The homotopy letter c cannot be direct (f^+ would be output at Step 1) nor inverse (g would not be a map). Similarly, d is not direct. Hence $v=wd\cdots$. If d is trivial then g is an isomorphism. If d is inverse then f_+ is output at Step 1′ and g clearly factors through f_+.

THE MAP $f^+=0$ IS OUTPUT AT STEP 7. One now applies the dual arguments for f_+ to get the required factorisation through f_+. □

7. APPLICATION: DISCRETE DERIVED CATEGORIES

For the definition and background on discrete derived categories we refer the reader to [11,13,22]. Derived-discrete algebras are derived equivalent to either path algebras of simply-laced Dynkin quivers or the bound path algebra $\Lambda(r,n,m)$ defined in Fig. 2; when we refer to discrete derived categories, we shall always mean $D^b(\Lambda(r,n,m))$.

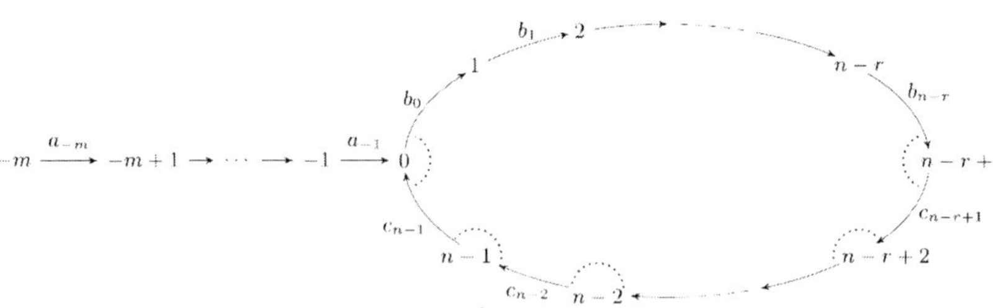

Figure 2. The bound quiver defining $\Lambda(r,n,m)$.

In this section, we recover the universal Hom-space dimension bound described in [13] when $\Lambda(r,n,m)$ is of finite global dimension and extend it to the case $\Lambda(r=n,n,m)$, which has infinite global dimension. Recall from [8] that $\Lambda(r,n,m)$ has no homotopy bands. From now on $\Lambda:=\Lambda(r,n,m)$ and $\mathrm{St}(\Lambda)$ will denote homotopy strings over Λ.

A *subword* of a homotopy string w is defined in the obvious fashion: the left- and rightmost homotopy letters of the subword may be (incomplete) substrings of the corresponding letters of w. We now describe all the homotopy strings for a discrete derived category. Note that, when $r=n$, by our labelling convention there are no 'b' arrows.

LEMMA 7.1. *Consider the following homotopy strings:*

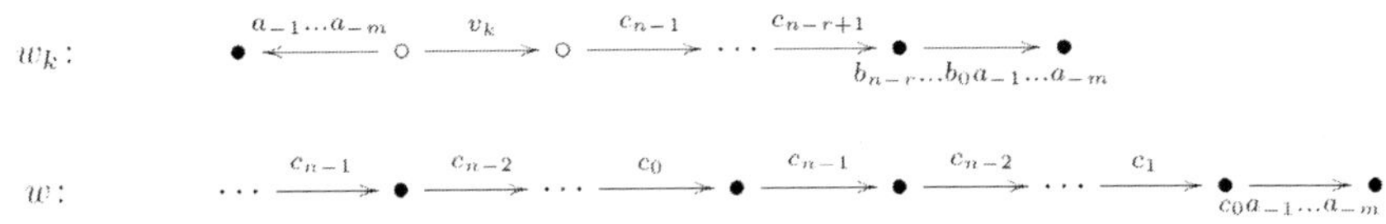

where v_k is the k-fold concatenation of $\bullet \xrightarrow{c_{n-1}} \cdots \xrightarrow{c_{n-r+1}} \bullet \xrightarrow{b_{n-r}\dots b_0} \bullet$. *Then:*

Where v_k is the k-fold concatenation of $\bullet \xrightarrow{c_{n-1}} \cdots \xrightarrow{c_{n-r+1}} \bullet \xrightarrow{b_{n-r}\dots b_0} \bullet$. *Then:*

(1) *if $r<n$, all homotopy strings are (shifted) copies of subwords of thewkfor $k\geq 0$;*

(2) *if $r=n$, all homotopy strings are (shifted) copies of subwords of w andwkfor $k\geq 0$.*

Lemma 7.2. *If $r>1$ we have*hom $\Lambda(P(i),P(j))\leq 1$ *for all*$-m\leq i,j<n$. *If $r=1$ we have additionally* $\hom_\Lambda(P(0),P(j))=2$ *for all* $-m\leq j\leq 0$.

Lemma 7.3. *Suppose $v,w\in$*St(Λ) *and consider the following unfolded diagram:*

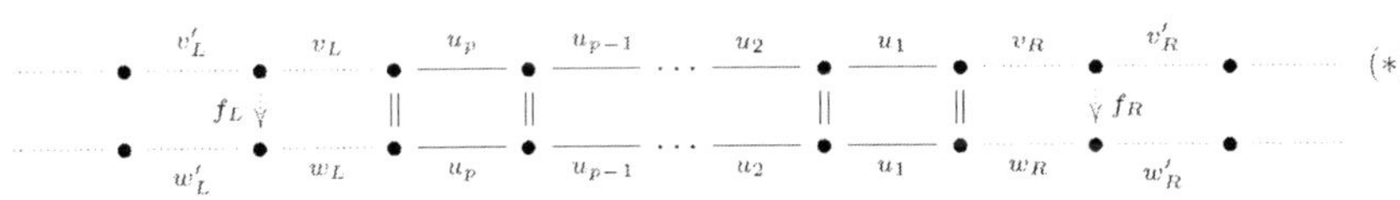

(1) *Suppose*(*)*represents a quasi-graph map* $P_v\to\Sigma^{-1}P_w$. *If v_L is non-zero, then one of v'_L or w'_L is zero.*

(2) *Suppose*(*)*represents a graph map* $P_v\to P_w$. *Then either*

(i) $f_L\neq 0$ *and one of v'_L or w'_L is zero; or*

(ii) $f_L=0$ *and if v_L is non-zero then* $v_L=(a_{-1}\dots a_{-i})-1$*for some* $1\leq i\leq m$.

Dual statements hold for the right-hand end of the diagram.

PROOF. Recall from Definition 3.11 that if v_L is not zero then $v_L\neq w_L$. We note that the only ways that vL and w_L can differ is if one of them is zero or if one (or both) of them is a subpath of $b_{n-r}\dots b_0a_{-1}\dots a_{-m}$, and it is clear that in all cases one of v'_L or w'_L is zero. This shows (1); (2) is similar. □

The upshot of Lemma 7.3 is that any graph map $P_v \to P_w$ or quasi-graph map $P_v \to \Sigma^{-1} P_w$ spans every degree where P_w and P_v are both non-zero.

THEOREM 7.4. *Suppose* $v,w \in \mathrm{St}(\Lambda)$. *If* $r>1$ *then* $\hom_{D^b(\Lambda)}(P_v, P_w) \leq 1$. *If* $r = 1$ *then* $\hom_{D^b(\Lambda)}(P_v, P_w) \leq 2$.

PROOF. For $r>1$, observe that Lemma 7.2 combines with Lemmas 4.3 and 7.3 to give $\hom_{D^b(\Lambda)}(P_v, P_w)$ in the cases that there is a graph map $P_v \to P_w$ or a non-singleton homotopy class $P_v \to P_w$.

For $r \geq 1$, we claim that if there is a single or double map $f: P_v \to P_w$ such that $H(f)$ is a singleton homotopy class, then $\hom_D(P_v, P_w)=1$. If f is a single map, a case analysis reveals that, of the options presented in Definition 3.7, only (*i*) could arise. Clearly, there can be no other basis maps $P_v \to P_w$. Similarly, by considering all of the possible cases where f is a double map, we find that either v or w is a homotopy string of length 1 and that in each of these cases there is no other possible basis map.

When $r=1$, the only way there can be more than one basis map $P_v \to P_w$ is if there is a graph map and a single map supported in the same degree, both the graph map and the homotopy class containing this single map will span the rest of the string, by Lemmas 4.3 and 7.3. □

EXAMPLE 7.5. The following example shows that the upper bound can be attained. Let $\Lambda = \Lambda(1,1,3)$ and consider the homotopy strings $v=(b_2b_1b_0,2,3)(b_2b_1b_0a_{-1},3,4)$ and $w=(a_{-1},2,1)(b_2b_1b_0,1,2)(b_2b_1b_0,2,3)(b_2b_1b_0a_{-1},3,4)$. Pictured below, from left to right: The algebra $\Lambda(1,1,3)$, a graph map $P_v \to P_w$, and a quasi-graph map $P_v \to \Sigma^{-1} P_w$.

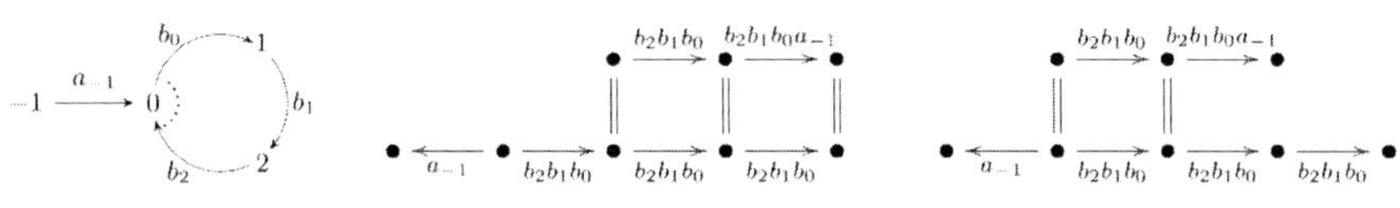

ACKNOWLEDGMENTS

We would like to thank Peter Jørgensen and Mike Prest for valuable comments. KA and DP gratefully acknowledge financial support of the EPSRC through grant EP/K022490/1, and KA would like to thank The University of Manchester for the kind hospitality during two research visits. RL and DP also acknowledge the kind hospitality of the Algebra Group at the Norwegian University of Science and Technology in Trondheim, and for financial support on a research visit there.

REFERENCES

1. F. Babaei Alitappeh, Special biserial cluster-tilted algebras and derived categories of cluster-tilted algebras of type A, PhD thesis, Norwegian University of Science and Technology, Department of Mathematical Sciences, 2014.
2. Kristin Krogh Arnesen, Yvonne Grimeland, The Auslander–Reiten components of Kb(projΛ)for a cluster-tilted algebra of type ˜A, J. Algebra Appl. 14(1) (2015) 1550005.
3. Assem, T. Brüstle, G. Charbonneau-Jodoin, P-G. Plamondon, Gentle algebras arising from surface triangulations, Algebra Number Theory 4(2) (2010) 201–229.
4. Assem, A. Skowroński, Iterated tilted algebras of type ˜**A**n, Math. Z. 195(2) (1987) 269–290.
5. D. Avella Alaminos, C. Geiß, Combinatorial derived invariants for gentle algebras, J. Pure Appl. Algebra 212(1) (2008) 228–243.
6. J. Bastian, Mutation classes of _An-quivers and derived equivalence classification of cluster tilted algebras of type _An, Algebra Number Theory 5(5) (2011) 567–594.
7. A.A. Be˘ılinson, Coherent sheaves on **P**nand problems in linear algebra, Funktsional. Anal. i Prilozhen. 12(3) (1978) 68–69.
8. V. Bekkert, H.A. Merklen, Indecomposables in derived categories of gentle algebras, Algebr. Rep-resent. Theory 6(3) (2003) 285–302.
9. G. Bobiński, The almost split triangles for perfect complexes over gentle algebras, J. Pure Appl. Algebra 215(4) (2011) 642–654.
10. G. Bobiński, The graded centers of derived discrete algebras, J. Algebra 333 (2011) 55–66.
11. G. Bobiński, C. Geiß, A. Skowroński, Classification of discrete derived categories, Cent. Eur. J. Math. 2(1) (2004) 19–49 (electronic).
12. Grzegorz Bobiński, Henning Krause, The Krull–Gabriel dimension of discrete derived categories, Bull. Sci. Math. 139(3) (2015) 269–282.
13. N. Broomhead, D. Pauksztello, D. Ploog, Discrete derived categories I: homomorphisms, autoe-quivalences and t-structures, Math. Z. (2016), http://dx.doi.org/10.1007/s00209-016-1690-1, online first; arXiv:1312.5203.
14. Nathan Broomhead, David Pauksztello, David Ploog, Discrete derived categories II: the silting pairs CW complex and the stability manifold, J. Lond. Math. Soc. (2) 93(2) (2016) 273–300.
15. Burban, Y. Drozd, Derived categories of nodal algebras, J. Algebra 272(1) (2004) 46–94.
16. M.C.R. Butler, C.M. Ringel, Auslander–Reiten sequences with few middle terms and applications to string algebras, Comm. Algebra 15(1–2) (1987) 145–179.
17. W.W. Crawley-Boevey, Maps between representations of zero-relation algebras, J. Algebra 126(2) (1989) 259–263.
18. W.W. Crawley-Boevey, Infinite-dimensional modules in the representation theory of finite-dimensional algebras, in: Algebras and Modules, I, Trondheim, 1996, in: CMS Conf. Proc., vol.23, Amer. Math. Soc., Providence, RI, 1998, pp.29–54.
19. D. Happel, Triangulated Categories in the Representation Theory of Finite-Dimensional Algebras, London Mathematical Society Lecture Note Series, vol.119, Cambridge University Press, Cam-bridge, 1988.
20. P.J. Hilton, U. Stammbach, A Course in Homological Algebra, second edition, Graduate Texts in Mathematics, vol.4, Springer-Verlag, New York, 1997.
21. H. Krause, Maps between tree and band modules, J. Algebra 137(1) (1991) 186–194.
22. D. Vossieck, The algebras with discrete derived category, J. Algebra 243(1) (2001) 168–176.
23. C.A. Weibel, An Introduction to Homological Algebra, Cambridge Studies in Advanced Mathemat-ics, vol.38, Cambridge University Press, Cambridge, 1994.

CHAPTER 5

On the Probability of Generating a Monolithic Group

Eloisa Detomi[a], Andrea Lucchini[a], Colva M. Roney-Dougal[b,*,1]

[a] *Università degli Studi di Padova, Dipartimento di Matematica, Via Trieste 63,35121 Padova, Italy*

[b] *University of St Andrews, Mathematical Institute, St Andrews, Fife KY16 9SS, Scotland, United Kingdom*

ABSTRACT

A group L is *primitive monolithic* if L has a unique minimal normal subgroup, N, and trivial Frattini subgroup. By $P_{L,N}(k)$ we denote the conditional probability that k randomly chosen elements of L generate L, given that they project onto generators for L/N. In this article we show that $P_{L,N}(k)$ is controlled by $P_{Y,S}(2)$, where $N \cong S^r$ and Y is a 2-generated almost simple group with socle S that is contained in the normalizer in L of the first direct factor of N. Information about $P_{L,N}(k)$ for L primitive monolithic yields various types of information about the generation of arbitrary finite and profinite groups.

Keywords: Finite group theory Random generation of finite groups Finite simple groups Crown-based power

1. INTRODUCTION

A group L is *primitive monolithic* if L has a unique minimal normal subgroup N, and trivial Frattini subgroup. By $P_{L,N}(k)$ we denote the conditional probability that k randomly chosen elements of L generate L, given that they project onto generators for L/N, and by d(L) we denote the cardinality of the smallest generating set for L.

Bounds on $P_{L,N}(k)$ were studied in [6]. The results presented there depend on a detailed analysis of the behavior of $P_{L,N}(d(L))$ when L is almost simple [14]. Although these bounds were strong enough to solve a series of open problems on the generation of finite groups, an interesting question arises from [14] and [6]: is it true that if $k \geq d(L)$ then $P_{L,N}(k)$ is bounded below by the conditional probability $P_{Y,S}(2)$? Here Y is a 2-generated group such that $S \trianglelefteq Y \leq N_L(S1)/C_L(S1) \leq \mathrm{Aut}S$, with $N \cong S1 \times \cdots \times Sr \cong Sr$. By [14], $P_{Y,S}(2) \geq 53/90$ and indeed $P_{Y,S}(2) \geq 0.9$ apart from very few exceptions, so a lower bound $P_{L,N}(k) \geq P_{Y,S}(2)$ can be considered quite satisfactory. In any case, a positive solution would allow a neater and more efficient formulation of the results in [6].

In fact, the only remaining case to prove is when $k=r=2$, which needs a deeper insight into the structure of almost simple groups. By [3], if X is almost simple with socle S, then $d(X)=\max(2,d(X/S)) \leq 3$. In Section 4 of [6] it is shown that in order to settle our conjecture, it would suffice to prove that every almost simple group X satisfies the following property P: *there exists a 2-generated subgroup Y, with* $S = \mathrm{soc}X \leq Y \leq X$ *and*

$$P_{X,S}(3) - P_{Y,S}(2) \geqslant \frac{|\operatorname{Out} S|}{|S|} \quad (1)$$

In the present paper we prove that indeed all the finite almost simple groups have this property:

Proposition 1.1

Every finite almost simple group has property P.

This immediately allows us to settle the conjecture and give a positive answer to our question.

Theorem 1.2

Let L be a primitive monolithic group with non-abelian socle $N \cong S_1 \times \cdots \times S_r$, *and identify S with*$S1$. *If* $k \geq d(L)$ *then there exists a 2-generated group Y with* $S \trianglelefteq Y \leq \mathrm{i}(S_1)/C_L(S_1) \leq \mathrm{Aut}\, S$ *such that* $P_{L,N}(k) \geq P_{Y,S}(2)$.

Let G be a finite group, and let L be monolithic with socle N. We define the *crown-based power* of L of *size t* to be

$Lt = \{(l_1,\ldots,l_t) \in L^t \mid l_1 N = \cdots = l_t N\}$.

In [4] it was proved that there exist a primitive monolithic group L and a positive integer t such that the crown-based power L of size t is an epimorphic image of G, and $d(G) = d(L_t)$. Clearly, $d(L_t)$ increases with t, hence for each $k \geq d(L)$ there exists a largest positive integer $fL(k)$ such that $d(L^{f_L(k)}) \leq k$. Information about the behavior of $f_L(k)$ is of relevance to several problems in the generation of finite and profinite groups: see, for example, [6].

Theorem 2.7 in [4] gives a formula for $f_L(k)$:

$$f_L(k) = \begin{cases} \frac{P_{L,N}(k)|N|^k}{|C_{\mathrm{Aut}\,N}(L/N)|} & \text{if } N' = N \\ \log_q\left(\frac{|N|^{k-1}}{|H^1(L/N,N)|}\right) & \text{if } N' = 1, \text{ where } q = |\mathrm{End}_{L/N} N| \end{cases}$$

Notice that when N is abelian, we can readily compute $f_L(k)$, but when $N \cong S^r$ is a direct product of non-abelian simple groups S, the value of $f_L(k)$ depends on $P_{L,N}(k)$.

We therefore get the following corollary to Theorem 1.2. Note that throughout, all logarithms are to base 2 unless otherwise specified.

Corollary 1.3

Let L be a monolithic primitive group with non-abelian socle $N=S_r$ and let $d=d(L)$. Then there exists a 2-generated almost simple group Y with socle S, with the property that

$$f_L(d) \geqslant f_Y(2)\frac{|N|^{d-1}}{r|S|}$$

In particular,

$$f_L(d) \geqslant \alpha\frac{|N|^{d-1}}{r\log|S|}$$

$$where\ \alpha = \tfrac{121}{1680}\log 20\,160 > 1.029$$

In Section 2 we prove Proposition 1.1 when $d(X)=2$. In Section 3 we collect some background material for the proof of Proposition 1.1 in the cases where $d(X) \neq 2$: in particular we show that if $d(X)=3$, then S is either a linear or an orthogonal group. In Section 4 we prove Proposition 1.1 when $S=L_n(q)$ and $d(X) = 3$, then in Section 5 we prove the result when $S = \mathrm{O}_n^{+}(q)$ and $d(X) = 3$. In the last section we prove Corollary 1.3.

2. THE 2-GENERATOR CASE

In this section we prove Proposition 1.1 for the case $d(X)=2$. Given a non-abelian simple group S and a prime divisor u of $|S|$, we define the number c_u as follows: $c_u=2u$ if u^2 does not divide $|S|$, otherwise $c_u=u^2$. Then we set

$$\alpha_u = \max\left\{c_u, \min_{t\in S, |t|=u}\{|C_S(t)|\}\right\}$$

Proposition 2.1

Let X be a 2-generated almost simple group with socle S. If

$\alpha_u \geq |\mathrm{Out}\, S|/P_{X,S}(2)$

for every prime divisor u of |S|, then X satisfies property P.

Proof

Say X= $\langle g_1, g_2\rangle$ and set

$\Phi=\{(s_1, s_2, s_3) \in S_3| \langle g_1s_1,g_2s_2,s_3=X\}=\{(s_1, s_2, s_3) \in S_3| \langle g_1s_1,g_2s_2,s_3\rangle \geq S\}$.

Note that, by Gaschütz [8], $P_{X,S}(3)=|\Phi|/|S|3$.

This set Φ contains the set

$\Phi_1=\{(s_1, s_2, s_3) \in S_3| \langle g_2s_2, s_3\rangle \geq S\}$

whose cardinality is $P_{Y,S}(2)|S^2|\cdot|S|=P_{Y,S}(2)|S|3$ for $Y=\langle g_2,S\rangle$ (the first component is free).

Moreover, Φ contains the set

$\Phi_2=\{(s_1, s_2, s_3) \in S_3| \langle g_1s_1,g_2s_2\rangle \geq S$ and $\langle g_2s_2,s_3\rangle \geq S\}$

where the second condition ensures that $\Phi1\cap\Phi2=\emptyset$. Say

$$\alpha = \min_{s_2\in S}\left|\left\{s_3 \in S \mid \langle g_2s_2, s_3\rangle \not\geq S\right\}\right|$$

then $|\Phi_2|\geq P_{X,S}(2)|S_2|\cdot\alpha$. Therefore

$$P_{X,S}(3) = \frac{|\Phi|}{|S|^3} \geqslant \frac{|\Phi_1|}{|S|^3} + \frac{|\Phi_2|}{|S|^3}$$

$$\geqslant P_{Y,S}(2) + \frac{P_{X,S}(2)\alpha}{|S|}$$

so we are left to prove that $\frac{P_{X,S}(2)\alpha}{|S|} \geqslant \frac{|\operatorname{Out} S|}{|S|}$, that is,

$P_{X,S}(2)\alpha \geq |\mathrm{Out}\, S|$.

Given $s_2 \in S$, we know that $C_S(g_2s_2) \neq \{1\}$, since by [15] a simple group cannot have a fixed-point-free automorphism. Therefore, there exists a non-trivial element $t \in S$ such that $g_2s_2 \in C_X(t)$, and we can assume t to be of prime order, say u, by taking suitable powers, if necessary.

If $s_3 \in N_S(\langle t \rangle)$, then $\langle g_2s_2, s_3 \rangle \geq S$, because $\langle t \rangle$ is centralized by g_2s_2 and normalized by s_3, and hence is normal in $\langle g_2s_2, s_3 \rangle$.

Thus

$$\alpha \geqslant \min_{t \in P} \left| N_S(\langle t \rangle) \right| \geqslant \min_{t \in P} \left| C_S(t) \right|$$

where P is a Sylow u-subgroup. If u^2 divides $|S|$, then $|NP(\langle t \rangle)| \geq u^2$ so u^2 divides $|NS(\langle t \rangle)|$. Otherwise, $P = \langle t \rangle$ is abelian and by Burnside's normal p-complement theorem, P is not self-normalizing, hence $|N_S(\langle t \rangle)| \geq 2u$. □

By [14], $P_{X,S}(2) \geq 1/2$, and actually $P_{X,S}(2) \geq 0.9$ for all but a finite set S of exceptions. The exceptions have socle one of: An (with $5 \leq n \leq 11$), $L_2(q)$ (with $7 \leq q \leq 11$), $L_3(q)$ (with $3 \leq q \leq 4$), $S_4(3)$, M_{11} or M_{12}.

Lemma 2.2

Let X be a 2-generated almost simple group with socle S. If there exists a 2-generator almost simple group H with socle S such that $P_{H,S}(2) < 0.9$, or if $|\mathrm{Out}\, S| \leq 4$, then X satisfies property P.

Proof

The probabilities $P_{H,S}(2)$ for $H \in S$ are listed in [14]. Furthermore, [14, Corollary 1.2] proves that $P_{X,S}(3) \geq 139/150$ for all almost simple groups X. Thus it is straightforward to check that if X contains a 2-generator almost simple group H with socle S such that $P_{H,S}(2)<0.9$, then X satisfies property P.

Now assume that $|\mathrm{Out}\ S|\leq 4$ and X has no 2-generated subgroup with socle S that belongs to S; in particular, $P_{X,S}(2)\geq 0.9$. By Corollary 4 in [10], all finite non-abelian simple groups containing an element t such that $|C_S(t)|<5$ lie in S. Therefore $|C_S(t)|\geqslant 5$ for every $t\in S$. Then $|C_S(t)|\geq 5\geq 10|\mathrm{Out}\ S|/9\geq |\mathrm{Out}\ S|/P_{X,S}(2)$ for every $t\in S$, hence by Proposition 2.1 the group X satisfies property P. □

Lemma 2.3

Let S be simple, with $|\mathrm{Out}\ S|>4$. *Then* $\alpha_2 \geq \frac{10}{9}|\mathrm{Out}\ S|$.

Proof

Since $|\mathrm{Out}\ S|>4$ we deduce that $|S|\geq |L_2(27)|=9828$. In [1, Remark 2] it is stated that for every non-abelian simple group S and any involution t we can bound $|S|\leq |C_S(t)|^3$. Since $|\mathrm{Out}\ S|\leq \frac{6}{7}\log|S|$ (see [14]), it follows that $|C_S(t)|\geq |S|^{1/3}\geq \frac{106}{77}\log|S|\geq \frac{10}{9}|\mathrm{Out}\ S|$ for every involution t. □

Fix a type of simple algebraic group X of rank r over the algebraic closure of a finite field. Let σ be an endomorphism of X with fixed point group $X\sigma$ of finite order. We will write $X(q)=X\sigma$ if q is the absolute value of the eigenvalues of σ on the character group of the maximal torus T of X.

Lemma 2.4

(See[7, Lemma 3.4].) Let C be the centralizer of an element in $X(q)$, *then* $|C|\geq(q-1)r$.

In the remainder of this section we consider the families of simple groups S with the property that $|\mathrm{Out}\ S|>4$ for at least one group in the family. In the majority of cases, it suffices to use Lemma 2.4 to bound the order of an element centralizer in the simple group, and hence to show that the conditions of Proposition 2.1 are satisfied; however for a small number of groups we must calculate various values of α_u explicitly.

Lemma 2.5

Let X be a 2-generated almost simple group with socle $S=L_n(q)$. *Then X satisfies property P.*

Proof

Recall that $|\mathrm{Out}\ S|=2dh$ for $n\geq 3$, and $|\mathrm{Out}\ S|=dh$ when $n=2$, where $q=ph$ and $d=(n,q-1)$. By Lemma 2.2, we can assume that $dh\geq 2$, and that $P_{X,S}(2)\geq 0.9$. In particular, the case $S=L_n(2)$ follows.

Let $t \in S=L_n(q)$. We apply the previous lemma, with $r=n-1$ and dividing by d to pass from $X(q)$ to the associated simple group, to deduce that $|C_S(t)| \geq (q-1)n-1d$. By Proposition 2.1 it suffices to prove that

$$(q-1)^{n-1} \geqslant \frac{20}{9} d^2 h \quad \text{if } n \geqslant 3$$

$$q-1 \geqslant \frac{10}{9} d^2 h \quad \text{if } n=2 \tag{2}$$

Suppose first that $n=2$. If $q=2h$ then $d=1$, and the result follows. Otherwise, if $q-1<10d^2h/9$ then $q \in \{5,9\}$ and $P_{X,L2(q)}(2)<0.9$, contrary to assumption.

Now let $n=3$, so that $d \leq 3$. Our assumption on $P_{X,S}(2)$ implies that $q \geq 5$. If $(q-1)^2<20h$ then $q \in \{5,8\}$, but $d=1$ for both of these q, so it suffices to verify that $(q-1)^2 \geq 20h/9$, which is clear.

Let $n \geq 4$. We use the fact that $d \leq q-1$, hence $(q-1)^{n-1}/d^2 \geq q-1$. For $p=2$, if $2^h-1<20h/9$ then $h<3$. Hence $q=22$, and so $(q-1)n-1>40d^2/9$ for $n \geqslant 4$ (if $n=4$ then $d=1$).
If $p=3$ then $3^h-1 \geqslant 20h/9$ whenever $h \geqslant 2$; if $h=1$, then $|\mathrm{Out}\ S| \leqslant 4$, contrary to assumption. If $p \geqslant 5$ then $q-1 \geqslant 5h-1 \geqslant 20h/9$ for every h. □

Lemma 2.6

Let X be an almost simple group with socle $S=U_n(q)$, $n \geqslant 3$. Then X satisfies property P.

Proof

Let $t \in S=Un(q)$. Applying Lemma 2.4 with $r=n-1$ and dividing by $d=(n,q+1)$ to pass to the associated simple group, we get $|C_S(t)| \geqslant \frac{(q-1)^{n-1}}{d}$. Since $|\mathrm{Out}\ S|=2dh$, with $q=ph$, by Proposition 2.1 and Lemma 2.2 it suffices to prove that

$$(q-1)^{n-1} \geqslant \frac{20}{9} d^2 h \tag{3}$$

for groups S with $dh>2$ and $P_{X,S} \geqslant 0.9$.

Now $d \leqslant n$, hence if $q>2$ then $(q-1)^{n-1}/d^2 \geqslant (q-1)^{n-1}/n^2 \geqslant (q-1)^{3-1}/3^2$. Then $(q-1)^{3-1}/3^2 \geqslant \frac{20}{9} h$ for $q=7$ or $q \geqslant 9$.

We now study the cases $q=2,4,8,3,5$. Let $q=2$. As $h=1$, we can assume that 3 divides n (and $n \neq 3$, as $U_3(2)$ is not simple). Then 3^2 divides $|S|=\frac{d}{9}2^{n(n-1)/2}(2^2-1)(2^3+1)(2^4-1)\prod_{i=5}^{n}(2^i-(-1)^i)$, hence $\alpha_3 \geqslant 9 \geqslant (10/9)6=(10/9)|\mathrm{Out}\ S|$.

The same holds for the other odd prime divisors u of $|S|$, since $2u \geqslant 9$, and we conclude from Proposition 2.1 and Lemma 2.3.

Let $q=4$. As $h=2$, we can assume that 5|n. If $S=U_5(4)$, then $|\mathrm{Out}\ S|=20$ and we can use GAP to check that $\alpha_u \geqslant (10/9)20$ for every odd prime divisor of S, and as before we conclude from Proposition 2.1 and Lemma 2.3. If $n \geqslant 10$, then (3) holds.

Let $q=8$. If $S=U_3(8)$ then $|\mathrm{Out}\ S|=18$, and we use GAP to check that $\alpha_u \geqslant 20$ for all odd prime divisors u of $|S|$. If $n \geqslant 4$, then $(8-1)^{n-1}/d^2 \geqslant 7^{n-1}/n^2 \geqslant 73/16 \geqslant (20/9)3$ and (3) holds.

Let $q=3$. As $h=1$, we can assume that 4 | n. If $S=U_4(3)$, then $|\mathrm{Out}\ S|=8$ and, as 3^2 divides |S|, we see that $\alpha_u \geqslant 9 \geqslant (10/9)8$ for every odd prime divisor u of $|S|$. If $n \geqslant 8$, then $2^7 \geqslant (20/9)4^2 \geqslant (10/9)2d^2$ and (3) holds.

Let $q=5$. As $h=1$, we can assume that 3 | n. If $S=U_3(5)$, then $|\mathrm{Out}\ S|=6$ and, as 3^2 divides $|S|$, we see that $\alpha_u \geqslant 7 \geqslant (10/9)6$ for every odd prime divisor u of $|S|$. If $n \geqslant 6$, then $4^5 \geqslant (20/9)6^2 \geqslant (10/9)2d^2$ and (3) is satisfied. □

Lemma 2.7

Let X be an almost simple group with socle $S=S_{2n}(q)(n \geqslant 2)$ or $S=O_{2n+1}(q)(q$ odd and $n \geqslant 3)$. Then X satisfies property P.

Proof

Let $t \in S$. Applying Lemma 2.4 and dividing by $d=(2,q-1)$ to pass to the associated simple group, we get $|C_S(t)| \geqslant \frac{(q-1)^n}{d}$. Here $d \leqslant 2$, and by Lemma 2.2 we may assume that $|\mathrm{Out}\ S|=dh>4$, so $h \geqslant 3$. It therefore suffices to prove that

$$\frac{(q-1)^n}{h} \geqslant \frac{10}{9}4 \quad \text{if } q \text{ is odd}$$

$$\frac{(2^h-1)^n}{h} \geqslant \frac{10}{9} \quad \text{if } q=2^h>2^4 \tag{4}$$

If q is odd, then $(q-1)^n/h \geqslant (q-1)^2/h \geqslant (3^3-1)^2/3>40/9$. Otherwise $q=2^h$ and $(2^h-1)^n/h \geqslant (2^3-1)^2/3 \geqslant 10/9$. Thus (4) is always satisfied. □

Lemma 2.8

Let X be a 2-generated almost simple group with socle $S=O^+_{2n}(q), n \geqslant 4$. Then X satisfies property P.

Proof

Let $t \in S$. Applying Lemma 2.4 and dividing by $d=(4,q^n-1)$ to pass to the associated simple group, we get $|C_S(t)| \geqslant \frac{(q-1)^n}{d}$. Since $|\mathrm{Out}\, S|=2dh$ for $n \geqslant 5$ and $|\mathrm{Out}\, S|=6dh$ for $n=4$, it suffices to prove that

$$\frac{(q-1)^n}{d^2h} \geqslant \frac{10}{9}2 \quad \text{if } n \geqslant 5$$
$$\frac{(2^h-1)^n}{d^2h} \geqslant \frac{10}{9}6 \quad \text{if } n = 4 \tag{5}$$

Consider first the case $n \geqslant 5$. If q is odd, then $\frac{(q-1)^n}{d^2h} \geqslant \frac{(p^h-1)^5}{4^2h} \geqslant \frac{2^{5h}}{16h} \geqslant \frac{20}{9}$ whenever $h \geqslant 2$; if $h=1$, then (5) is satisfied for either $p \geqslant 5$ or $p=3$ and $n \geqslant 6$; finally, for $p=3$ and $n=5$ we get $d=(4,3^5-1)=2$ and thus $2^5 \geqslant \frac{20}{2}2^2$ and (5) is satisfied. If $n \geqslant 5$ and q is even, then $d=1$ and $|\mathrm{Out}\, S|=2h$, so we can assume $h \geqslant 3$. Then $(2^h-1)^n/h \geqslant (2^h-1)^5/h \geqslant (2^3-1)^5/3 > 20/9$.

Let $n=4$. If q is odd and $h \geqslant 2$ then $\frac{(q-1)^4}{d^2h} \geqslant \frac{(p^h-1)^4}{4^2h} \geqslant \frac{2^{4h}}{16h} \geqslant \frac{20}{3}$. If $h=1$, then (5) is satisfied for $p \geqslant 5$. If $S=O_8^+(3)$, then $P_{X,S}(2) \geqslant 139/150$, by [14, Remark 4.1], so $|\mathrm{Out}\, S|/P_{X,S}(2)<26$. A straightforward GAP calculation shows that $N_S(\langle t \rangle) \geqslant 26$ whenever $|t|$ is an odd prime, and the result follows from Lemma 2.3 and Proposition 2.1.

Finally, if $n=4$ and $q=2^h$, then $d=1$ and (5) is satisfied whenever $h \geqslant 2$. If $S=O^+_8(2)$, then $|S|=2^{12}\cdot 3^5\cdot 5^2\cdot 7$ and $(10/9)\ |\mathrm{Out}\, S|<7$; thus $\alpha_u \geqslant (10/9)\ |\mathrm{Out}\, S|$ for every odd prime u and we conclude as before. □

Lemma 2.9

Let X be a 2-generated almost simple group with socle $S=O^-_{2n}(q)$, with $n \geqslant 4$. Then X satisfies property P.

Proof

Let $t \in S$. Applying Lemma 2.4 and dividing by $d=(4,q^n+1)$ to pass to the associated simple group, we get $|C_S(t)| \geqslant (q-1)$nd. Since $|\mathrm{Out}\, S|=2dh$, it suffices to prove that

$$\frac{(q-1)^n}{d^2h} \geqslant \frac{10}{9}2 \tag{6}$$

and by Lemma 2.2 we may assume that $dh>2$ and $P_{X,S} \geqslant 0.9$.

If q is odd, then $\frac{(q-1)^n}{d^2h} \geqslant \frac{(p^h-1)^4}{4^2h} \geqslant \frac{2^{4h}}{16h} \geqslant \frac{20}{9}$ whenever $h\geqslant 2$. If $h=1$, then (6) is satisfied for either $p\geqslant 5$ or $p=3$ and $n\geqslant 6$. If $p=3$ and $n=4$ then $d=2$, so $2^4 \geqslant \frac{20}{9}2^2$ and again (6) is satisfied. If $S=O^-_{10}(5)$, then $(10/9)\,|\mathrm{Out}\,S|<9$; since 3^2 divides $|S|$ and $2\cdot 5>9$, for every odd prime u we see that $\alpha_u\geqslant 9$ for every odd prime u. The result follows from Proposition 2.1.

Finally, if $q=2^h$ then $d=1$, so by assumption $h\geqslant 3$, and (6) holds. □

Lemma 2.10

Let X be a 2-generated almost simple group with socle S. If S is a simple group of exceptional type, then X satisfies property P.

Proof

Let $t\in S$. Applying Lemma 2.4 and dividing by d to pass to the associated simple group, we get $|C_S(t)| \geqslant \frac{(q-1)^n}{d}$. By Lemma 2.2 it suffices to prove that

$$\frac{(q-1)^n}{d|\,\mathrm{Out}\,S|} \geqslant \frac{10}{9}$$

and we may assume that $|\mathrm{Out}\,S|>4$.

Let $S=G_2(q)$. Then $d=1$ and $|\mathrm{Out}\,S|\leqslant 2h$, so $h\geqslant 3$. Thus $\frac{(q-1)^2}{2h} \geqslant \frac{(2^h-1)^2}{2h} \geqslant (2^3-1)^2/6 \geqslant 10/9$ and (7) is satisfied. The same argument holds for $S=F_4(q)$, since d=1, $|\mathrm{Out}\,S|\leqslant 2h$ and $n=4$.

Let $S={}^2B_2(q)$. Then $d=1$ and $|\mathrm{Out}\,S|=h$, so $h\geqslant 5$. Thus $(q^{1/2}-1)2/h\geqslant(2^{h/2}-1)^2/h\geqslant(2^{5/2}-1)^2/5\geqslant 10/9$ and (7) is satisfied. The same argument holds for $S={}^2G_2(q)$ and for $S={}^2F_4(q)$.

Let $S={}^3D_4(q)$. Since $d=1$ and $|\mathrm{Out}\,S|\leqslant 3h$, we can assume that $h\geqslant 2$. Then $\frac{(q-1)^6}{2\cdot d^2h} \geqslant \frac{(q-1)^5}{18} \geqslant (3-1)^5/18 \geqslant 10/9$ and (7) is satisfied.

Let $S=E_n(q)$, $n=6,7,8$. Since $d\leqslant(q-1)$, $h\leqslant(q-1)$ and $|\mathrm{Out}\,S|\leqslant 2dh$, if $q\geqslant 3$ then $\frac{(q-1)^n}{2d^2h} \geqslant (q-1)^{6-3}/2 \geqslant (3-1)^3/2 \geqslant 10/9$ and (7) is satisfied. If $q=2$, then $|\mathrm{Out}\,S|\leqslant 2$, contradicting our assumptions.

Finally, let $S={}^2E_6(q)$. Then $d=(3,q+1)$ and $|\mathrm{Out}\,S|=2dh$. As $d\leqslant 3$ and $h\leqslant q-1$, if $q\geqslant 3$ then $\frac{(q-1)^6}{2\cdot d^2h} \geqslant \frac{(q-1)^5}{18} \geqslant (3-1)^5/18 \geqslant 10/9$ and (7) is satisfied. If $S={}^2E_6(2)$ then $(10/9)\,|\mathrm{Out}\,S|<7$: since 3^2 divides $|S|$ and $2\cdot 5>7$, for every odd prime u dividing $|S|$ we get $\alpha_u\geqslant 7$. The result follows from Lemma 2.3 and Proposition 2.1. □

3. BACKGROUND MATERIAL FOR THE CASE $D(X)=3$

In this section we collect some results and prove some elementary lemmas that will be helpful in both of the next two sections.

Theorem 3.1

(See Häsä[9].) Let G be a finite almost simple classical group, of dimension n over F_q. Let m(G)be the number of conjugacy classes of maximal subgroups of G that do not contain soc *G. Then*

$m(G)\leqslant 2_n^{5.2}+n\log\log q.$

Lemma 3.2

Let I be a finite set of positive integers, and let $q\geqslant 3$. Then $\prod_{i\in I}(q^i-1)>q-1+\sum_{i\in I}i.$

Proof

We see that

$$\prod_{i\in\mathcal{I}}(q^i-1)=q^{\sum_{i\in\mathcal{I}}i}\prod_{i\in\mathcal{I}}(1-q^{-i})>q^{\sum_{i\in\mathcal{I}}i}\prod_{i=1}^{\infty}(1-q^{-i})$$

This product converges to a limit c_q with $1/q<cq<1$□ .

The following result will be used extensively, without repeated citations, throughout Sections 4 and 5.

Proposition 3.3

Let $q>2$.

(1) *Let $G=SL_n(q)$. Then $q^{n2-2}<|G|<q^{n2-1}$.*

(2) *Let $G=SU_n(q)$. Then $|G|<qn^2-1$.*

(3) *Let $G=Sp_n(q)$. Then $|G|<q(n^2+n)/2$.*

(4) *Let $G_+=SO^+_n(q)$. Then $q^{n2/2-n/2-1}<|G+|<q(n^2-n)/2$.*

(5) *Let $G_-=SO^-_n(q)n-q$. Then $q^{n2/2-n/2-1}<|G-|<2q(^{n2-n})/2$.*

(6) *Let $G_\circ=SO^\circ_n(q)$. Then $|G_\circ|<q(n^2-n)/2$.*

Proof

(1) We use the expression $|G|=q^{n(n-1)/2}\prod_{i=2}^{n}(q^i-1)$. The upper bound is clear, and the lower bound follows from Lemma 3.2.

(2) We use $|G|=q^{n(n-1)/2} \prod_{i=2}^{n}(q^i-(-1)^i)$. Note that $(q^{2k}-1)(q^{2k+1}+1)<q^{4k}+1$ for all k. Thus if n is odd then $|G|<q^{n(n-1)/2} \prod_{i=1}^{(n-1)/2} q^{4i+1} = q^{(n^2-n)/2+(n^2+n-2)/2}$. The result for n even is similar.

(3) This follows easily from $|G|=q^{n^2/4} \prod_{i=1}^{n/2}(q^{2i}-1)$.

(4) and (5) We use $|G_\pm| = q^{n(n-2)/4}(q^{n/2} \mp 1)\prod_{i=1}^{n/2-1}(q^{2i}-1)$. For the upper bound for G_-, note that $q^{n/2}+1<2q^{n/2}$. For the lower bound, use Lemma 3.2.

(6) This follows easily from $|G_\circ|=q^{(n-1)^2/4} \prod_{i=1}^{(n-1)/2}(q^{2i}-1)$. □

The following lemma follows straightforwardly from [3].

Lemma 3.4

Let X be almost simple with socle S a classical group. If d(X)=3, then n is even and at least four, and q is an even power of an odd prime. Furthermore, either $S=L_n(q)$ and $X \ntrianglelefteq P\Gamma L_n(q)$, or $n \geqslant 8$ and $S=O^+_n(q)$.

Lemma 3.5

Let X be almost simple with socle S. Let M be the set of maximal subgroups of X that supplement S, up to X-conjugacy. Then

$$P_{X,S}(l) \geqslant 1 - \sum_{M\in\mathcal{M}} \frac{1}{|S:M\cap S|^{l-1}}.$$

Furthermore, if $M=M1\cup M2$ and $\sum_{M\in M1}|M|\leqslant m$, then

$$P_{X,S}(l) \geqslant 1 - \frac{m^{l-1}}{|X|^{l-1}} - \sum_{M\in\mathcal{M}_2} \frac{1}{|S:M\cap S|^{l-1}}.$$

Proof

The proof of the first claim is a routine calculation, using the fact that if l elements of X fail to generate X, but do project onto generators for X/S, then they must lie in some maximal subgroup of X that supplements S. For the second, rewrite $1/[S:M\cap S]$ as $|M|/|X|$. □

4. THE LINEAR GROUPS *X* WITH *D*(*X*)=3

In this section we prove that if X is almost simple with socle $S=L_n(q)$, and $d(X)=3$ (so that $q\geqslant 9$), then X has property P. We fix this notation and these assumptions throughout this section.

The following is a straightforward calculation.

Lemma 4.1 *The value of* $|\mathrm{Out}\ S|/|S|$ *is less than* $q^{-(n2-5)}$.

Lemma 4.2

The probability $P_S(2)$*is bounded above by*

$$P_S(2) \leqslant 1 - \frac{(q-1)^2}{(q^n-1)^2} \leqslant 1 - \frac{4}{q^{2n-1}}$$

Proof

The group S has a permutation action on (qn−1)/(q−1) points, and hence a maximal subgroup of the same index. For the final inequality, use (qn−1+⋯+1)2<2q2n−2□ .

Lemma 4.3

The group X has at most n(5+log n+log log q)conjugacy classes of geometric maximal subgroups.

Proof

The group $X \not\leqslant P\Gamma L_n(q)$, by Lemma 3.4. We use [12, Table 3.5.A] to count the geometric maximal subgroups of X, up to X-conjugacy. There are at most $n-1$ C_1-subgroups, $n/2$ C_2-subgroups, $\log n$ C_3-subgroups, $(\sqrt{n}-1)^2$ $C4$-subgroups, $n \log\log q$ $C5$-subgroups, no C_6-subgroups (since q is an odd square), $n \log n$ C_7-subgroups, and $5n/2$ C_8-subgroups. Since $n\geq 4$ we may simplify using $\log n\leq\sqrt{n}$. □

Lemma 4.4

Assume that n≥8. Then the largest maximal subgroup of X that supplements S is geometric and has index $(q^n-1)(q^{n-1}-1)/(q-1)^2$. *Furthermore, the largest non-geometric maximal subgroup of X has index at least* $q^{n2-3n-3}$.

Proof

We take the classification of geometric subgroups of X from [12, Table 3.5.A]. We will work in $SL_n(q)$, since the formulae for the subgroup orders are more straightforward.

Consider first the C_1-subgroups. By Lemma 3.4, the group $X \nleq P\Gamma Ln(q)$, so the parabolic subgroups Pi are not maximal in X unless $i=n/2$, in which case they have order less than $q^{n2/4}|SL_{n/2}(q)|^2(q-1)<q^{3n2/4-1}<q^{n2-2n}$. The subgroup of type $P_{k,n-k}$ has order $q^{k(2n-3k)}|SL_k(q)|2|SL_{n-2k}(q)|(q-1)^2$, so the index of $P_{1,n-1}$ follows from Proposition 3.3, and $|P_{k,n-k}|<q^{n2-2nk+3k2-1}$. Since $n\geq 2k$, $|P_{k,n-k}|\leq|P_{1,n-1}|$. The completely reducible subgroup of type $GL_1(q)\oplus GL_{n-1}(q)$ has order $q(n^{2-3n+2})/2 \prod_{i=1}^{n-1}(q^i-1) < |P_{1,n-1}|$, whilst for $k\geq 2$ the subgroups of type $GL_k(q) \oplus GL_{n-k}(q)$ have order $|SL_k(q)||SL_{n-k}(q)|(q-1)<q^{n2-2nk+2k2-1}<q^{n2-2n}$.

A $C2$-subgroup has order $|SL_{n/t}(q)|^t(q-1)^{t-1}t!<q(^{n2/t})-1_t t$ for some divisor $t>1$ of n. It suffices to show that $n^2(t-1)/t+1\geq 2n+t\ log_q t$. Now $t\geq 2$, so $n^2(t-1)/t\geq n^2/2$. Furthermore, $t\leq n$ and $q\geq 9$ so $2n+t log_q t<n log_2 n$, whilst $n^2/2+1> n\ logn$ for all $n\geq 8$.

The C3-subgroups have order at most $|GL_{n/r}(q^r)|.r<q^{n2/r+r}<q^{n2-2n}$, for some prime divisor r of n. The C4-subgroups have order less than $|GL_k(q)||GL_n/k(q)|<q^{n2/k2+k2}<q^{n2/4+n}<q^{n2-2n}$, since $k\geq 2$. The $C5$-subgroups have order at most $|SL_n(q_0)|(q-1)<q^{n2/2+1}<q^{n2-2n}$, since $q^r{}_0=q$ with $r>1$. Class $C6$ is empty as q is a square.

The $C7$-subgroups have order less than $|GL_k(q)|^r r!<q^{rk2}r^r$. Using $k=n^{1/r}$ and $2\leq r<n$, it suffices to prove that $rn^{2/r}+r\log_q r\leq n^2-2n$. Now $rlog_q r<(1/3)rlog_2 r$, so $rn^{2/r}+rlog_q r<n^{3/2}+(1/6)n^{1/2}logn<2n^{3/2}$, and the result follows.

In class $C8$, the symplectic groups have order less than $q^{n2/2+n/2+1}$, and are larger than the orthogonal groups. The unitary groups have order at most $|SU_n(q^{1/2})|(q-1)<q^{n2-2n}$.

Finally, by [13], if a subgroup H of X is not geometric, then either H has socle A_{n+1} or A_{n+2} (which would imply that H preserves a non-degenerate form, and hence is non-maximal: see for instance [11, Table 2(a)]), or $|H|$ is at most q^{3n}. Since $n\geq 8$ this is less than $q^{n2-2n-1}$. □

Theorem 4.5

Let $n\geq 8$. Then X satisfies property P.

Proof

We use Lemma 3.5: group together first the geometric maximal subgroups of S, using Lemma 4.3 and the first part of Lemma 4.4, and then the non-geometric maximal subgroups, using Theorem 3.1 and the second part of Lemma
4.4.

$$\begin{aligned} P_{X,S}(3) &\geqslant 1 - \frac{n(5+\log n+\log\log q)(q-1)^4}{(q^n-1)^2(q^{n-1}-1)^2} - \frac{2n^{5.2}+n\log\log q}{q^{2n^2-6n-6}} \\ &\geqslant 1 - \frac{q^{\log n}(q+\log(q^{\log n})+q)}{(q^{n-1}+\cdots+1)^2(q^{n-2}+\cdots+1)^2} - \frac{2q^{5.2\log_9 n}+q^{\log_9 n+1}}{q^{2n^2-6n-6}} \\ &\geqslant 1 - \frac{q^{2\log n}}{q^{2(2n-3)}} - \frac{2q^{2\log n}}{q^{2n^2-6n-6}} \\ &\geqslant 1 - q^{-2(2n-\log n-3)} - q^{-(2n^2-7n-6)}. \end{aligned}$$

Then by Lemma 4.2,

$$P_{X,S}(3) - P_S(2) \geqslant \frac{4}{q^{2n-1}} - \frac{1}{q^{2(2n-\log n-3)}} - \frac{1}{q^{2n^2-7n-6}} > \frac{1}{q^{n^2-5}}$$

The result then follows from Lemma 4.1□ .

We finish this section by considering the cases not covered by Theorem 4.5.

Proposition 4.6

Let n=4. Then X satisfies property P.

Proof

First we show that

$$P_{X,S}(3) \geqslant 1 - \frac{5(q-1)^2}{(q^2+1)^2(q^3-1)^2} \tag{8}$$

By Lemma 3.4, the group $X \nleqslant \Gamma L_4(q)$, and $q \geq 9$. We work in $SL_4(q)$. We shall divide the conjugacy classes of maximal subgroups of *X* that supplement *S* into five sets, each of combined order at most $q^6(q-1)(q^2-1)^2$, and then (8) will follow from Lemma 3.5. We take the list of maximal subgroups of *X* from [2, Tables 8.8 and 8.9].

There is one conjugacy class of *C*1-subgroups of each type. The subgroups of type P_2 have order $q^6(q-1)(q^2-1)^2>q^{10}$, and hence index $(q^3-1)(q^2+1)/(q-1)$. The subgroups of type $P_{1,3}$ have order less than q^{10}. The subgroups of type $GL_1(q)\oplus GL_3(q)$ have order less than *q*9.

There are two classes of *C*2-subgroups, of combined order less than *q*8. There is one class of *C*3-subgroups, of order less than *q*8.

There are at most 4*log log q* classes of C_5-subgroups, of combined order at most $16(\log\log q)q^3(q-1)(q^{3/2}-1)(q^2-1)<q^{10}$.

There are at most two classes of symplectic groups, of combined order at most $4q^4(q^2-1)(q^4-1)<|P_2|$, since $q \geq 9$. There are at most four classes of unitary groups, of combined order at most $16(q^{1/2})^{15}<q^9$, by Proposition 3.3. There are at most four classes of orthogonal groups, with total order sum less than $32q^6<q^8$ by Proposition 3.3. Since q is not prime, there are no further maximal subgroups.

Our first set of subgroups contains only $P2$; our second contains $P_{1,3}$; our third contains the subgroup of type $GL_1(q)\oplus GL_3(q)$, the C_2- and the $C3$-subgroups, the unitary and the orthogonal subgroups; our fourth is the $C5$-subgroups; and our fifth is the symplectic subgroups. Eq. (8) is now proved.

By Lemma 4.2

$$P_{X,S}(3)-P_S(2)\geqslant\frac{(q-1)^2((q^3-1)^2-5(q^2-1)^2)}{(q^3-1)^2(q^4-1)^2}>\frac{(q-1)^2(q^6-q^5)}{q^{14}}$$

The result now follows from Lemma 4.1□ .

Proposition 4.7

Let n=6. Then X satisfies property P.

Proof

First we show that

$P_{X,S}(3)\geq 1-3_q{}^{-16}$ (9)

We take the lists of maximal subgroups of X from [2, Tables 8.24 and 8.25]. As before, we shall work in $SL_6(q)$. We shall divide the conjugacy class representatives of the maximal subgroups into three sets, such that the sum of the orders of the subgroups in each set is at most q^{26}. Proposition 3.3 then implies (9).

There is one class of each type of $C1$-subgroup. The subgroup P_3 has order less than q^{26}, and forms the first set on its own. The subgroup $P_{1,5}$ has order less than q^{26}, and forms the second set on its own.

The remaining groups all form one set. The subgroup $P_{2,4}$ has order less than q^{23}. The subgroup of type $GL_1(q)\oplus GL_5(q)$ has order less than q^{25}. The subgroup of type $GL_2(q)\oplus GL_4(q)$ has order less than q^{19}.

Next we consider the C_2-, C_3- and C_4-subgroups, of which there is one class of each type. The C_2-subgroups of types $GL_1(q)\wr S6$, $GL_2(q)\wr S3$ and $GL_3(q)\wr S_2$ have order less than q^8, q^{12} and q^{18}, respectively, since $q\geq 9$. The $C3$-subgroups of types $GL_3(q^2)$ and $GL_2(q^3)$ have order less than q^{18} and q^{12}, respectively. The $C4$-subgroups have order less than q^{11}.

Now consider the C_5-subgroups. There are at most 6*log log q* classes, each of order at most $6q^{35/2}$. Thus we bound the sum of their orders by q^{20}.

Next we consider the C_8- and C_9-subgroups. There are at most six classes of orthogonal groups, of total order less than q^{17}. There are at most three classes of symplectic groups, of total order at most q^{22}. There are at most six classes of unitary groups, of total order at most q^{20}. Since $d(X)=3$, the C^9-subgroups have order sum less than $4|SL_3(4)|<q^{10}$, so (9) now follows.

The result is now immediate from Lemmas 4.1 and 4.2. □

5. The orthogonal groups X with $d(X)=3$

Throughout this section, we let $S=\Omega^+_n(q)$ for $n\geq 8$, and let $\bar{S}=S/Z(S)$ be the corresponding simple group. We let X be almost simple with socle $\bar{S}$, and assume throughout that $d(X)=3$. Note that this implies that $q\geq 9$. We shall show that X has property P. Information about the maximal subgroups of X is taken from [12] and [2, Chapter 2].

The following is a straightforward calculation, using Proposition 3.3.

Lemma 5.1

The value of $|\mathrm{Out}\,\bar{S}|/|\bar{S}|$ *is less than* $q^{-(n2/2-n/2-3)}$.

Lemma 5.2 *The probability* $P_{\bar{S}}(2)$ *is bounded above by*

$$P_{\bar{S}}(2) \leqslant 1 - \frac{2(q-1)^2}{(q^{n-1}+q^{n/2}-q^{n/2-1}-1)^2} + \frac{1}{q^{4n-10}}$$

Proof

The group S has a permutation action on $(q^{n/2}-1)(q^{n/2-1}+1)/(q-1)$ points, on the cosets of the parabolic $P1$. Here $P1$ is the stabilizer of a totally singular 1-space $\langle v1\rangle$, and has shape $[q^{n-2}].(\frac{q-1}{2}\times\Omega^+_{n-2}(q)).2$.

Let $g\in S$ send $v1$ to $v2\in\langle v1\rangle\perp\backslash\langle v1\rangle$. Then $K:=P_1\cap P^{g}{}_1$ has shape $[q^{1+2(n-4)}].(((q-1)/2)^2\times\Omega^+_{n-4}(q)).22$. Thus

$[S{:}K]=q(q^{n/2}-1)(q^{(n-4)/2}+1)(q^{n-2}-1)/(q-1)2\geq q^{2n-5}$,

since $n\geq 8$ and $q\geq 9$. If two elements of S both lie in $P1$, or in $P^{g}{}_1$, then they do not generate S. Thus $P_{\bar{S}}(2)\leq 1-[S{:}P_1]^{-2}-[S{:}P^{g}{}_1]^{-2}+[S{:}K]^{-2}$, and the result follows. □

In Lemmas 5.3 to 5.9 we find upper bounds for the sum of the orders of the preimages in S of the intersection of various classes of maximal subgroups of X with $\bar{S}$, up to X-conjugacy. We describe this as their 'contribution to the subgroup order sum'.

Lemma 5.3

The contribution to the subgroup order sum of the maximal subgroups P_2, P_3,…,$P_{n/2}$ is at most $(1/2)q^{n^2/2-39n/16+7}$ *when* $n\geq 14$, *and at most* q^{51} *when* $n=12$.

Proof

There is one class of each P_k except when $k\geq n/2-1$. When $k=n/2$ there are two or no classes, depending on $X/\overline{S}$, and when $k=n/2-1$ there are no classes or one class, with the same conditions on $X/\overline{S}$.

If $k<n/2$, then

$$|P_k| = q^{k\left(n-\frac{1+3k}{2}\right)}|\mathrm{GL}_k(q)||\Omega^+_{n-2k}(q)| < 1/2q^{(n^2-n-2nk+k+3k^2)/2},$$

(10)

whilst $|P_{n/2}|$ is half of this.

For $n=12$ the result now follows from (10) and a routine calculation, so assume for the rest of the proof that $n\geq 14$. There are at most $n/2-1$ classes, and we claim that each group has order less than $(1/2)q^{n^2/2-5n/2+7}$. Since $n/2-1<q^{n/16}$ for these n and q, the result will follow.

For $k=2$ the claim follows from (10). For other k, it suffices to show

$(n^2-5n+14)-(n^2-n-2nk+k+3k^2)=-3k^2+k(2n-1)-4n+14\geq 0$.

The left hand side is a quadratic in k with negative k^2 coefficient, which is positive when $k=3$, and non-negative when $k=n/2$, so it must be positive for all values of k in between. □

Lemma 5.4

Let K be the groups of type $\mathrm{GO}^{\epsilon}_m(q)\perp\mathrm{GO}^{\epsilon}_{n-m}(q)$, *with* $m\geq 2$. *The contribution of K to the subgroup order sum is at most* $q^{n^2/2-9n/4+4}$.

Proof

If m is odd then there are up to two conjugacy classes, whilst if m is even then there is one class with $\epsilon=+$ and one with $\epsilon=-$.

Each group has size at most $|\mathrm{SO}^{\epsilon}_m(q)||\mathrm{SO}^{\epsilon}_{n-m}(q)|\leq 4q^{\frac{1}{2}(n^2+2m^2-n(2m+1))}$. Substituting $m+1$ in place of m, and using $2m\leq n-2$, shows that this decreases with increasing m, so is bounded above by $4q^{\frac{1}{2}(n^2-5n+8)}$. Bounding $4(n-4)$ by $q^n/4$ gives the result. □

Lemma 5.5

The contribution of the C2-subgroups to the subgroup order sum is less than $q^{8n2/25+n+1}$ *when* n≥14, *and less than* q^{37} *when* n=12.

Proof

The result for n=12 is a straightforward calculation, using [2, Table 8.82], so assume n≥14.

Since $d(X)$=3, there are no groups of type $GO_1(p)\wr Sn$ or $GO_{n/2}(q)^2$. There are at most two classes of type $GL_{n/2}(q)$, each of order at most $2q^{n2/4}$.

This leaves only type $GO^{\epsilon}_m(q)\wr S_t$, where n=mt with m>1. For each m there are at most two classes, and the group order is bounded above by

$$\begin{aligned}|SO^{\epsilon}_m(q)|^t 2^{2(t-2)}t! &< q^{n(m-1)/2}2^{2(t-2)}t! \\ &< q^{n(n/t-1)/2}t^{2t} = q^{n(n/t-1)/2+2t\log_q t} \\ &< q^{n(n/4-1/2+2\log_9 n)} < q^{n(8n/25+15/16)}.\end{aligned}$$

The number of choices for t is at most $n/2-1\leq q^{n/16}$. Thus the total contribution of these groups is $2q^{8n2/25+n}$. The result follows. □

Lemma 5.6

The contribution of the C_3*-subgroups to the subgroup order sum is less than* $q^{n2/4+1}$.

Proof

If $n\equiv 0$ mod 4, then there are up to two classes of groups of order at most $(q+1)|SU_{n/2}(q)|<2q^{n2/4}$. Conversely, if $n\equiv 2$ mod 4, then there are up to two classes of groups of order at most $2|SO^0_{n/2}(q^2)|<2q^{n2/4-n/2}$. Thus these two types contribute at most $4q^{n2/4}$ between them.

The remaining type is $GO^+_{n/s}(q^s)$. For s=2 there are up to two classes, of order at most $2|SO^+_{n/2}(q^2)|<2q^{n2/4-n/2}$. For odd s there are *log* $n<\frac{2}{3}q^{n/2}$ classes, each of order at most $3|\Omega^+_{n/3}(q^3)|<\frac{3}{2}q^{n2/6-n/2}$. So for s odd the order sum is at most $q^{n2/6}$. Thus the $C3$-groups contribute less than $5q^{n2/4+qn2/6}$, as required. □

Lemma 5.7

The contribution of the $C_4\cup C_7$*-subgroups to the subgroup order sum is at most* $q^{n2/8+n/2+3}$.

Proof

We work our way through the types of C_4- and C_7-subgroups. Consider first type $Sp_m(q) \otimes Sp_{n/m}(q)$. When these groups are maximal, there are two classes. Since both m and n/m are even, and $m<\sqrt{n}$, there are at most $n/8$ possible m. Each group has order at most $|Sp_m(q)||Sp_{n/m}(q)|<q^{\frac{1}{2}(m2+m+n2/m2+n/m)}$. Substitute $m+2$ for m, and use $m(m+2)\leq n$, to see that this is non-increasing with m, and so is maximal when $m=2$. Thus the sum of the group orders is less than $(n/4)q^{n2/8+n/4+3}<(1/2)q^{n2/8+n/2+3}$.

Consider next type $GO^{\epsilon_1}_{n_1}(q) \otimes GO^{\epsilon_2}_{n_2}(q)$. Here $n1\geq 4$ with n1 even, $n_2\geq 3$ and $n_1n_2=n$. If n_2 is odd then $\epsilon_1=+$ and there is a unique conjugacy class, so there are at most $n/4$ groups. Otherwise (ϵ_1,ϵ_2) is (+,+), (+,−) or (−,−). If $\epsilon_1=\epsilon_2$ then without loss of generality $n_1<\sqrt{n}$, and since n≥24 for this type to exist we deduce n1<n/4. For each such n_1 we get up to two classes of each type. If $(\epsilon_1,\epsilon_2)=(+,-)$ then n_1 and n_2 are even, so there are at most $n/4$ possible n_1, and for each n_1 we get up to two classes. So we have found at most $7n/4$ classes of groups. Each group has order at most $2|SO^{\epsilon_1}_m(q)||SO^{\epsilon_2}_{n/m}(q)| < 8q^{m^2/2-m/2+n^2/2m^2-n/2m}$.

Substituting $m+1$ for m gives a non-increasing function for $m<\sqrt{n}$, so the maximum of this bound is when $m=3$. Thus we bound the sum of the group orders by $7n/4 \cdot 8q^{n2/18-n/6+3}<q^{n2/18+n/6+3}$.

Consider next the C_7-subgroups, which are only maximal when $t\leq 3$. For type $Sp_m(q)\wr S_t$, we require $t=2$ for maximality, so we find one class, of order at most $|Sp_{\sqrt{n}}(q)|2<q^{n+\sqrt{n}}$. For type $GO^+_m(q)\wr$St, when $t=2$ there is one class, of order $|SO^+_{\sqrt{n}}(q)|^2 \cdot 2 < 2q^{n-\sqrt{n}}$ and when $t=3$ there are two classes, of total order at most $96|SO^+_{n1/3}(q)|^3<q^{3/2n2/3-3/2n1/3+2}$. For type $GO^-_m(q)\wr$St, when $t=2$ there are at most two classes, of total order at most $8|SO^-_{\sqrt{n}}(q)|2<q^{n-\sqrt{n}+2}$, and when $t=3$ there are at most two classes, of total order at most $96|SO^-_{n1/3}(q)|^3<q^{3/2n2/3-3/2n1/3+4}$. The order sum of the non-symplectic groups is less than $12q^{n+\sqrt{n}}$, so the total contribution from $C7$-subgroups is less than $q^{n+\sqrt{n}+1}$. The result follows. □

Lemma 5.8

The contribution of the C_5-subgroups of X to the subgroup order sum is at most $q^{n2/4-n/4+2}$.

Proof

There are at most two classes of types $GO^-_n(q^{1/2})$ and $GO^+_n(q^{1/2})$, and *log log q* of type $GO^+_n(q^{1/r})$. Groups of type $GO^-_n(q^{1/2})$ have order at most $2|SO^-_n(q^{1/2})|<4q^{n2/4-n/4}$, whilst groups of type $GO^+_n(q^{1/2})$ have order at most $2|SO^+_n(q^{1/2})|<2q^{n2/4-n/4}$. Type $GO^+_n(q^{1/r})$ has group order at most $q^{n2/6-n/6}$, since $r\geq 3$. Thus we bound the sum by $12q^{n2/4-n/4}+(\log\log q)q^{n2/6-n/6}\leq q^{n2/4-n/4+3/2}+q^{n2/6-n/6+1}$. □

Lemma 5.9

The contribution of the $C_2\cup\cdots\cup C_9$*-subgroups to the subgroup order sum is at most* $2q^{8n2/25+n+1}$ *when* $n\geq 14$*, and at most* $2q^{44}$ *when* $n=12$.

Proof

We first consider $n\neq 12$. By Lemma 5.5 the sum of the orders of the maximal $C2$-subgroups is at most $N:=q^{8n2/25+n+1}$. We divide the remaining maximal subgroups under consideration into four sets, which in total sum to N.

The first is the maximal $C3$-subgroups, which by Lemma 5.5 have order summing to $q^{n2/4+1}$. The second is the maximal $C_4\cup C_7$-subgroups, which by Lemma 5.7 have order summing to $q^{n2/8+n/2+3}$. The fourth is the maximal C_5-subgroups, which by Lemma 5.8 have order summing to $q^{n2/4-n/4+2}$.

The final set is the maximal C_9-subgroups. By [13] any such subgroup of X either has order at most q^{3n} (in X) or is almost simple, with socle A_{n+1} or A_{n+2}. The groups with socle A_{n+1} or A_{n+2} arise only over prime fields. By Theorem 3.1 there are at most $2n^{5/2}+n \log\log q$ classes of subgroups, and $2n^{5/2}+n\log\log q<q^{2n/3/2}$ since $n\geq 12$ and $q\geq 9$. So the sum of the orders of the maximal C_9-subgroups is at most $q^{3n+2n/3}$.

Classes C_6 and C_8 are empty, so the result for $n\geq 14$ follows. For $n=12$ we perform a similar calculation, but using the more precise value in Lemma 5.5, and noting that the largest contribution is from Class C_9. □

Theorem 5.10

Let $n\geq 12$*. Then* X *satisfies property* P.

Proof

To bound $P_{X,\bar{S}}(3)$ we consider two individual conjugacy classes of groups, namely the groups of type $P1$ and $GO1(q)\perp GO_{n-1}(q)$, and three sets of groups, namely the remaining parabolic subgroups, the remaining completely reducible subgroups, and all other maximal subgroups.

If $n \geq 18$, then $2q^{8n^2/25+n+1}+\frac{1}{2}q^{n^2/2-39n/16+7}<q^{n^2/2-9n/4+4}$, so we use Lemmas 5.3, 5.4 and 5.9 to bound the sum of the three sets of groups by $2q^{n^2/2-9n/4+4}$. If $n \in \{12,14,16\}$ then we bound the sum of the orders of these groups by $2q^{51}$, $3q^{78/2}$ and q^{99}, respectively.

The parabolic group $P1$ has order $q^{n-2}(q-1)|\Omega^+_{n-2}(q)|$, and there are two classes of groups of type $GO_1(q) \perp GO_{n-1}(q)$ each of order $|SO_{n-1}(q)|$.

Thus, using Proposition 3.3, for $n \geq 18$ we calculate

$$P_{X,S}(3) \geqslant 1-\left(\frac{q^{n-2}(q-1)|\Omega^+_{n-2}(q)|}{|\Omega^+_n(q)|}\right)^2-2\left(\frac{|SO_{n-1}(q)|}{|\Omega^+_n(q)|}\right)^2-\left(\frac{2q^{n^2/2-9n/4+4}}{\frac{1}{2}q^{n^2/2-n/2-1}}\right)^2$$

$$\geqslant 1-\frac{(q-1)^2}{(q^{n-1}+q^{n/2}-q^{n/2-1}-1)^2}-\frac{8}{q^{n-2}(q^n-2q^{n/2}+1)}-\frac{16}{q^{7n/2-10}}.$$

The result then follows from Lemmas 5.1 and 5.2. For $n \in \{12,14,16\}$ we carry out a similar calculation. □

Proposition 5.11

Let $n=10$. Then X satisfies property P.

Proof

We use [2, Tables 8.66 and 8.67]. Consider first the parabolic subgroups, of which there is one class of each. The group $P1$ has order less than $\frac{1}{2}q^{37}$, whilst the orders of P_2, P_3 and P_4 are less than $\frac{1}{2}q^{32}$, $\frac{1}{2}q^{30}$, and $\frac{1}{2}q^{31}$, respectively. The group P_5 is non-maximal, since $d(X)=3$.

Consider next the completely reducible subgroups. We shall multiply the order of the stabilizers of odd-dimensional subspaces by 2, since there are at most two classes. Thus the sum of the orders of the groups of type $GO_1(q) \perp GO_9(q)$, $GO^{\pm}_2(q) \perp GO^{\pm}_8(q)$, $GO_3(q) \perp GO_7(q)$ and $GO^{\pm}_4(q) \perp GO^{\pm}_6(q)$ are less than $2|SO_9(q)|$, $5q^{29}$, $2q^{24}$ and $5q^{21}$, respectively. We bound the sum of the last three of these by q^{30}.

Moving on to the C_2-subgroups, type $GO^+_2(q) \wr S_5$ contributes less than $960|SO^+_2(q)|5<q9$. The order sum for types $GO_5(q) \wr S_2$ and $GO_5(q)^2$ is less than $4|SO_5(q)|^2<4q^{20}$. There is one class of groups of type $GL_5(q).2$, of order at most q^{25}. There are at most two classes of C_3-subgroups, of combined order less than $4q^{20}$. There are at most *log log q* classes of C_5-subgroups of type $\Omega^+_{10}(q_0)$, of combined order at most q^{16}. There are two classes of groups of each of the types $SO^{\pm}10(q^{1/2})$, of combined order at most $2q^{45/2}<q^{23}$.

These complete the maximal subgroups of X. Thus the sum of the C_2-, C_3- and C_5-subgroups of X is at most $2q^{25}$, and so the sum of the orders of all maximal subgroups of X, up to X-conjugacy, is $|P_1|+2|SO_9(q)|+q^{32}$. The result now follows from Lemmas 5.1 and 5.2. □

We finish with n=8. Detailed maximal subgroup information is taken from [2, Table 8.50], and it may help the reader to consult this table.

Lemma 5.12

Let n=8. Then

$$P_{\bar{S}}(2) \leqslant 1 - \frac{4(q-1)^2}{(q^3+1)^2(q^4+1)^2} + \frac{6}{q^{22}}$$

Proof

Let the basis of the natural vector space be $e1,\ldots,e4,f4,\ldots,f1$, so that the bilinear form has antidiagonal matrix with all nonzero entries 1. The group S has a permutation action on $(q^3+1)(q^4-1)/(q-1)$ points, on the cosets of $H_1=S\ \langle_{e1}\rangle$, of shape $[q^6].(\frac{q-1}{2}\times\Omega_6^+(q)).2$. For $i\in\{2,3,4\}$, let H_i be the conjugate of H that stabilizes $\langle e_i\rangle$.

If $i\neq j$ then $H_{ij}:=H_i\cap H_j$ has shape

$[q^9].(((q-1)/2)^2\times\Omega^+_4(q)).2^2$.

Thus $[S:H_{ij}]=q(q^2+1)(q^3+q^2+q+1)(q^5+\ldots+1)>q11$. Furthermore, if i, j, k are distinct then $H_{ijk}:=H_i\cap H_j\cap H_k$ has shape

$[q^9].(((q-1)/2)^3\times\Omega^+_2(q)).23$

and so has index $q^3(q+1)(q^3+q^2+q+1)^2(q^5+\ldots+1)<q^{16}$ in S. The group $H_{1234}=H_{123}\cap H_4$ has shape $[q^6].((q-1)/2)^4.2^3$ and so has index $q^6(q+1)(q^3+q^2+q+1)^2(q^5+\ldots+1)>q^{18}$ in S.

Hence

$$\begin{aligned}P_{\bar{S}}(2) &\leqslant 1-4[S:H_1]^{-2}+6[S:H_{12}]^{-2}-4[S:H_{123}]^{-2}+[S:H_{1234}]^{-2}\\ &\leqslant 1-4(q-1)^2\left(q^3+1\right)^{-2}\left(q^4-1\right)^{-2}+6q^{-22}-4q^{-32}+q^{-36}\\ &\leqslant 1-4(q-1)^2\left(q^3+1\right)^{-2}\left(q^4-1\right)^{-2}+6q^{-22}\end{aligned}$$

and the result follows.

Proposition 5.13

Let n=8. Then X satisfies property P.

Proof

We shall work through the 42 rows of [2, Table 9.50], and divide the maximal subgroups of X that complement $\bar{S}$ into three families: the groups that are Aut $\bar{S}$-conjugate to P_1, the groups that are Aut $\bar{S}$-conjugate to the C_1-subgroups of type $GO_1(q) \oplus GO7(q)$, and the rest. The result will then follow from Lemmas 3.5, 5.1 and 5.12.

Rows 3, 9, 10, 11, 14, 17, 20 to 24, 30, 32, and from 35 onwards, do not occur for $d(X)=3$, so can be excluded from our calculations.

Rows 1 and 2 describe the groups that are Aut $\bar{S}$-conjugate to P_1. The sum of the squares of the reciprocals of their indices is $3(q-1)^2/((q^3+1)^2(q^4-1)^2)$.

Rows 7 and 8 describe the groups of type $GO_1(q) \oplus GO_7(q)$. The sum of the squares of the reciprocals of their indices is less than $24/(q^6(q^4-1)^2)$.

Rows 4 to 6 describe C_1-subgroups, of total order less than $2q^{19}$. Rows 12 and 13 describe C_1-subgroups, of total order less than $3(q-1)q^{15}$. Rows 15 and 16 describe C_1-subgroups, of total order less than $3(q+1)q^{15}$. These four rows describe groups of total order less than $6q^{16}$. Rows 18 and 19 describe C_1-subgroups, of total order less than $6q^{13}$.

Rows 25 and 26 describe C_2-subgroups of total order less than $1732q^4<q^8$. Rows 27 to 29 describe C_2-subgroups of total order less than $8q^{12}<q^{13}$.

Row 31 is C_5-subgroups, of total order less than *log log q* $(q^{1/3})^{28}/2<q^{10}$. Rows 33 and 34 describe C_5-subgroups, of total order less than $12q^{14}$.

We bound the sum of the orders of the class representatives of all of the maximal subgroups of X that supplement $\bar{S}$, other than those that are Aut $\bar{S}$-conjugate to groups of type P_1 or $GO_1(q) \oplus GO_7(q)$, by $3q^{19}$. Thus the sum of the squares of the reciprocals of their indices is at most $36/_q{}^{16}$, and the result follows. □

6. Proof of Corollary 1.3

Proof of Corollary 1.3

Let $N=S_1\times\ldots\times S_r$, set $X=N_L(S_1)/CL(S_1)$, and identify S with soc(X). By [4, Theorem 2.7] we know that $f_L(d) = \frac{P_{L,N}(d)|N|^d}{|C_{\mathrm{Aut}\,N}(L/N)|}$, where $d=d(L)$. By Proposition 1.1, there exists a 2-generated group Y with $S \trianglelefteq Y \leq X$ such that $P_{L,N}(d) \geq P_{Y,S}(2)$. In [5, proof of Lemma 1] it is shown that

$$|C_{\mathrm{Aut}\ N}(L/N)| \leqslant r|S|^{r-1}|C_{\mathrm{Aut}\ S}(X/S)|.$$

Thus $|C_{\mathrm{Aut}\ N}(L/N)| \leqslant r|S|^{r-1}|C_{\mathrm{Aut}\ S}(Y/S)|$ and

$$f_L(d) = \frac{P_{L,N}(d)|N|^d}{|C_{\mathrm{Aut}\ N}(L/N)|} \geqslant \frac{P_{Y,S}(2)|N|^d}{r|S|^{r-1}|C_{\mathrm{Aut}\ S}(Y/S)|}. \quad (11)$$

Now $f_Y(2) = \frac{P_{Y,S}(2)|S|^2}{|C_{\mathrm{Aut}\ S}(Y/S)|}$ by [4, Theorem 2.7], so that (11) becomes

$$f_L(d) \geqslant f_Y(2)\frac{|N|^{d-1}}{r|S|}$$

and the first part of the corollary is proved.

The second part of the corollary follows from (11), an explicit calculation using [14, Table 1] of $P_{Y,S}(2)/|\mathrm{Out}\ S|$ for those groups Y with $P_{Y,S}(2)\leq 0.9$, and the bound $|\mathrm{Out}\ S|\leq 6/7\ log|S|$ given in [14, Lemma 2.1] for the general case. □

REFERENCES

1. B. Amberg, L. Kazarin**Large subgroups of a finite group of even order**Proc. Amer. Math. Soc., 140 (1) (2012), pp. 65-68
2. John N. Bray, Derek F. Holt, Colva M. Roney-Dougal**The Maximal Subgroups of the Low Dimensional Finite Classical Groups**London Math. Soc. Lecture Note Ser., vol. 407, Cambridge University Press, Cambridge (2013)
3. F. Dalla Volta, A. Lucchini**Generation of almost simple groups**J. Algebra, 178 (1) (1995), pp. 194-223
4. F. Dalla Volta, A. Lucchini**Finite groups that need more generators than any proper quotient**J. Aust. Math. Soc. A, 64 (1) (1998), pp. 82-91
5. F. Dalla Volta, A. Lucchini**The smallest group with non-zero presentation rank**J. Group Theory, 2 (2) (1999), pp. 147-155
6. E. Detomi, A. Lucchini**Probabilistic generation of finite groups with a unique minimal normal subgroup**J. Lond. Math. Soc., 87 (3) (2013), pp. 689-706
7. J. Fulman, R. Guralnick**Derangements in simple and primitive groups**A. Ivanov, *et al.* (Eds.), Groups, Combinatorics, and Geometry, Durham, 2001, World Sci. Publ., River Edge, NJ (2003), pp. 99-121
8. W. Gaschütz**Die Eulersche Funktion endlicher auflösbarer Gruppen**Illinois J. Math., 3 (1959), pp. 469-476
9. Jokke Häsä**Growth of cross-characteristic representations of finite quasisimple groups of Lie type**arXiv:1112.3941v1 (2011)
10. M. Herzog**On centralizers of involutions**Proc. Amer. Math. Soc., 22 (1969), pp. 170-174
11. Gerhard Hiss, Gunter Malle**Low-dimensional representations of quasi-simple groups**LMS J. Comput. Math., 4 (2001), pp. 22-63

12. Peter Kleidman, Martin Liebeck**The Subgroup Structure of the Finite Classical Groups**London Math. Soc. Lecture Note Ser., vol. 129, Cambridge University Press, Cambridge (1990)
13. Martin W. Liebeck**On the orders of maximal subgroups of the finite classical groups**Proc. Lond. Math. Soc. (3), 50 (1985), pp. 426-446
14. Nina E. Menezes, Martyn Quick, Colva M. Roney-Dougal**The probability of generating a finite simple group**Israel J. Math., 198 (1) (2013), pp. 371-392
15. P. Rowley**Finite groups admitting a fixed-point-free automorphism group**J. Algebra, 174 (2) (1995), pp. 724-727

CHAPTER 6

Singular Equivalence and The (Fg) Condition

Øystein Skartsæterhagen.

Institutt for matematiske fag, NTNU, N-7491 Trondheim, Norway

ABSTRACT

We show that singular equivalences of Morita type with level between finite-dimensional Gorenstein algebras over a field preserve the **(Fg)** condition.

Keywords: Finite generation condition Hochschild cohomology Singular equivalences

1. INTRODUCTION

Throughout the paper, we let k be a fixed field.

Support varieties for modules over a group algebra kG were introduced by J.F. Carlson in [4], using the group cohomology ring $H^*(G,k)$. Later, Snashall and Solberg [18] defined support varieties for modules over an arbitrary finite-dimensional k-algebra Λ, using the Hochschild cohomology ring $HH^*(\Lambda)$.

We say that a finite-dimensional k-algebra Λ satisfies the **(Fg)** condition if the Hochschild cohomology ring $HH^*(\Lambda)$ of Λ is Noetherian and the Yoneda algebra $\mathrm{Ext}^*_\Lambda(M,M)$ is a finitely generated $HH^*(\Lambda)$-module for every finitely generated Λ-module M (for more details, see the definition in Section 7). It was shown in [9] that many of the results for support varieties over a group algebra also hold for support varieties over a selfinjective algebra which satisfies the **(Fg)** condition. We can thus think of the **(Fg)** condition as a criterion for deciding whether a given algebra has a nice theory of support varieties.

It is therefore interesting to investigate whether the **(Fg)** condition holds for various algebras, and to find out which relations between algebras preserve the **(Fg)** condition. This question has been considered in [16] for algebras whose module categories are related by a recollement of abelian categories, and in [12] for derived equivalence of algebras. In this paper, we consider singular equivalence of algebras.

The *singularity category* $D_{sg}(\Lambda)$ of a k-algebra Λ, introduced by Buchweitz in [3], is defined as the Verdier quotient

$D_{sg}(\Lambda)=D^b(\mathrm{mod}\Lambda)/\mathrm{perf}(\Lambda)$

of the bounded derived category $D_b(\mathrm{mod}\Lambda)$ by the subcategory of perfect complexes. This is a triangulated category. We say that two k-algebras Λ and Σ are *singularly equivalent* if there exists a triangle equivalence $f:D_{sg}(\Lambda)\rightarrow D_{sg}(\Sigma)$ between their singularity categories, and the functor f is then called a *singular equivalence* between the algebras Λ and Σ.

The purpose of this paper is to investigate to what extent singular equivalences preserve the **(Fg)** condition. Since arbitrary singular equivalences are hard to work with and do not necessarily have nice properties, we restrict our attention to special classes of singular equivalences.

A *singular equivalence of Morita type* (introduced by Chen and Sun in [7]) between k-algebras Λ and Σ is a singular equivalence

$$D_{sg}(\Lambda) \xrightarrow{N\otimes_\Lambda -} D_{sg}(\Sigma)$$

which is induced by a tensor functor $N\otimes_\Lambda-$, where N is a Σ–Λ bimodule subject to some technical requirements. Wang [19] has introduced a generalized version of singular equivalence of Morita type called *singular equivalence of Morita type with level*. We recall the definitions of these two types of singular equivalences in Section 2. The question we want to answer in this paper is: Do singular equivalences of Morita type with level preserve the **(Fg)** condition?

All algebras that satisfy the **(Fg)** condition are Gorenstein algebras (see Theorem 7.2), and singular equivalences of Morita type with level do not preserve Gorensteinness. Moreover, even if one of the algebras involved in a singular equivalence of Morita type with level satisfies the **(Fg)** condition, the other algebra does not need to be a Gorenstein algebra (see Example 7.5). This means that the **(Fg)** condition is in general not preserved under singular equivalence of Morita type with level.

However, we can consider the question of whether it is only when one of the algebras is non-Gorenstein that such counterexamples arise. In other words, if we require all our algebras to be Gorenstein, is it then true that singular equivalences of Morita type with level preserve the **(Fg)** condition? The main result of this paper, Theorem 7.4, answers this question affirmatively: A singular equivalence of Morita type with level between finite-dimensional Gorenstein algebras over a field preserves the **(Fg)** condition. As a consequence of this, we obtain a similar statement for *stable* equivalence of Morita type (Corollary 7.6), where we do not need the assumption of Gorensteinness.

A result similar to the main result of this paper has been shown independently, and in a different way, in a recent preprint by Yiping Chen.

The content of the paper is structured as follows.

In Section 2, we state the definitions of singular equivalence of Morita type and singular equivalence of Morita type with level, and look at some easily derived consequences.

In Section 3, we begin to look at what more we can deduce from a singular equivalence of Morita type with level when the assumption of Gorensteinness is added. We recall the well-known result stating that the singularity category of a Gorenstein algebra is equivalent to the stable category of maximal Cohen–Macaulay modules. This implies that a singular equivalence

$$f\colon \mathsf{D}_{\mathsf{sg}}(\Lambda) \xrightarrow{\simeq} \mathsf{D}_{\mathsf{sg}}(\Sigma)$$

between Gorenstein algebras Λ and Σ gives an equivalence

$$g\colon \underline{\mathrm{CM}}(\Lambda) \xrightarrow{\simeq} \underline{\mathrm{CM}}(\Sigma)$$

between their stable categories of maximal Cohen–Macaulay modules. We show that if the singular equivalence f is of Morita type with level, and thus induced by a tensor functor, then the equivalence g is induced by the same tensor functor.

In Section 4, we consider certain maps of the form

$$\mathrm{Ext}^n_\Lambda(U,V) \to \mathrm{Ext}^n_\Lambda(\Omega^i_\Lambda(U),\Omega^i_\Lambda(V))$$

which we call rotation maps. We show that these maps are isomorphisms if the algebra Λ is Gorenstein and $n>\mathrm{id}_\Lambda\Lambda$. This means that in extension groups of sufficiently high degree over a Gorenstein algebra, we can replace both modules by syzygies. This result is used in the following three sections.

In Section 5, we show that if we have a singular equivalence of Morita type with level

$$\mathsf{D}_{\mathsf{sg}}(\Lambda) \xrightarrow[N\otimes_\Lambda -]{\simeq} \mathsf{D}_{\mathsf{sg}}(\Sigma)$$

between two Gorenstein algebras Λ and Σ, then we have isomorphisms

$$\operatorname{Ext}^n_\Lambda(A, B) \xrightarrow[N\otimes_\Lambda -]{\cong} \operatorname{Ext}^n_\Sigma(N \otimes_\Lambda A, N \otimes_\Lambda B)$$

(for A and B in modΛ) (1.1)

between extension groups over Λ and extension groups over Σ, in all sufficiently large degrees n. In the terminology of [16], this implies that a tensor functor inducing a singular equivalence of Morita type with level between Gorenstein algebras is an eventually homological isomorphism. The proof of this result builds on the result about stable categories of Cohen–Macaulay modules from Section 3.

In Section 6, we show that a singular equivalence of Morita type with level between Gorenstein algebras preserves Hochschild cohomology in almost all degrees. That is, if two Gorenstein algebras Λ and Σ are singularly equivalent of Morita type with level, then there are isomorphisms

$HH^n(\Lambda)\cong HH^n(\Sigma)$ (1.2)

for almost all n, and these isomorphisms respect the ring structure of the Hochschild cohomology.

In Section 7, we show the main result of the paper: A singular equivalence of Morita type with level between finite-dimensional Gorenstein algebras over a field preserves the **(Fg)** condition. The main ingredients in the proof of this result are the isomorphism (1.1) of extension groups from Section 5 and the isomorphism (1.2) of Hochschild cohomology groups from Section 6.

2. SINGULAR EQUIVALENCES OF MORITA TYPE WITH LEVEL

In this section, we recall the definitions we need regarding singular equivalences. We begin with the concept of singularity categories.

Definition 2.1 Let Λ be a k-algebra. The *singularity category*$\mathsf{D}_{sg}(\Lambda)$ of Λ is a triangulated category defined as the Verdier quotient

$D_{sg}(\Lambda)=D^b(\text{mod}\Lambda)/\text{perf}(\Lambda)$

of the bounded derived category Db(modΛ) by the subcategory of perfect complexes. We say that two algebras Λ and Σ are *singularly equivalent* if their singularity categories $D_{sg}(\Lambda)$ and $D_{sg}(\Sigma)$ are equivalent as triangulated categories. A triangle equivalence between $D_{sg}(\Lambda)$ and $D_{sg}(\Sigma)$ is called a *singular equivalence* between the algebras Λ and Σ.

The singularity category of an algebra was first defined by Buchweitz in [3, Definition 1.2.2]. In his definition, the singularity category is called the *stabilized derived category*, and it is denoted by $D^b(\Lambda)$. Later, Orlov [15] used the same construction in algebraic geometry to define the *triangulated category of singularities* of a scheme *X*, denoted $D_{Sg}(X)$. We follow the recent convention of using Orlov's terminology and notation for algebras as well. The term *singular equivalence* was introduced by Chen [6].

Analogously to the special type of stable equivalences called *stable equivalences of Morita type*, Chen and Sun have defined a special type of singular equivalences called *singular equivalences of Morita type* in their preprint [7]. This concept was further explored by Zhou and Zimmermann in [20].

Definition 2.2 Let Λ and Σ be finite-dimensional *k*-algebras, and let *M* be a Λ–Σ bimodule and *N* a Σ–Λ bimodule. We say that *M* and *N* induce a *singular equivalence of Morita type* between Λ and Σ (and that Λ and Σ are *singularly equivalent of Morita type*) if the following conditions are satisfied:

1.*M* is finitely generated and projective as a left Λ-module and as a right Σ-module.

2.*N* is finitely generated and projective as a left Σ-module and as a right Λ-module.

3.There is a finitely generated Λ^e-module *X* with finite projective dimension such that $M_{\otimes\Sigma}N\cong\Lambda_{\oplus}X$ as Λ^e-modules.

4.There is a finitely generated Σe-module *Y* with finite projective dimension such that $N_{\otimes\Lambda}M\cong\Sigma_{\oplus}Y$ as Σ^e-modules.

Notice that the definition is precisely the same as the definition of stable equivalence of Morita type, except that the modules *X* and *Y* are not necessarily projective, but only have finite projective dimension. Thus stable equivalences of Morita type occur as a special case of singular equivalences of Morita type.

The following proposition describes how a singular equivalence of Morita type is a singular equivalence, thus justifying the name.

Proposition 2.3 *(See[20, Proposition 2.3].) Let ${}_{\Lambda}M_{\Sigma}$ and ${}_{\Sigma}N_{\Lambda}$ be bimodules which induce a singular equivalence of Morita type between two k-algebras Λ and Σ. Then the functors*

$N\otimes_{\Lambda}-:D_{sg}(\Lambda)\to D_{sg}(\Sigma)$ and $M\otimes_{\Sigma}-:D_{sg}(\Sigma)\to D_{sg}(\Lambda)$

are equivalences of triangulated categories, and they are quasi-inverses of each other.

Inspired by the notion of singular equivalence of Morita type, Wang [19] has defined a more general type of singular equivalence called *singular equivalence of Morita type with level.*

Definition 2.4 Let Λ and Σ be finite-dimensional *k*-algebras, and let *M* be a Λ–Σ bimodule and *N* a Σ–Λ bimodule. Let *l* be a nonnegative integer. We say that *M* and *N* induce a *singular equivalence of Morita type with level l* between Λ and Σ (and that Λ and Σ are *singularly equivalent of Morita type with level l*) if the following conditions are satisfied:

1.*M* is finitely generated and projective as a left Λ-module and as a right Σ-module.

2.*N* is finitely generated and projective as a left Σ-module and as a right Λ-module.

3.There is an isomorphism $M\otimes_{\Sigma}N\cong\Omega^{l}_{\Lambda e}(\Lambda)$ in the stable category mod_Λ^{e}.

4.There is an isomorphism $N\otimes_{\Lambda}M\cong\Omega^{l}_{\Sigma e}(\Sigma)$ in the stable category mod_Σe.

Just as in the case of singular equivalence of Morita type, the conditions in the definition of singular equivalence of Morita type with level are designed to ensure that the functors $N\otimes_{\Lambda}-$ and $M\otimes_{\Sigma}-$ induce singular equivalences.

Proposition 2.5 *(See[19, Remark 2.2].) Let ${}_{\Lambda}M_{\Sigma}$ and ${}_{\Sigma}N_{\Lambda}$ be bimodules which induce a singular equivalence of Morita type with level l between two k-algebras Λ and Σ. Then the functors*

$N\otimes_{\Lambda}-:D_{sg}(\Lambda)\to D_{sg}(\Sigma)$ and $M\otimes_{\Sigma}-:D_{sg}(\Sigma)\to D_{sg}(\Lambda)$

are equivalences of triangulated categories. The compositions

$M\otimes_{\Sigma}N\otimes_{\Lambda}-:D_{sg}(\Lambda)\to D_{sg}(\Lambda)$ and $N\otimes_{\Lambda}M\otimes_{\Sigma}-:Dsg(\Sigma)\to Dsg(\Sigma)$

are isomorphic to the shift functor$[-l]$*on the respective categories* $D_{sg}(\Lambda)$ *and* $D_{sg}(\Sigma)$.

We now show that the notion of singular equivalence of Morita type with level generalizes the notion of singular equivalence of Morita type, in the sense that any equivalence of the latter type is also of the former type. This is mentioned without proof in [19].

Proposition 2.6 *Let* Λ *and* Σ *be finite-dimensional k-algebras. If a functor* $f:D_{sg}(\Lambda)\to D_{sg}(\Sigma)$ *is a singular equivalence of Morita type, then it is also a singular equivalence of Morita type with level.*

Proof Let M, N, X and Y be bimodules satisfying the requirements of a singular equivalence of Morita type, such that $f=(N\otimes_{\Lambda}-)$. Let $l=\max\{pd_{\Lambda^e}X, pd_{\Sigma^e}Y\}$. Let M' be an l-th syzygy of M as Λ–Σ-bimodule, and let

$$0\to M'\to P_{l-1}\to\cdots\to P_0\to M\to 0 \tag{2.1}$$

be the beginning of a projective resolution of M. We show that the bimodules M' and N induce a singular equivalence of Morita type with level l.

If we consider the bimodules in sequence (2.1) as one-sided modules (left Λ-modules or right Σ-modules), then M and the modules $P_0,\ldots,P_{l-1}$ are projective, and thus M' must be projective as well. Thus condition (1) in Definition 2.4 is satisfied. Condition (2) in Definition 2.4 is trivially satisfied, since it is the same as condition (2) in Definition 2.2.

Tensoring sequence (2.1) with N gives the sequence

$$0\to M'\otimes_{\Sigma}N\to P_{l-1}\otimes_{\Sigma}N\to\cdots\to P_0\otimes_{\Sigma}N\to M\otimes_{\Sigma}N\to 0.$$

This sequence is exact since N is projective as left Σ-module, and the modules $P_i\otimes_{\Sigma}N$ are projective Λ^e-modules since N is projective as right Λ-module. The Λ^e-module $M'\otimes_{\Sigma}N$ is therefore an l-th syzygy of $M\otimes_{\Sigma}N$. Since $M\otimes_{\Sigma}N$ is isomorphic to $\Lambda\oplus X$ and the projective dimension of X is at most l, this means that $M'\otimes_{\Sigma}N$ is an l-th syzygy of Λ as Λ^e-module. Similarly, we can show that $N\otimes_{\Lambda}M'$ is an l-th syzygy of Σ as Σ^e-module. This means that conditions (3) and (4) in Definition 2.4 are satisfied. □

In the rest of the paper we work with singular equivalences of Morita type with level. By the above proposition, all results where we assume such an equivalence are also applicable to singular equivalences of Morita type.

As seen above, if ${}_{\Lambda}M_{\Sigma}$ and ${}_{\Sigma}N_{\Lambda}$ are bimodules which induce a singular equivalence of Morita type with level, then the functors $N\otimes_{\Lambda}-$ and $M\otimes_{\Sigma}-$ are equivalences between the singularity categories of Λ and Σ. We end this section by examining some properties of these tensor functors when viewed as functors between the module categories mod Λ and mod Σ.

Lemma 2.7 *Let* ${}_{\Lambda}M_{\Sigma}$ *and* ${}_{\Sigma}N_{\Lambda}$ *be bimodules which induce a singular equivalence of Morita type with level between two k-algebras* Λ *and* Σ. *Then the functors*

$$N\otimes_{\Lambda}-:\mathrm{mod}\Lambda\to\mathrm{mod}\Sigma \text{ and } M\otimes_{\Sigma}-:\mathrm{mod}\Sigma\to\mathrm{mod}\Lambda$$

are exact and take projective modules to projective modules. In particular, this means that they take projective resolutions to projective resolutions.

Proof Consider the functor $N \otimes_\Lambda -$. This functor is exact since N is projective as right Λ-module, and it takes projective modules to projective modules since N is projective as left Σ-module. □

Let Λ, Σ, M and N be as in the above lemma. Since the functor

$$N \otimes_\Lambda - : \text{mod}\Lambda \to \text{mod}\Sigma$$

is exact, it induces homomorphisms of extension groups. By abuse of notation, we denote these maps by $N \otimes_\Lambda -$ as well. More precisely, for Λ-modules U and V and an integer $n \geq 0$, we define a map

$$N \otimes_\Lambda - : \text{Ext}^n_\Lambda (U,V) \to \text{Ext}^n_\Sigma (N \otimes_\Lambda U, N \otimes_\Lambda V). \tag{2.2}$$

For $n=0$, the map $N \otimes_\Lambda -$ simply sends a homomorphism $f: U \to V$ to the homomorphism $N \otimes_\Lambda f: N \otimes_\Lambda U \to N \otimes_\Lambda V$. For $n>0$, the map $N \otimes_\Lambda -$ sends the element represented by the extension

$$0 \to V \to E_n \to \cdots \to E_1 \to U \to 0$$

to the element represented by the extension

$$0 \to N \otimes_\Lambda V \to N \otimes_\Lambda E_n \to \cdots \to N \otimes_\Lambda E_1 \to N \otimes_\Lambda U \to 0$$

obtained by applying the functor $N \otimes_\Lambda -$ to all objects and maps.

The maps (2.2) play an important role later in the paper. In Section 5, we show that if Λ and Σ are Gorenstein algebras, then these maps are isomorphisms for almost all n. This fact is used in the proof of the main theorem (Theorem 7.4).

3. GORENSTEIN ALGEBRAS AND MAXIMAL COHEN–MACAULAY MODULES

So far, we have considered the situation of two k-algebras Λ and Σ, together with bimodules ${}_\Lambda M_\Sigma$ and ${}_\Sigma N_\Lambda$ inducing a singular equivalence of Morita type with level between Λ and Σ. From now on, we restrict our attention to the special case where both Λ and Σ are Gorenstein algebras. In this section, we prove our first result under this assumption, namely Proposition 3.7, which states that the tensor functors $N \otimes_\Lambda -$ and $M \otimes_\Sigma -$ induce triangle equivalences between the stable categories of maximal Cohen–Macaulay modules over Λ and Σ.

We begin by recalling the definition of Gorenstein algebras.

Definition 3.1 A k-algebra Λ is a *Gorenstein algebra* if the injective dimension of Λ as a left Λ-module is finite and the injective dimension of Λ as a right Λ-module is finite:

$$\mathrm{id}_\Lambda({}_\Lambda\Lambda)<\infty \text{ and } \mathrm{id}_{\Lambda^{op}}(\Lambda_\Lambda)<\infty.$$

If Λ is a Gorenstein algebra, then idΛ(ΛΛ) and idΛop(ΛΛ) are the same number, and this number is called the *Gorenstein dimension* of Λ. In later sections, we need the following result about Gorenstein algebras.

Lemma 3.2 *(See[2, Lemma 2.1].) If Λ is a Gorenstein k-algebra with Gorenstein dimension d, then its enveloping algebra Λ^e is a Gorenstein algebra with Gorenstein dimension at most* 2*d.*

We continue by recalling the definition of maximal Cohen–Macaulay modules.

Definition 3.3 Let Λ be a k-algebra. A finitely generated Λ-module C is a *maximal Cohen–Macaulay module* if $\mathrm{Ext}^n_\Lambda(C,\Lambda)=0$ for every positive integer n. We denote the subcategory of mod Λ consisting of all maximal Cohen–Macaulay modules by CM(Λ), and the corresponding stable category modulo projectives by CM (Λ).

In the following lemma, we recall some characterizations of maximal Cohen–Macaulay modules over Gorenstein algebras.

Lemma 3.4 *Let Λ be a finite-dimensional Gorenstein k-algebra and C a finitely generated Λ-module. The following are equivalent.*

1.*C is a maximal Cohen–Macaulay module.*

2.*C has a projective coresolution. That is, there exists an exact sequence*

$0\to C\to P_{-1}\to P_{-2}\to\cdots$

where every P_i is a projective Λ-module.

3.*For every n>0, there is a Λ-module A such that C is an n-th syzygy of A.*

4.*For some $n\geq \mathrm{id}_\Lambda\Lambda$, there is a Λ-module A such that C is an n-th syzygy of A.*

Proof We only need to show that statement (1) implies statement (2); the implications (2)⇒(3)⇒(4) are obvious, and the implication (4)⇒(1) follows directly from Definition 3.3.

We use Theorem 5.4 (b) from [1]. We first describe the notation used in [1] for certain subcategories of a module category.

For a Λ-module T with the property that $\mathrm{Ext}^i_\Lambda(T,T)=0$ for every $i>0$, we define the subcategories ${}^\perp T$ and $\mathcal{X}_T$ of mod Λ. The category ${}^\perp T$ is the subcategory of mod Λ consisting of all modules A such that $\mathrm{Ext}^i_\Lambda(A,T)=0$ for every $i>0$. The category $\mathscr{X}_T$ is the subcategory of ${}^{\perp}T$ consisting of all modules A such that there is an exact sequence

$$0 \to A \to T_0 \xrightarrow{f_0} T_1 \xrightarrow{f_1} T_2 \xrightarrow{f_2} \cdots$$

where T_i is in add T and im f_i is in ${}^{\perp}T$ for every $i \geq 0$.

Theorem 5.4 (b) in [1] says that if T is a cotilting module, then the categories ${}^{\perp}T$ and $\mathscr{X}_T$ are equal.

Now consider the case T=Λ. Since Λ is a Gorenstein algebra, it is a cotilting module, and then by the above we have ${}^{\perp}\Lambda = \mathscr{X}_\Lambda$. Furthermore, ${}^{\perp}\Lambda$ is the category CM(Λ) of maximal Cohen–Macaulay modules. Therefore, every maximal Cohen–Macaulay module is in the category $\mathscr{X}_\Lambda$, and thus it has a resolution of the form

$0 \to C \to P_{-1} \to P_{-2} \to \cdots$

where every Pi is a projective Λ-module. □

We now recall the theorem by Buchweitz which provides the connection we need between singularity categories and stable categories of maximal Cohen–Macaulay modules.

Theorem 3.5 *(See[3, Theorem 4.4.1].) Let* Λ *be a finite-dimensional Gorenstein algebra. Then there is an equivalence of triangulated categories*

$$\underline{\mathrm{CM}}(\Lambda) \xrightarrow{\simeq} \mathsf{D}_{\mathsf{sg}}(\Lambda)$$

*given by sending every object in*CM(Λ)*to a stalk complex concentrated in degree* 0.

A direct consequence of Theorem 3.5 is that if two finite-dimensional Gorenstein algebras Λ and Σ are singularly equivalent, then the categories CM(Λ) and CM(Σ) are triangle equivalent. If the algebras are not only singularly equivalent, but singularly equivalent of Morita type (with level), then there are tensor functors $N \otimes_\Lambda -$ and $M \otimes_\Sigma -$ that induce equivalences between the singularity categories $\mathrm{D_{sg}}(\Lambda)$ and $\mathrm{D_{sg}}(\Sigma)$. What we aim to prove now is that these tensor functors also induce equivalences between the stable categories CM(Λ) and CM(Σ) of maximal Cohen–Macaulay modules. We first show that these functors preserve the property of being a maximal Cohen–Macaulay module.

Lemma 3.6 *Let* ${}_\Lambda M_\Sigma$ *and* ${}_\Sigma N_\Lambda$ *be bimodules which induce a singular equivalence of Morita type with level between two finite-dimensional Gorenstein k-algebras* Λ *and* Σ. *Then the functors*

$N \otimes_\Lambda - : \mathrm{mod}\Lambda \to \mathrm{mod}\Sigma$ and $M \otimes_\Sigma - : \mathrm{mod}\Sigma \to \mathrm{mod}\Lambda$

send maximal Cohen–Macaulay modules to maximal Cohen–Macaulay modules.

Proof Let $n = \max\{\mathrm{id}_\Lambda \Lambda, \mathrm{id}_\Sigma \Sigma\}$ (this is finite since the algebras Λ and Σ are Gorenstein). Let C be a maximal Cohen–Macaulay module over Λ. Then by Lemma 3.4, there is a Λ-module A such that C is an n-th syzygy of A. By Lemma 2.7, the Σ-module $N \otimes_\Lambda C$ is an n-th syzygy of $N \otimes_\Lambda A$, and therefore by Lemma 3.4 it is a maximal Cohen–Macaulay module. □

Finally, we are ready to prove the main result of this section.

Proposition 3.7 *Let ${}_\Lambda M_\Sigma$ and ${}_\Sigma N_\Lambda$ be bimodules which induce a singular equivalence of Morita type with level between two finite-dimensional Gorenstein k-algebras Λ and Σ. Then the functors*

$N \otimes_\Lambda - : \underline{\mathrm{CM}}(\Lambda) \to \underline{\mathrm{CM}}(\Sigma)$ and $M \otimes_\Sigma - : \underline{\mathrm{CM}}(\Sigma) \to \underline{\mathrm{CM}}(\Lambda)$

are equivalences of triangulated categories.

Proof We first check that $N \otimes_\Lambda -$ actually gives a functor from $\underline{\mathrm{CM}}(\Lambda)$ to $\underline{\mathrm{CM}}(\Sigma)$. We know from Lemma 3.6 that it gives a functor from CM(Λ) to CM(Σ). By Lemma 2.7, we see that if f is a map of Λ-modules that factors through a projective module, then the map $N \otimes_\Lambda f$ also factors through a projective module. Thus $N \otimes_\Lambda -$ gives a well-defined functor from $\underline{\mathrm{CM}}(\Lambda)$ to $\underline{\mathrm{CM}}(\Sigma)$.

Consider the diagram

$$\begin{array}{ccc} \underline{\mathrm{CM}}(\Lambda) & \xrightarrow{N \otimes_\Lambda -} & \underline{\mathrm{CM}}(\Sigma) \\ \downarrow \simeq & & \downarrow \simeq \\ \mathsf{D}_{\mathsf{sg}}(\Lambda) & \xrightarrow[\simeq]{N \otimes_\Lambda -} & \mathsf{D}_{\mathsf{sg}}(\Sigma) \end{array}$$

of categories and functors, where the vertical functors are the equivalences from Theorem 3.5, and the functor $N \otimes_\Lambda -$ in the bottom row is an equivalence by Proposition 2.5. The diagram commutes, and therefore the functor $N \otimes_\Lambda -$ in the top row is also an equivalence. □

4. ROTATIONS OF EXTENSIONS

If U and V are modules over an algebra Λ, then dimension shift gives isomorphisms $\mathrm{Ext}^n_\Lambda(U, V) \cong \mathrm{Ext}^{n-i}_\Lambda(\Omega^i_\Lambda(U), V)$ for integers n and i with $n > i > 0$. If the algebra Λ is Gorenstein, then all projective Λ-modules have finite injective dimension. This means that for sufficiently large n (more precisely, $n > \mathrm{id}_\Lambda \Lambda$), we can use projective resolutions to do dimension shifting in the second argument of Ext as well. That is, we have isomorphisms $\mathrm{Ext}^n_\Lambda(U,V) \cong \mathrm{Ext}^{n+i}_\Lambda(U,\Omega^i_\Lambda(V))$. By dimension shifting in both arguments, we then get isomorphisms

$$\mathrm{Ext}^n_\Lambda(U,V) \cong \mathrm{Ext}^n_\Lambda(\Omega^i_\Lambda(U),\Omega^i_\Lambda(V))$$

where we stay in the same degree *n*, but replace both arguments to Ext by their *i*-th syzygies. In this section, we describe such isomorphisms, which we call rotation maps, and which are going to be used several times in later sections.

For defining the rotation maps, we do not need to assume that we are working over a Gorenstein algebra. This however means that the maps are not necessarily isomorphisms. We first define the maps in a general setting, and then in Lemma 4.2 describe the conditions we need for ensuring that they are isomorphisms.

Definition 4.1 Let Λ be a finite-dimensional *k*-algebra, and let *U* and *V* be finitely generated Λ-modules. Choose projective resolutions $\pi:\cdots\to P_1\to P_0\to U\to 0$ and $\tau:\cdots\to Q_1\to Q_0\to V\to 0$ of the modules *U* and *V*. Let *i* and *n* be integers with $i<n$, and let

$\pi_i:0\to\Omega^i_\Lambda(U)\to P_{i-1}\to\cdots\to P_0\to U\to 0,$

$\tau_i:0\to\Omega^i_\Lambda(V)\to Q_i-1\to\cdots\to Q_0\to V\to 0$

be truncations of the chosen projective resolutions. We define the *i-th rotation* of the extension group $\mathrm{Ext}^n_\Lambda(U,V)$ with respect to the resolutions π and τ to be the map

$$\rho_i:\ \mathrm{Ext}^n_\Lambda(U,V) \longrightarrow \mathrm{Ext}^n_\Lambda(\Omega^i_\Lambda(U),\Omega^i_\Lambda(V))$$

$$\mathrm{Ext}^n_\Lambda(U,V) \xrightarrow{(\tau_i)_*} \mathrm{Ext}^{n+i}_\Lambda(U,\Omega^i_\Lambda(V)) \xrightarrow{(\pi_i^*)^{-1}} \mathrm{Ext}^n_\Lambda(\Omega^i_\Lambda(U),\Omega^i_\Lambda(V))$$

given by $\rho_i=(\pi^*_i)^{-1}(\tau_i)_*$.

Consider the situation in the above definition. If the algebra Λ is Gorenstein and $n>\mathrm{id}_\Lambda\Lambda$, then for each of the projective modules Q_j, we have $\mathrm{id}_\Lambda Q_j\le\mathrm{id}_\Lambda\Lambda<n$, and thus the map $(\tau_i)^*$ is an isomorphism. This gives the following result.

Lemma 4.2 *Let Λ be a finite-dimensional Gorenstein k-algebra, and let U and V be finitely generated Λ-modules. For every $n>\mathrm{id}_\Lambda\Lambda$ and every $i<n$, the i-th rotation*

$$\mathrm{Ext}^n_\Lambda(U,V) \to \mathrm{Ext}^n_\Lambda(\Omega^i_\Lambda(U),\Omega^i_\Lambda(V))$$

(*with respect to any projective resolutions of U and V*) *is an isomorphism.*

If we look at a rotation map of an extension group $\mathrm{Ext}^n_\Lambda(U,U)$ with the same module in both arguments, then the action of the map can be viewed as a concrete "rotation" of the extensions, as we will now see. Let $\pi: \cdots \to P_1 \to P_0 \to U \to 0$ be a projective resolution of U, and consider the i-th rotation map

$$\rho_i : \mathrm{Ext}^n_\Lambda(U,U) \to \mathrm{Ext}^n_\Lambda(\Omega^i_\Lambda(U), \Omega^i_\Lambda(U))$$

with respect to the resolution π. Every element of ExtΛn(U,U) can be represented by an exact sequence of the form

$$0 \to U \to E \to P_{n-2} \to \cdots \to P_i \longrightarrow P_{i-1} \to \cdots \to P_0 \to U \to 0 \quad \text{(with } P_i \searrow \Omega^i_\Lambda(U) \nearrow P_{i-1}\text{)}$$

Applying the map ρi to the element represented by this sequence produces the element represented by the following sequence:

$$0 \to \Omega^i_\Lambda(U) \to P_{i-1} \to \cdots \to P_0 \longrightarrow E \to P_{n-2} \to \cdots \to P_i \to \Omega^i_\Lambda(U) \to 0 \quad \text{(with } P_0 \searrow U \nearrow E\text{)}$$

We have thus rotated the sequence by removing an i-fold sequence from the right side and moving it to the left side.

5. ISOMORPHISMS BETWEEN EXTENSION GROUPS

In this section, we show that if Λ and Σ are Gorenstein algebras which are singularly equivalent of Morita type with level, then we have isomorphisms between extension groups over Λ and extension groups over Σ in sufficiently high degrees. More precisely, if ${}_\Lambda M_\Sigma$ and ${}_\Sigma N_\Lambda$ are bimodules which induce a singular equivalence of Morita type with level between the algebras Λ and Σ, then the functor $N \otimes_\Lambda -$ induces an isomorphism

$$\mathrm{Ext}^n_\Lambda(A,B) \cong \mathrm{Ext}^n_\Sigma(N \otimes_\Lambda A, N \otimes_\Lambda B) \tag{5.1}$$

for every $n \geq \max\{\mathrm{id}_\Lambda \Lambda, \mathrm{id}_\Sigma \Sigma\}$ and for any Λ-modules A and B. This is stated as Proposition 5.4.

To prove this result, we use maximal Cohen–Macaulay modules and the results from Section 3, as well as the rotation maps from Section 4. By Proposition 3.7, we know that in the setting described above, we have isomorphisms

$$\underline{\mathrm{Hom}}_\Lambda(C, C') \cong \underline{\mathrm{Hom}}_\Sigma(N \otimes_\Lambda C, N \otimes_\Lambda C') \quad (5.2)$$

between stable Hom groups over Λ and Σ for maximal Cohen–Macaulay Λ-modules C and C'. Lemma 5.2 below relates stable Hom groups to extension groups. Using this and isomorphism (5.2), we show (Proposition 5.3) that there are isomorphisms

$$\mathrm{Ext}^n_\Lambda(C, C') \cong \mathrm{Ext}^n_\Sigma(N \otimes_\Lambda C, N \otimes_\Lambda C')$$

for all maximal Cohen–Macaulay modules C and C' and every positive integer n. Finally, to arrive at isomorphism (5.1) for any Λ-modules A and B in Proposition 5.4, we use Proposition 5.3 together with two facts about Gorenstein algebras from earlier sections: all syzygies of sufficiently high degree are maximal Cohen–Macaulay modules, and by using a rotation map, we can replace the modules A and B by their syzygies.

We begin this section by showing, in the following two lemmas, how extension groups between maximal Cohen–Macaulay modules can be described as stable Hom groups. If C and C' are maximal Cohen–Macaulay modules over an algebra Λ, then we get (Lemma 5.2) an isomorphism

$$\mathrm{Ext}^n_\Lambda(C, C') \cong \underline{\mathrm{Hom}}_\Lambda(K_n, C')$$

for every positive integer n, with K_n an n-th syzygy of C.

In fact, it turns out that the conditions on C and C' can be relaxed somewhat. Recall that C being a maximal Cohen–Macaulay module means that $\mathrm{Ext}^i_\Lambda(C,\Lambda)=0$ for every positive integer i. To get the above isomorphism in degree n, it is sufficient to assume that $\mathrm{Ext}^n_\Lambda(C,\Lambda)=0$, and we do not need to put any assumptions on the module C'. We use this weaker assumption in the lemmas.

The following notation is used in the two lemmas. Given two modules A and B over an algebra Λ, we write $\mathscr{P}_\Lambda(A,B) \subseteq \mathrm{Hom}_\Lambda(A,B)$ for the subspace of $\mathrm{Hom}_\Lambda(A,B)$ consisting of morphisms that factor through a projective module; then the stable Hom group is $\underline{\mathrm{Hom}}_\Lambda(A,B)=\mathrm{Hom}_\Lambda(A,B)/\mathscr{P}_\Lambda(A,B)$.

In the first lemma, we consider the special case $n=1$.

Lemma 5.1 *Let Λ be a finite-dimensional k-algebra, and let A and C be finitely generated Λ-modules such that* $\mathrm{Ext}^1_\Lambda(C,\Lambda)=0$. *Let*

$$\eta\colon 0 \to K \xrightarrow{\alpha} P \xrightarrow{\beta} C \to 0$$

be a short exact sequence of Λ-modules with P projective. Then the sequence

$$0 \to \mathscr{P}_\Lambda(K, A) \hookrightarrow \mathrm{Hom}_\Lambda(K, A) \xrightarrow{\eta^*} \mathrm{Ext}^1_\Lambda(C, A) \to 0$$

of k-vector spaces is exact.

Proof By applying the functor $\mathrm{Hom}_\Lambda(-,A)$ to the sequence η, we get the exact sequence

$$0 \to \mathrm{Hom}_\Lambda(C, A) \xrightarrow{\beta^*} \mathrm{Hom}_\Lambda(P, A) \xrightarrow{\alpha^*} \mathrm{Hom}_\Lambda(K, A) \xrightarrow{\eta^*} \mathrm{Ext}^1_\Lambda(C, A) \to 0$$

From this we obtain the short exact sequence

$$0 \to \mathrm{im}\,\alpha^* \hookrightarrow \mathrm{Hom}_\Lambda(K, A) \xrightarrow{\eta^*} \mathrm{Ext}^1_\Lambda(C, A) \to 0$$

Now we only need to show that $\mathrm{im}\alpha^* = \mathscr{P}_\Lambda(K,A)$. If a homomorphism $f{:}K \to A$ lies in $\mathrm{im}\alpha^*$, then it factors through the map $\alpha{:}K \to P$, and since the module P is projective, this means that f lies in $\mathscr{P}_\Lambda(K,A)$. We thus have $\mathrm{im}\alpha^* \subseteq \mathscr{P}_\Lambda(K,A)$.

For the opposite inclusion, let Q be a projective Λ-module. Since we have assumed that $\mathrm{Ext}^1_\Lambda(C,\Lambda)=0$, we also have $\mathrm{Ext}^1_\Lambda(C,Q)=0$. Then from the long exact sequence obtained by applying the functor $\mathrm{Hom}_\Lambda(-,Q)$ to the short exact sequence η, we see that every homomorphism $g{:}K \to Q$ factors through the homomorphism $\alpha{:}K \to P$. Thus every homomorphism which starts in K and factors through some projective module, also factors through α, and we get $\mathscr{P}_\Lambda(K,A) \subseteq \mathrm{im}\alpha^*$. □

Now we continue to extension groups in arbitrary degree by using the above lemma and dimension shifting.

Lemma 5.2 *Let* Λ *be a finite-dimensional k-algebra, let A and C be finitely generated* Λ*-modules, and let n be a positive integer. Assume that* $\mathrm{Ext}^n_\Lambda(C,\Lambda)=0$. *Let*

$\pi_n{:}0 \to K_n \to P_{n-1} \to P_{n-2} \to \cdots \to P_1 \to P_0 \to C \to 0$

be the beginning of a projective resolution of C with K_n *as the n-th syzygy. Then the sequence*

$$0 \to \mathscr{P}_\Lambda(K_n, A) \hookrightarrow \mathrm{Hom}_\Lambda(K_n, A) \xrightarrow{\pi_n^*} \mathrm{Ext}^n_\Lambda(C, A) \to 0$$

of k-vector spaces is exact, and thus the map π_n^* *induces an isomorphism*

$$\overline{\pi_n^*} : \underline{\mathrm{Hom}}_\Lambda(K_n, A) \xrightarrow{\cong} \mathrm{Ext}^n_\Lambda(C, A)$$

Proof Decompose the sequence π_n into two exact sequences

$\eta{:}0 \to K_n \to P_{n-1} \to K_{n-1} \to 0$

and $\pi_{n-1}: 0 \to K_{n-1} \to P_{n-2} \to P_{n-3} \to \cdots \to P_1 \to P_0 \to C \to 0$,

such that $\pi_n = \eta \circ \pi_{n-1}$. By dimension shifting, we have an isomorphism

$$\pi_{n-1}^*: \operatorname{Ext}_\Lambda^1(K_{n-1}, A) \xrightarrow{\cong} \operatorname{Ext}_\Lambda^n(C, A)$$

We observe that $\pi_n^* = \pi_{n-1}^* \circ \eta^*$, so the following diagram is commutative.

$$\begin{array}{ccccccccc} 0 & \longrightarrow & \mathscr{P}_\Lambda(K_n, A) & \hookrightarrow & \operatorname{Hom}_\Lambda(K_n, A) & \xrightarrow{\eta^*} & \operatorname{Ext}_\Lambda^1(K_{n-1}, A) & \longrightarrow & 0 \\ & & \| & & \| & & \downarrow{\scriptstyle \pi_{n-1}^*}\ \cong & & \\ 0 & \longrightarrow & \mathscr{P}_\Lambda(K_n, A) & \hookrightarrow & \operatorname{Hom}_\Lambda(K_n, A) & \xrightarrow{\pi_n^*} & \operatorname{Ext}_\Lambda^n(C, A) & \longrightarrow & 0 \end{array}$$

By Lemma 5.1□ , the top row of this diagram is exact. Since all the vertical maps are isomorphisms, the bottom row is also exact.

We now show that we get the isomorphisms we want between extension groups in the special case where the involved modules are maximal Cohen–Macaulay modules. In this case, we get isomorphisms between extension groups in all positive degrees, while in the general case which is considered afterwards (Proposition 5.4), we only get isomorphisms in almost all degrees.

Proposition 5.3 *Let Λ and Σ be finite-dimensional Gorenstein algebras which are singularly equivalent of Morita type with level, and let ${}_\Lambda M_\Sigma$ and ${}_\Sigma N_\Lambda$ be bimodules which induce a singular equivalence of Morita type with level between Λ and Σ. Let C and C'be maximal Cohen–Macaulay modules over Λ. Then for every positive integer n, the map*

$$\operatorname{Ext}_\Lambda^n(C, C') \xrightarrow{N \otimes_\Lambda -} \operatorname{Ext}_\Sigma^n(N \otimes_\Lambda C, N \otimes_\Lambda C')$$

is an isomorphism.

Proof The idea is to translate the two Ext groups to stable Hom groups by using Lemma 5.2, and then use the equivalence of stable categories of Cohen–Macaulay modules from Proposition 3.7.

Let

$\pi_n: 0 \to K_n \to P_{n-1} \to \cdots \to P_0 \to C \to 0$

be the beginning of a projective resolution of C with K_n as n-th syzygy. By Lemma 2.7, the sequence $N \otimes_\Lambda \pi_n$, which is obtained by applying the functor $N \otimes_\Lambda -$ to all objects and maps in π_n, is the beginning of a projective resolution of the Σ-module $N \otimes_\Lambda C$, with $N \otimes_\Lambda K_n$ as the n-th syzygy.

Since C and C' are maximal Cohen–Macaulay modules, we deduce that $N\otimes_\Lambda C$, $N\otimes_\Lambda C'$, K_n and $N\otimes_\Lambda K_n$ are also maximal Cohen–Macaulay modules, by using Lemma 3.4 and Lemma 3.6. We form the following commutative diagram of k-vector spaces.

$$\begin{array}{ccc} \underline{\mathrm{Hom}}_\Lambda(K_n, C') & \xrightarrow[\cong]{N\otimes_\Lambda -} & \underline{\mathrm{Hom}}_\Sigma(N\otimes_\Lambda K_n, N\otimes_\Lambda C') \\ \pi_n^* \downarrow \cong & & \cong \downarrow (N\otimes_\Lambda \pi_n)^* \\ \mathrm{Ext}^n_\Lambda(C, C') & \xrightarrow{N\otimes_\Lambda -} & \mathrm{Ext}^n_\Sigma(N\otimes_\Lambda C, N\otimes_\Lambda C') \end{array}$$

The vertical maps are isomorphisms by Lemma 5.2, and the map in the top row is an isomorphism by Proposition 3.7. Therefore the map in the bottom row is also an isomorphism, and this concludes the proof. □

Finally, we come to the main result of this section, where we show that if two Gorenstein algebras Λ and Σ are singularly equivalent of Morita type with level, then for every extension group (of sufficiently high degree) over Λ, there is an isomorphic extension group over Σ.

Proposition 5.4 *Let Λ and Σ be finite-dimensional Gorenstein k-algebras which are singularly equivalent of Morita type with level, and let* ${}_\Lambda M_\Sigma$ *and* ${}_\Sigma N_\Lambda$ *be bimodules which induce a singular equivalence of Morita type with level between Λ and Σ. Let*

$d=\max\{\mathrm{id}_\Lambda \Lambda, \mathrm{id}_\Sigma \Sigma\}$

be the maximum of the injective dimensions of Λ and Σ. Then for every integer n>d, we have k-vector space isomorphisms

$$\mathrm{Ext}^n_\Lambda(A, B) \xrightarrow[N\otimes_\Lambda -]{\cong} \mathrm{Ext}^n_\Sigma(N\otimes_\Lambda A, N\otimes_\Lambda B)$$ for Λ

modules A and B,

$$\mathrm{Ext}^n_\Sigma(A', B') \xrightarrow[M\otimes_\Sigma -]{\cong} \mathrm{Ext}^n_\Lambda(M\otimes_\Sigma A', M\otimes_\Sigma B')$$ for Σ-

modules A′ and B′.

Proof Let A and B be Λ-modules, and let

$\pi: \cdots \to P_1 \to P_0 \to_A \to 0$ and $\tau: \cdots \to Q_1 \to Q_0 \to B \to 0$

be projective resolutions. Then by Lemma 2.7, the sequences $N\otimes_\Lambda \pi$ and $N\otimes_\Lambda \tau$ are projective resolutions of the Σ-modules $N\otimes_\Lambda A$ and $N\otimes_\Lambda B$. We form the following commutative diagram, where ρ_d is the d-th rotation map with respect to the resolutions π and τ, and ρ'_d the d-th rotation map with respect to the resolutions $N\otimes_\Lambda \pi$ and $N\otimes_\Lambda \tau$. These maps are isomorphisms by Lemma 4.2.

$$\begin{array}{ccc} \operatorname{Ext}^n_\Lambda(A,B) & \xrightarrow{N\otimes_\Lambda -} & \operatorname{Ext}^n_\Sigma(N\otimes_\Lambda A, N\otimes_\Lambda B) \\ \rho_d \downarrow \cong & & \cong \downarrow \rho'_d \\ \operatorname{Ext}^n_\Lambda(\Omega^d_\Lambda(A),\Omega^d_\Lambda(B)) & \xrightarrow[\cong]{N\otimes_\Lambda -} & \operatorname{Ext}^n_\Sigma(N\otimes_\Lambda \Omega^d_\Lambda(A), N\otimes_\Lambda \Omega^d_\Lambda(B)) \end{array}$$

By Lemma 3.4, the syzygies $\Omega^d{}_\Lambda$ (A) and $\Omega^d{}_\Lambda$ (B) are maximal Cohen–Macaulay modules, and then by Proposition 5.3, the map $N\otimes_\Lambda-$ in the bottom row is an isomorphism. It follows that the map $N\otimes_\Lambda-$ in the top row is an isomorphism. This gives the first of the two isomorphisms we want. The second isomorphism follows by symmetry. □

6. HOCHSCHILD COHOMOLOGY RINGS

In this section, we define the Hochschild cohomology ring $\mathrm{HH}^*(\Lambda)$ of an algebra Λ, and we show that if two Gorenstein k-algebras are singularly equivalent of Morita type with level, then their Hochschild cohomology rings are isomorphic in almost all degrees.

We first introduce some notation for rings of extensions. If Λ is a k-algebra and A a Λ-module, then we define

$$\mathcal{E}^*_\Lambda(A) = \operatorname{Ext}^*_\Lambda(A,A) = \bigoplus_{n\geq 0} \operatorname{Ext}^n_\Lambda(A,A)$$

That is, $\mathcal{E}^*_\Lambda$ (A) denotes the graded k-algebra which is the direct sum of all extension groups of A by itself, with multiplication given by Yoneda product.

We are interested in the "asymptotic" behaviour of such graded rings of extensions; that is, we want to find isomorphisms which hold in all degrees above some finite bound. Given an extension ring $\mathcal{E}^*_\Lambda$ (A), we therefore consider the graded ideals of the form

$$\mathcal{E}^{>d}_\Lambda(A) = \bigoplus_{n>d} \operatorname{Ext}^n_\Lambda(A,A)$$

for some integer d. Such an ideal is a graded nonunital k-algebra. When we want to say that two extension rings $\mathcal{E}^*_\Lambda$ (A) and $\mathcal{E}^*_\Sigma$ (B) (for k-algebras Λ and Σ with modules A_Λ and B_Σ) are the same in almost all degrees, the appropriate kind of isomorphism to look for, in order to preserve all the relevant structure, is an isomorphism of graded nonunital k-algebras between $\mathcal{E}_{\Lambda}^{>d}(A)$ and $\mathcal{E}_{\Sigma}^{>d}(B)$, for some integer d. This section is mainly concerned with finding isomorphisms of this type.

We define the Hochschild cohomology of an algebra as the extension ring of the algebra over its enveloping algebra.

Definition 6.1 Let Λ be a finite-dimensional k-algebra. The *Hochschild cohomology ring* of Λ is the extension ring $\mathrm{HH}^*(\Lambda) = \mathcal{E}_{\Lambda^e}^{*}(\Lambda)$.

Hochschild cohomology was first defined by G. Hochschild in [11]. The original definition uses the bar resolution. We follow the definition in [5], where Hochschild cohomology is given by extension groups. Since we have assumed that k is a field, this definition is equivalent to the original one. More generally, the two definitions are equivalent whenever Λ is projective over k (see [5, IX, §6]).

We now turn to the problem of showing that singular equivalences of Morita type with level between Gorenstein algebras preserve Hochschild cohomology in almost all degrees. We need the following diagram lemma, known as the "3×3 splice".

Lemma 6.2 *(See[14, Lemma VIII.3.1].) Let R be a ring, and let*

$$
\begin{array}{cccccc}
 & \eta': & & \eta: & & \eta'': \\
\eta_A: & A' & \longrightarrow & A & \longrightarrow & A'' \\
 & \downarrow & & \downarrow & & \downarrow \\
\eta_B: & B' & \longrightarrow & B & \longrightarrow & B'' \\
 & \downarrow & & \downarrow & & \downarrow \\
\eta_C: & C' & \longrightarrow & C & \longrightarrow & C''
\end{array}
$$

be a commutative diagram of R-modules, where the three rows η_A, η_B *and* η_C, *as well as the three columns* η', η *and* η'', *are short exact sequences. Then the elements in the extension group* $\mathrm{Ext}^1_R(C'', A')$ *represented by the composition* $\eta_A \circ \eta''$ *and by the composition* $\eta' \circ \eta_C$ *are the additive inverses of each other:*

$$[\eta_A \circ \eta''] = -[\eta' \circ \eta_C].$$

If two bimodules ${}_{\Lambda}M_{\Sigma}$ and ${}_{\Sigma}N_{\Lambda}$ induce a singular equivalence of Morita type with level between algebras Λ and Σ, then the Λ^e-module $M \otimes_{\Sigma} N$ is a syzygy of Λ. In the following lemma, we use Lemma 6.2 to show that under certain assumptions, the tensor functors $(M \otimes_{\Sigma} N) \otimes_{\Lambda} -$ and $- \otimes_{\Lambda} (M \otimes_{\Sigma} N)$ induce isomorphisms of Ext groups in almost all degrees. This is afterwards used in the proof of Theorem 6.4.

Lemma 6.3 *Let Λ be a finite-dimensional Gorenstein k-algebra, and let U be a Λ^e-module which is projective as a left Λ-module and as a right Λ-module. Let $d \geq 2 \cdot \mathrm{id}_\Lambda \Lambda$. Let K be an i-th syzygy of Λ as Λ^e-module, for some $i<d$. Then the maps*

$$K \otimes_\Lambda - : \mathcal{E}_{\Lambda^e}^{>d}(U) \to \mathcal{E}_{\Lambda^e}^{>d}(K \otimes_\Lambda U) \quad \text{and} \quad - \otimes_\Lambda K : \mathcal{E}_{\Lambda^e}^{>d}(U) \to \mathcal{E}_{\Lambda^e}^{>d}(U \otimes_\Lambda K)$$

are isomorphisms of graded nonunital k-algebras.

Proof We show that the map $K \otimes_\Lambda -$ is an isomorphism; the proof for $- \otimes_\Lambda K$ is similar. Let

$\pi : \cdots \to P_1 \to P_0 \to \Lambda \to 0$

be a projective resolution of Λ as Λ^e-module, with K as the i-th syzygy, and let

$\sigma : \cdots \to P1 \otimes_\Lambda U \to P_0 \otimes_\Lambda U \to U \to 0$

be the result of applying the functor $- \otimes_\Lambda U$ to the sequence π and identifying $\Lambda \otimes_\Lambda U$ with U in the last term. This sequence is exact since U is projective as left module, and every $P_j \otimes_\Lambda U$ is projective since U is projective as right module. Thus, σ is a projective resolution of U, and $K \otimes_\Lambda U$ is an i-th syzygy of U.

By Lemma 3.2, the enveloping algebra Λ^e of Λ is Gorenstein, and we have $\mathrm{id}_{\Lambda^e} \Lambda^e \leq 2 \cdot \mathrm{id}_\Lambda \Lambda \leq d$. Then by Lemma 4.2, the i-th rotation map

$$\rho_i : \mathcal{E}_{\Lambda^e}^{>d}(U) \to \mathcal{E}_{\Lambda^e}^{>d}(K \otimes_\Lambda U)$$

(with respect to the resolution σ) is an isomorphism of graded nonunital k-algebras. We show that the map $K \otimes_\Lambda -$ is an isomorphism by showing that it is equal to the map ρ_i, up to sign. More precisely, we show that for any homogeneous element $[\eta] \in \mathcal{E}_{\Lambda^e}^{>d}(U)$ of degree $n>d$, we have

$K \otimes_\Lambda [\eta] = (-1)^{in} \cdot \rho_i([\eta])$.

Let $[\eta] \in \mathcal{E}_{\Lambda^e}^{>d}(U)$ be a homogeneous element of degree $n>d$ represented by an exact sequence

$\eta : 0 \to U \to E_n \to \cdots \to E_1 \to U \to 0$.

We can assume without loss of generality that all the modules E_j are projective as left Λ-modules and as right Λ-modules. Let

$\pi_i : 0 \to K \to P_{i-1} \to \cdots \to P_0 \to \Lambda \to 0$

$\sigma_i : 0 \to K \otimes_\Lambda U \to P_{i-1} \otimes_\Lambda U \to \cdots \to P_0 \otimes_\Lambda U \to U \to 0$

be truncations of the projective resolutions π and σ. We construct the commutative diagram in Fig. 1 by tensoring π_i with η over Λ and identifying $\Lambda_{\otimes\Lambda}-$ with the identity in the last row. The rows and columns of the diagram are exact sequences.

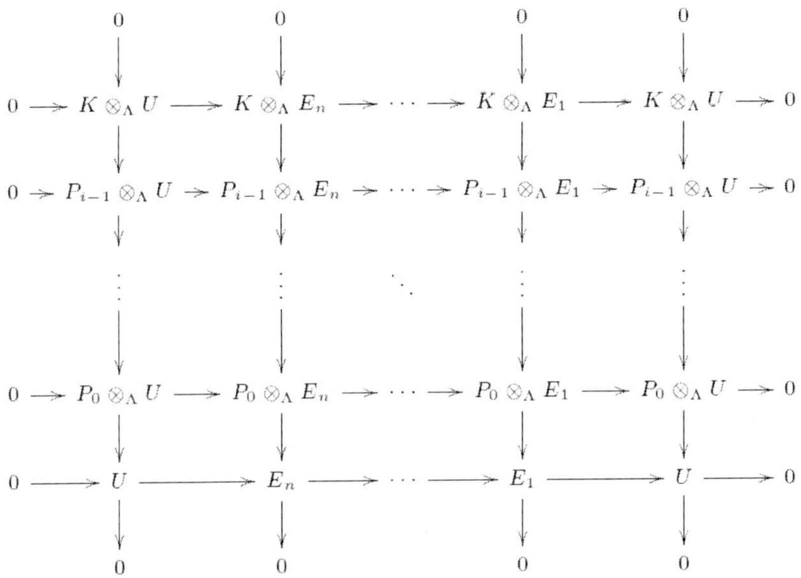

Figure. 1. Commutative diagram used in the proof of Lemma 6.3.

The bottom row in the diagram is the sequence η, the top row is the sequence $K\otimes\Lambda\eta$, and the first and the last column are both equal to the sequence σi. By using Lemma 6.2 repeatedly, we get the equality

$$[(K_{\otimes\Lambda}\eta)\circ\sigma_i]=(-1)^{in}[\sigma_i\circ\eta]$$

in the extension group $\mathrm{Ext}^{n+i}_{\Lambda^e}(U, K \otimes_\Lambda U)$. By the definition of the rotation map ρ_i (see Definition 4.1), we then get

$$K_{\otimes\Lambda}[\eta]=[K_{\otimes\Lambda}\eta]=(-1)^{in}\cdot\rho_i([\eta]).$$

Since the map ρi is an isomorphism, this means that the map $K_{\otimes\Lambda}-$ is an isomorphism as well. □

We now show that a singular equivalence of Morita type with level between Gorenstein k-algebras preserves the Hochschild cohomology in almost all degrees. A weaker form of this result, stating that a singular equivalence of Morita type preserves Hochschild cohomology groups in almost all degrees (but not necessarily the ring structure of the cohomology), appears in [20, Remark 4.3].

Theorem 6.4 *Let Λ and Σ be finite-dimensional Gorenstein k-algebras which are singularly equivalent of Morita type with level. Then we have the following.*

1.*The Hochschild cohomology rings* $\mathrm{HH}^*(\Lambda)$ *and* $\mathrm{HH}^*(\Sigma)$ *are isomorphic in almost all degrees, with isomorphisms that respect the ring structure.*

2.*Let ${}_\Lambda M_\Sigma$ and ${}_\Sigma N_\Lambda$ be bimodules which induce a singular equivalence of Morita type with level $l\geq 1$ (seeRemark 6.5) between Λ and Σ, and let $d=\max\{l, 2\cdot \mathrm{id}_\Lambda \Lambda, 2\cdot \mathrm{id}_\Sigma \Sigma\}$. Then there are isomorphisms*

$$\begin{array}{ccc} \mathrm{HH}^{>d}(\Lambda) & \xrightarrow[\cong]{\rho_l} & \mathcal{E}^{>d}_{\Lambda^e}(M\otimes_\Sigma \Sigma \otimes_\Sigma N) \\ {\scriptstyle N\otimes_\Lambda - \otimes_\Lambda M}\downarrow\cong & & \cong\uparrow{\scriptstyle M\otimes_\Sigma - \otimes_\Sigma N} \\ \mathcal{E}^{>d}_{\Sigma^e}(N\otimes_\Lambda \Lambda\otimes_\Lambda M) & \xleftarrow[\rho'_l]{\cong} & \mathrm{HH}^{>d}(\Sigma) \end{array}$$

of graded nonunital k-algebras, where the maps ρ_l and $\rho'_l$are rotation maps.

Proof We show part (2). Part (1) then follows directly.

Since M and N induce a singular equivalence of Morita type with level l, the module $M\otimes_\Sigma\Sigma\otimes_\Sigma N\cong M\otimes_\Sigma N$ is an l-th syzygy of Λ as a Λ^e-module. Let

$$\rho_l:\ \mathrm{HH}^{>d}(\Lambda)\rightarrow \mathcal{E}^{>d}_{\Lambda^e}(M\otimes_\Sigma \Sigma\otimes_\Sigma N)$$

be the l-th rotation map with respect to a projective resolution of Λ with $M\otimes_\Sigma\Sigma\otimes_\Sigma N$ as the l-th syzygy. By Lemma 3.2, the enveloping algebras Λ^e and Σ^e are Gorenstein algebras, and we have $\mathrm{id}_{\Lambda^e}\Lambda^e\leq 2\cdot\mathrm{id}_\Lambda\Lambda$ and $\mathrm{id}_{\Sigma^e}\Sigma^e\leq 2\cdot\mathrm{id}_\Sigma\Sigma$. By Lemma 4.2, the rotation map ρ_l is an isomorphism, since

$\max\{l,\mathrm{id}_{\Lambda^e}\Lambda^e,\mathrm{id}_{\Sigma^e}\Sigma^e\}\leq\max\{l,2\cdot\mathrm{id}_\Lambda\Lambda,2\cdot\mathrm{id}_\Sigma\Sigma\}=d.$

We can similarly define the rotation map ρ'_l and show that it is an isomorphism.

We now show that the maps $N\otimes_\Lambda-\otimes_\Lambda M$ and $M\otimes_\Sigma-\otimes_\Sigma N$ are isomorphisms. For any $n>d$, we can make the following diagram:

$$\begin{array}{ccc} \mathrm{HH}^{n}(\Lambda) & \xrightarrow[\cong]{\rho_l} & \mathcal{E}^{n}_{\Lambda^e}(M\otimes_\Sigma \Sigma \otimes_\Sigma N) \\ {\scriptstyle N\otimes_\Lambda - \otimes_\Lambda M}\downarrow & & \uparrow{\scriptstyle M\otimes_\Sigma - \otimes_\Sigma N} \\ \mathcal{E}^{n}_{\Sigma^e}(N\otimes_\Lambda \Lambda\otimes_\Lambda M) & \xleftarrow[\rho'_l]{\cong} & \mathrm{HH}^{n}(\Sigma) \end{array} \tag{6.1}$$

Consider the map $N{\otimes_\Lambda}-{\otimes_\Lambda}M$ in this diagram. We construct the following commutative diagram with this map at the top:

$$\begin{array}{ccc}
\mathrm{HH}^n(\Lambda) & \xrightarrow{N\otimes_\Lambda - \otimes_\Lambda M} & \mathcal{E}^n_{\Sigma^e}(N\otimes_\Lambda \Lambda \otimes_\Lambda M) \\
{\scriptstyle (M\otimes_\Sigma N)\otimes_\Lambda -}\downarrow \cong & & \downarrow {\scriptstyle M\otimes_\Sigma - \otimes_\Sigma N} \\
\mathcal{E}^n_{\Lambda^e}(M\otimes_\Sigma N\otimes_\Lambda \Lambda) & \xrightarrow[-\otimes_\Lambda (M\otimes_\Sigma N)]{\cong} & \mathcal{E}^n_{\Lambda^e}(M\otimes_\Sigma N\otimes_\Lambda \Lambda\otimes_\Lambda M\otimes_\Sigma N)
\end{array}$$

By Lemma 6.3, the maps $(M{\otimes_\Sigma}N){\otimes_\Lambda}-$ and $-{\otimes_\Lambda}(M{\otimes_\Sigma}N)$ in this diagram are isomorphisms, since $M{\otimes_\Sigma}N$ is an l-th syzygy of Λ as Λ^e-module. Therefore, the map $N{\otimes_\Lambda}-{\otimes_\Lambda}M$ in diagram (6.1) is a monomorphism. By a similar argument, the map $M{\otimes_\Sigma}-{\otimes_\Sigma}N$ in diagram (6.1) is a monomorphism. Since $\mathrm{HH}^n(\Lambda)$ and $\mathrm{HH}^n(\Sigma)$ are finite-dimensional over k, it follows that these monomorphisms must be isomorphisms. □

Remark 6.5 In Theorem 6.4 (2), we assumed that the level l is positive. The reason for this is that if we had allowed l=0, then we could not have made the rotation maps ρ_l and ρ'_l. This assumption does not strongly affect the applicability of the theorem, since any equivalence with level 0 implies the existence of an equivalence with level 1. In general, if two bimodules ${}_\Lambda M_\Sigma$ and ${}_\Sigma N_\Lambda$ induce a singular equivalence of Morita type with level l between algebras Λ and Σ, then the bimodules $\Omega^1_{\Lambda\otimes k\Sigma op}(M)$ and N induce a singular equivalence of level l+1 between Λ and Σ.

7. FINITE GENERATION

Support varieties for modules over artin algebras were defined by Snashall and Solberg in [18], using the Hochschild cohomology ring. In [9], Erdmann, Holloway, Snashall, Solberg and Taillefer defined two finite generation conditions **Fg1** and **Fg2** for the Hochschild cohomology ring of an algebra. These conditions ensure that the support varieties for modules over the given algebra have good properties. In [10], these conditions were reformulated as a new condition called **(Fg)** which is equivalent to the combination of **Fg1** and **Fg2**. We use the definition from [10].

In this section, we describe the finite generation condition **(Fg)**. We then show the main result of this paper (Theorem 7.4): A singular equivalence of Morita type with level between finite-dimensional Gorenstein k-algebras preserves the **(Fg)** condition.

In order to define the **(Fg)** condition, we first describe a way to view extension rings over an algebra as modules over the Hochschild cohomology ring. Let Λ be a finite-dimensional k-algebra and A a Λ-module. We define a graded ring homomorphism

$$\varphi_A : \mathrm{HH}^*(\Lambda) \to \mathcal{E}^*_\Lambda(A)$$

as follows. A homogeneous element of $\mathrm{HH}^*(\Lambda)$ can be represented by an exact sequence

$\eta{:}0\to\Lambda\to E\to P_n\to\cdots\to P_0\to\Lambda\to 0$

of Λ^e-modules, where each P_i is projective. Viewed as a sequence of right Λ-modules, this sequence splits. The complex

$\eta\otimes_\Lambda A{:}0\to\Lambda\otimes_\Lambda A\to E\otimes_\Lambda A\to P_n\otimes_\Lambda A\to\cdots\to P_0\otimes_\Lambda A\to\Lambda\otimes_\Lambda A\to 0$

is therefore an exact sequence. By composition with the isomorphism $\mu_A{:}\Lambda\otimes_\Lambda A\to A$ and its inverse, we get an extension

$\mu A\circ(\eta\otimes_\Lambda A)\circ\mu^{-1}{}_A{:}0\to A\to E\otimes_\Lambda A\to P_n\otimes_\Lambda A\to\cdots\to P_0\otimes_\Lambda A\to A\to 0$

of A by itself, and thus a representative of a homogeneous element in the extension ring $\mathcal{E}^*_\Lambda$ (A). The map φ_A is defined by the action

$\varphi_A([\eta])=[\mu_A\circ(\eta\otimes_\Lambda A)\circ\mu^{-1}{}_A]$

on homogeneous elements. By the map φ_A, the graded ring $\mathcal{E}^*_\Lambda(A)$ becomes a graded $\mathrm{HH}^*(\Lambda)$-module.

Definition 7.1 Let Λ be a finite-dimensional k-algebra. We say that Λ satisfies the **(Fg)** condition if the following holds.

1.The ring HH^* (Λ) is Noetherian.

2.The HH^* (Λ)-module $\mathcal{E}^*_\Lambda$ $(\Lambda/\mathrm{rad}\Lambda)$ is finitely generated. (The module structure is given by the map $\varphi_\Lambda/\mathrm{rad}_\Lambda$, as described above.)

By [17, Proposition 5.7], the **(Fg)** condition as defined here is equivalent to the combination of the conditions **Fg1** and **Fg2** defined in [9].

The following result describes why Gorenstein algebras are important in connection with the **(Fg)** condition.

Theorem 7.2 *(See[9, Theorem 1.5 (a)].) If an algebra satisfies the***(Fg)***condition, then it is a Gorenstein algebra.*

Our aim is to show that if two Gorenstein k-algebras are singularly equivalent of Morita type with level, then the **(Fg)** condition holds for one of the algebras if and only if it holds for the other. We use the following result, which describes a relation between two algebras ensuring that **(Fg)** for one of the algebras implies **(Fg)** for the other.

Proposition 7.3 *Let Λ and Σ be finite-dimensional k-algebras. Let $A=\Lambda/\mathrm{rad}\Lambda$, and assume that we have a commutative diagram*

$$\begin{array}{ccc} \mathrm{HH}^{>d}(\Lambda) & \xrightarrow{\varphi_A} & \mathcal{E}_\Lambda^{>d}(A) \\ \cong \downarrow f & & g \downarrow \cong \\ \mathrm{HH}^{>d}(\Sigma) & \xrightarrow{\varphi_B} & \mathcal{E}_\Sigma^{>d}(B) \end{array} \qquad (7.1)$$

of graded nonunital k-algebras, for some Σ-module B and some positive integer d, where the vertical maps f and g are isomorphisms. Assume that Σ satisfies the **(Fg)** *condition. Then Λ also satisfies* **(Fg)**.

Proof This follows from Proposition 6.3 in [16]. □

We are now ready to prove the main result of this paper.

Theorem 7.4 *Let Λ and Σ be finite-dimensional Gorenstein algebras over the field k. Assume that Λ and Σ are singularly equivalent of Morita type with level. Then Λ satisfies* **(Fg)** *if and only if Σ satisfies* **(Fg)**.

Proof We show that if Σ satisfies **(Fg)**, then Λ satisfies **(Fg)**. The opposite implication then follows by symmetry. Let ${}_\Lambda M_\Sigma$ and ${}_\Sigma N_\Lambda$ be bimodules which induce a singular equivalence of Morita type with level $l\geq 1$ (see Remark 6.5) between Λ and Σ. Let $d=\max\{l,2\cdot \mathrm{id}_\Lambda \Lambda, 2\cdot \mathrm{id}_\Sigma \Sigma\}$. Let A be the Λ-module $\Lambda/\mathrm{rad}\Lambda$.

The Λ^e-module $M\otimes_\Sigma \Sigma \otimes_\Sigma N \cong M \otimes_\Sigma N$ is an l-th syzygy of Λ as Λ^e-module. Let π be a projective resolution of Λ with $M\otimes_\Sigma \Sigma \otimes_\Sigma N$ as the l-th syzygy. Then the complex $\pi \otimes_\Lambda A$ is a projective resolution of $\Lambda \otimes_\Lambda A$, with $M\otimes_\Sigma \Sigma \otimes_\Sigma N \otimes_\Lambda A$ as the l-th syzygy. We construct the commutative diagram in Fig. 2, where the maps ρl and $\rho l'$ are the l-th rotation maps with respect to the resolutions π and $\pi\otimes_\Lambda A$, respectively. These maps are isomorphisms by Lemma 4.2. The map $M\otimes_\Sigma - \otimes_\Sigma N$ in the diagram is an isomorphism by Theorem 6.4, and the map $M\otimes_\Sigma -$ is an isomorphism by Proposition 5.4. The isomorphisms f and g are defined to be the appropriate compositions of the other isomorphisms in the diagram. By Proposition 7.3, this diagram shows that if the algebra Σ satisfies **(Fg)**, then Λ also satisfies **(Fg)**. □

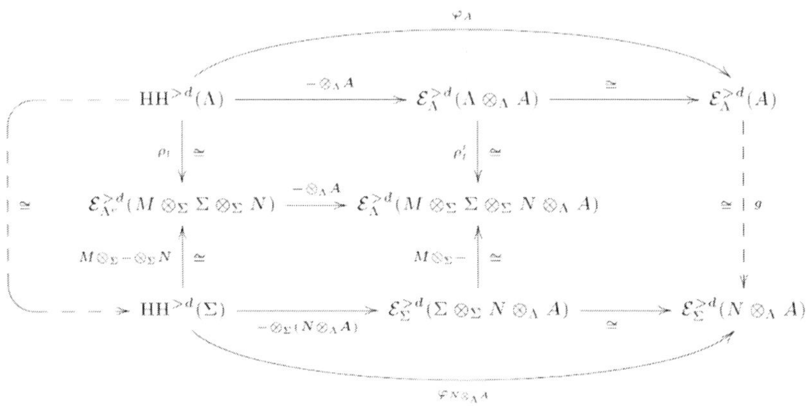

Figure. 2. Commutative diagram used in the proof of Theorem 7.4. We now show that the assumption of both algebras being Gorenstein is necessary in the above theorem. Example 5.5 in [16] contains two singularly equivalent algebras where one algebra satisfies **(Fg)** and the other is not Gorenstein. We use the same algebras, and show that there exists a singular equivalence of Morita type with level between them.

Example 7.5 Let $\Lambda=kQ/\langle\rho\rangle$ and $\Sigma=kR/\langle\sigma\rangle$ be k-algebras given by the following quivers and relations:

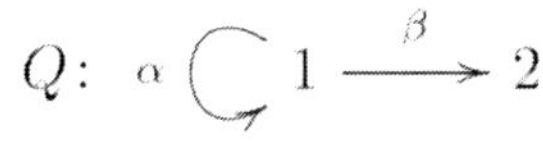

$$\rho=\{\alpha^2,\beta\alpha\}$$

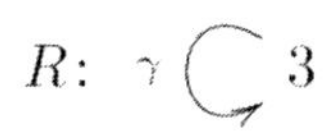

$$\sigma=\{\gamma^2\}$$

The tensor algebra $\Lambda\otimes_k\Sigma^{op}$ has the following quiver and relations:

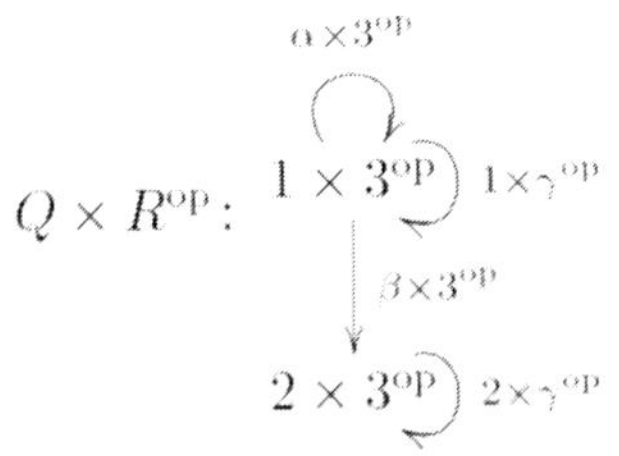

$$\left\{\begin{array}{c}(\alpha\times 3^{op})^2,(\beta\times 3^{op})(\alpha\times 3^{op}),\\ (1\times\gamma^{op})^2,(2\times\gamma^{op})^2,\\ (\alpha\times 3^{op})(1\times\gamma^{op})-(1\times\gamma^{op})(\alpha\times 3^{op}),\\ (\beta\times 3^{op})(1\times\gamma^{op})-(2\times\gamma^{op})(\beta\times 3^{op})\end{array}\right\}$$

The tensor algebra $\Sigma\otimes_k\Lambda^{op}$ has the following quiver and relations:

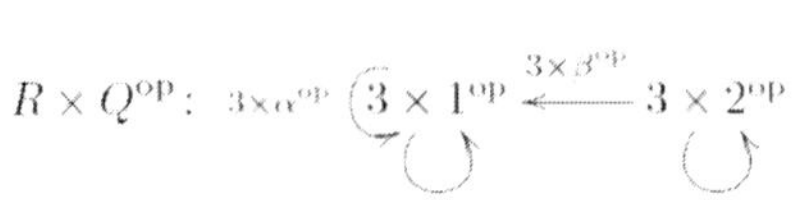

$$\left\{\begin{array}{c}(3\times\alpha^{op})^2,(3\times\alpha^{op})(3\times\beta^{op}),\\ (\gamma\times 1^{op})^2,(\gamma\times 2^{op})^2,\\ (3\times\alpha^{op})(\gamma\times 1^{op})-(\gamma\times 1^{op})(3\times\alpha^{op}),\\ (3\times\beta^{op})(\gamma\times 2^{op})-(\gamma\times 1^{op})(3\times\beta^{op})\end{array}\right\}$$

Let ${}_\Lambda M_\Sigma$ and ${}_\Sigma N_\Lambda$ be bimodules given by the following representations over $Q\times R^{op}$ and $R\times Q^{op}$, respectively:

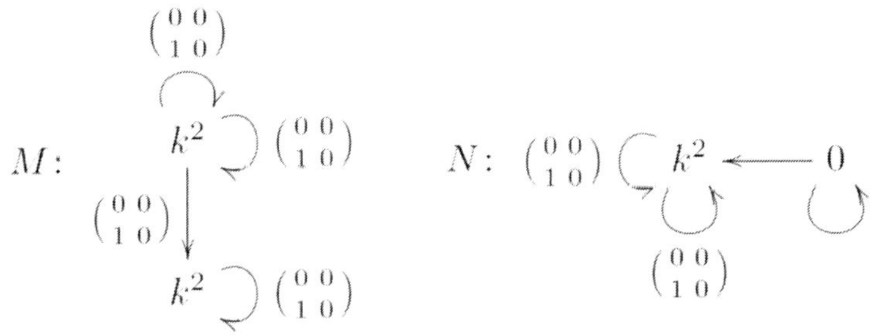

We show that the bimodules M and N induce a singular equivalence of Morita type with level 1 between the algebras Λ and Σ. We first check that these bimodules satisfy the first two conditions in Definition 2.4. Considering only the left or right structure of M and N, we have the following four isomorphisms:

$$ {}_\Lambda M \cong \Lambda \qquad N_\Lambda \cong e_2\Lambda $$

$$ {}_\Sigma N \cong \Sigma^2 \qquad M_\Sigma \cong \Sigma $$

Thus the bimodules M and N are projective when viewed as one-sided (left or right) modules.

To check the last two conditions in Definition 2.4, we compute the tensor products $M\otimes_\Sigma N$ and $N\otimes_\Lambda M$ as representations of quivers, and check that they are syzygies of Λ and Σ, respectively. The enveloping algebra Λ^e has the following quiver and relations:

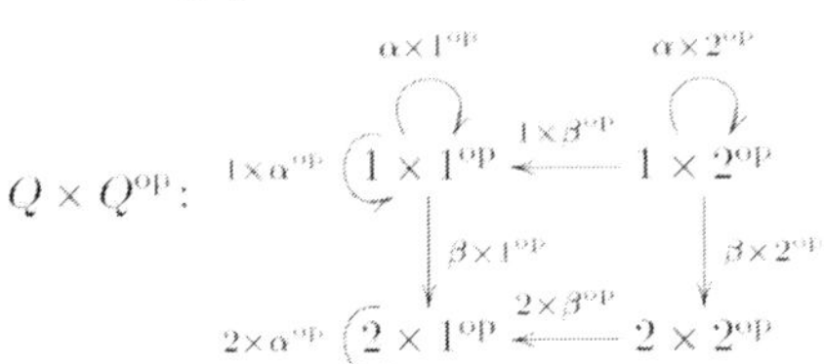

$$\left\{\begin{array}{c} (\alpha\times 1^{op})^2, (\beta\times 1^{op})(\alpha\times 1^{op}),\\ (\alpha\times 2^{op})^2, (\beta\times 2^{op})(\alpha\times 2^{op}),\\ (1\times\alpha^{op})^2, (1\times\alpha^{op})(1\times\beta^{op}),\\ (2\times\alpha^{op})^2, (2\times\alpha^{op})(2\times\beta^{op}),\\ (\alpha\times 1^{op})(1\times\alpha^{op}) - (1\times\alpha^{op})(\alpha\times 1^{op}),\\ (\alpha\times 1^{op})(1\times\beta^{op}) - (1\times\beta^{op})(\alpha\times 2^{op}),\\ (\beta\times 1^{op})(1\times\alpha^{op}) - (2\times\alpha^{op})(\beta\times 1^{op}),\\ (\beta\times 1^{op})(1\times\beta^{op}) - (2\times\beta^{op})(\beta\times 2^{op}) \end{array}\right\}$$

The tensor product $M\otimes_\Sigma N$ is the Λe-module given by the following representation over $Q\times Q^{op}$:

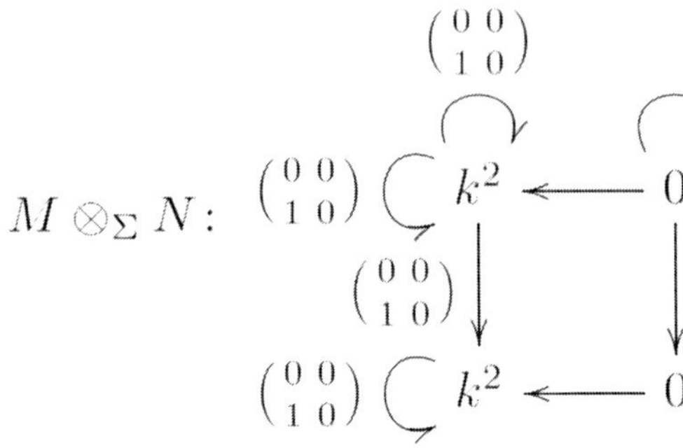

The algebra Λ considered as a Λ^e-module has the following representation over $Q\times Q^{op}$:

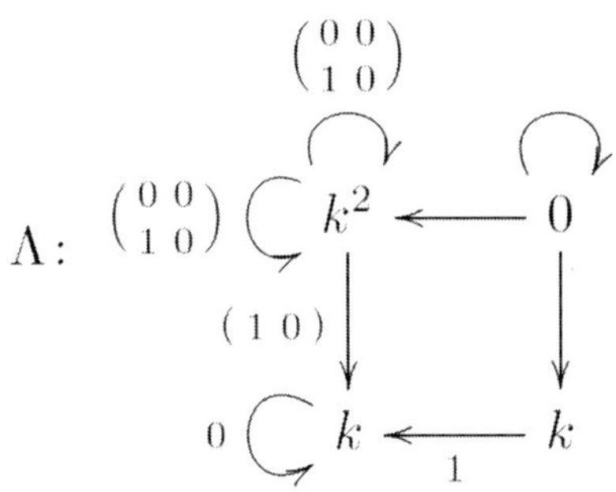

There is an exact sequence

$$0 \to M \otimes_\Sigma N \to \Lambda^e e_{1\times 1^{op}} \oplus \Lambda^e e_{2\times 2^{op}} \to \Lambda \to 0$$

of Λ^e-modules, and thus $M_{\otimes\Sigma}N$ is a first syzygy of Λ.

The enveloping algebra Σ^e has the following quiver and relations:

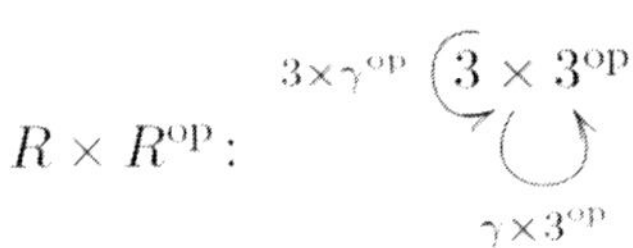

$$\left\{ \begin{matrix} (\gamma\times 3^{op})^2, (3\times\gamma^{op})^2, \\ (\gamma\times 3^{op})(3\times\gamma^{op}) - (3\times\gamma^{op})(\gamma\times 3^{op}) \end{matrix} \right\}$$

The algebra Σ considered as a Σ^e-module has the following representation over $R\times R^{op}$:

Its minimal projective resolution is

$\cdots \to \Sigma^e \to \Sigma^e \to \Sigma \to 0,$

with Σ itself as every syzygy. The tensor product $N \otimes_\Lambda M$ is isomorphic to Σ as Σ^e-module; in particular, it is a first syzygy of Σ.

We have now shown that the bimodules *M* and *N* induce a singular equivalence of Morita type with level 1 between the algebras Λ and Σ. The algebra Σ satisfies the **(Fg)** condition, but Λ does not, and is not even a Gorenstein algebra. This shows that the assumption of both algebras being Gorenstein can not be removed in Theorem 7.4.

For stable equivalences of Morita type (which are singular equivalences of Morita type with level 0), we can, under some conditions, remove the assumption of Gorensteinness.

Corollary 7.6

Let $_\Lambda M_\Sigma$ and $_\Sigma N_\Lambda$ be indecomposable bimodules that induce a stable equivalence of Morita type between two finite-dimensional k-algebras Λ and Σ. Assume that Λ and Σ have no semisimple blocks and that $\Lambda/\mathrm{rad}\Lambda$ and $\Sigma/\mathrm{rad}\Sigma$ are separable. Then Λ satisfies **(Fg)** *if and only if Σ satisfies* **(Fg)**.

Proof By [8, Corollary 3.1 (2)], the assumptions in the statement of the result imply that $(M\otimes_\Sigma -, N\otimes_\Lambda -)$ and $(N\otimes_\Lambda -, M\otimes_\Sigma -)$ are adjoint pairs. Then, by [13, Corollary 4.6], it follows that Λ is a Gorenstein algebra if and only if Σ is a Gorenstein algebra. The result now follows from Theorem 7.4. □

ACKNOWLEDGMENTS

I would like to thank Øyvind Solberg and Chrysostomos Psaroudakis for helpful discussions and suggestions. I would also like to thank Yiping Chen for informing me about the result stated in Corollary 7.6 and how it follows from the main result of this paper.

REFERENCES

1. Maurice Auslander, Idun Reiten**Applications of contravariantly finite subcategories**Adv. Math., 86 (1) (1991), pp. 111-152
2. Petter Andreas Bergh, David A. Jorgensen**Tate-Hochschild homology and cohomology of Frobenius algebras**J. Noncommut. Geom., 7 (4) (2013), pp. 907-937
3. Ragnar-Olaf Buchweitz, Maximal Cohen–Macaulay modules and Tate-cohomology over Gorenstein rings, 1987, unpublished manuscript.

4. Jon F. Carlson**The complexity and varieties of modules**Integral Representations and Applications, Lecture Notes in Math., Oberwolfach, 1980, vol. 882, Springer, Berlin–New York (1981), pp. 415-422
5. Henri Cartan, Samuel Eilenberg**Homological Algebra**Princeton University Press, Princeton, NJ (1956)
6. Xiao-Wu Chen**Singular equivalences induced by homological epimorphisms**Proc. Amer. Math. Soc., 142 (8) (2014), pp. 2633-2640
7. Xiao-Wu Chen, Longgang Sun, Singular equivalence of Morita type, 2012, unpublished manuscript.
8. Alex S. Dugas, Roberto Martínez-Villa**A note on stable equivalences of Morita type**J. Pure Appl. Algebra, 208 (2) (2007), pp. 421-433
9. Karin Erdmann, Miles Holloway, Rachel Taillefer, Nicole Snashall, Øyvind Solberg**Support varieties for selfinjective algebras**K-Theory, 33 (1) (2004), pp. 67-87
10. Karin Erdmann, Øyvind Solberg**Radical cube zero weakly symmetric algebras and support varieties**J. Pure Appl. Algebra, 215 (2) (2011), pp. 185-200
11. G. Hochschild**On the cohomology groups of an associative algebra**Ann. of Math. (2), 46 (1945), pp. 58-67
12. J. Külshammer, C. Psaroudakis, Ø. Skartsæterhagen**Derived invariance of support varieties**arXiv:1502.04880 (2015)
13. Yuming Liu, Changchang Xi**Constructions of stable equivalences of Morita type for finite-dimensional algebras. III**J. Lond. Math. Soc. (2), 76 (3) (2007), pp. 567-585
14. Saunders Mac Lane**Homology**Classics Math., Springer-Verlag, Berlin (1995)reprint of the 1975 edition
15. D.O. Orlov**Triangulated categories of singularities and D-branes in Landau–Ginzburg models**Tr. Mat. Inst. Steklova, 246 (2004), pp. 240-262English translation inProc. Steklov Inst. Math., 246 (2004), pp. 227-248
16. Chrysostomos Psaroudakis, Øystein Skartsæterhagen, Øyvind Solberg**Gorenstein categories, singular equivalences and finite generation of cohomology rings in recollements**Trans. Amer. Math. Soc. Ser. B, 1 (2014), pp. 45-95
17. Øyvind Solberg**Support varieties for modules and complexes**Trends in Representation Theory of Algebras and Related Topics, Contemp. Math., vol. 406, Amer. Math. Soc., Providence, RI (2006), pp. 239-270
18. Nicole Snashall, Øyvind Solberg**Support varieties and Hochschild cohomology rings**Proc. Lond. Math. Soc. (3), 88 (3) (2004), pp. 705-732
19. Zhengfang Wang**Singular equivalence of Morita type with level**J. Algebra, 439 (2015), pp. 245-269
20. Guodong Zhou, Alexander Zimmermann**On singular equivalences of Morita type**J. Algebra, 385 (2013), pp. 64-79

CHAPTER 7

A Lie–Rinehart Algebra with No Antipode

Ulrich Krähmer & Ana Rovi

School of Mathematics and Statistics, University of Glasgow, Glasgow, UK

KeyWords: Hopf algebroids, Lie algebroids, Lie–Rinehart algebras

INTRODUCTION

The enveloping algebra of a Lie algebra is a classical example of a Hopf algebra. Hence it is natural to ask whether the enveloping algebra of a Lie algebroid [12] or more generally of a Lie–Rinehart algebra [13] carries the structure of a Hopf algebroid. It turns out that they always are *left bialgebroids* (introduced under the name $\times_R$-bialgebras by Takeuchi [15]), see [16], and in fact *left Hopf algebroids* (introduced under the name $\times_R$-Hopf algebras by Schauenburg [14]), see [6, Example 2]; see also [5, 11].

However, there is also the definition of a Hopf algebroid due to Lu [9] and the one due to Bohm and Szlachanyi [2] (the latter will be called *full Hopf algebroids* from now on), which both assume the existence of an antipode satisfying certain axioms. The aim of the present paper is to communicate a concrete example of a Lie–Rinehart algebra whose universal enveloping algebra is not a Hopf algebroid in either of these two settings.

This clarifies further the relation between the three concepts: it is well-known and easily seen that every full Hopf algebroid is a left Hopf algebroid, see [2]. In [1], an example of a full Hopf algebroid was given that is not a Hopf algebroid in the sense of [9], but to the best of our knowledge, it is not known whether every Hopf algebroid in the sense of Lu satisfies the axioms of a full or at least of a left Hopf algebroid, and whether every left Hopf algebroid admits an antipode satisfying either of the definitions of [2,

9] (a counterexample announced in [8, Remark 3.12] did not appear in print).

In the light of [8, Proposition 3.11], it is known that the enveloping algebras of Lie algebroids [3] and of the Lie–Rinehart algebras associated to Poisson algebras [4, Section (3.2)] are full Hopf algebroids. However, here we prove the following theorem.

Theorem 1.1. *Let* K *be a field,* $R \_= K[x,y]/\{x.y,x^2,y2\}$,*L be the 1-dimensional Lie algebra with basis* {α}, *and* $E \in \mathrm{Der}_K(R)$*be the derivation with* E(*x*) = y, E(*y*) = 0.

1. *There is a Lie–Rinehart algebra structure on R,L) with R-module structure on* L *given by* x.α= *y.α* = 0 *and anchor map given by p*(α) = *E.*

2. *There is no right* V(R, L)*-module structure on* R *that extends multiplication in* R.

3. V(R, L) *is neither a full Hopf algebroid, nor a Hopf algebroid in the sense of [9].*

The note is structured as follows. In Section 2, we recall some basic definitions and prove the implication 2. ⇒ 3. of Theorem 1.1. In Section 3, we provide a construction method of Lie Rinehart algebras whose enveloping algebras satisfy part 2, for which the Lie–Rinehart algebra from 1 is a basic example.

2. BACKGROUND

This section contains background on Lie–Rinehart algebras [13], see also [4, 7, 8, 11] for more information. For the corresponding differential geometric notion of a Lie algebroid see [12] and for example [10] for further details.

We fix a field *K*. An unadorned ⊗ denotes the tensor product of *K*-vector spaces.

Definition 2.1. A Lie–Rinehart algebra consists of the following elements:

1. A commutative *K*-algebra (*R*,.);

2. A Lie algebra (L,[-,-]$_L$) over *K;*

3. A left *R*-module structure *R*⊗*L*?*L*,*r*⊗?→*r* . ∈, *R*, ?∈ L;and

4. An R-linear Lie algebra homomorphism *P*:L→Der$_K$(*R*) satisfying

$$[\xi, r \cdot \zeta]_L = r \cdot [\xi, \zeta]_L + \rho(\xi)(r) \cdot \zeta, \quad r \in R, \xi, \zeta \in L \text{.} \quad (2.1)$$

The map *P* is referred to as the anchor map.

There are two fundamental examples: if *R* is any commutative algebra, one can take *L* to be Der_K(R)with its usual Lie algebra and *R*-module structure, and *P* = id. The other extreme is *R* = *K* and *P* = 0, *L* being any Lie algebra.

In his paper [13], Rinehart generalised the construction of the universal enveloping algebra of a Lie algebra to Lie–Rinehart algebras, see Section 2 therein for the precise construction. The result is an associative *K*-algebra *V(R, L)* that is generated by the (sum of the) images of a *K*-algebra map

$R \rightarrow V \quad (R, L)$

and a Lie algebra map

$$(L, [-, -]_L) \longrightarrow (V(R, L), [-, -]), \quad \xi \longmapsto \bar{\xi}$$

where [-,-]denotes the commutator in *V (R,L)*. As Rinehart, we do not distinguish between an element in *R* and its image in *V (R L)* which is justified as the first map is always injective. The construction is such that in V R_ L one has

for all $r \in R$, �� $\in L$

$$[\bar{\xi}, r] = \rho(\xi)(r), \quad r\bar{\xi} = \overline{r \cdot \xi} \quad (2.2)$$

where the product in *V(R, L)* is denoted by concatenation.

As indicated in the introduction, *V(R, L)* has the structure of a left Hopf algebroid. Its counit endows R with the structure of a left *V(R L)*-module, in such a way that the induced action of r ∈ *R* is given by multiplication, and the induced action of � $\in L$ is given by the anchor map. The following fact is well known and yields the implication 2. ⇒ 3. in Theorem 1.1.

Lemma 2.2. *If* H *is either a full Hopf algebroid or a Hopf algebroid in the sense*

of Lu, with antipode S : H → H, *left counit �: H → R, and source and target maps*

s_t : R → H, *then defining for h ∈H, r ∈R*

$$rh := \varepsilon(S(h)s(r)) \quad (2.3)$$

yields a right H-module structure on R for which the underlying left R-action on R is

given by left multiplication.

Proof. The canonical left action of a left bialgebroid *H* on the base algebra *R*

is given by $hr := \varepsilon(hs(r)) = \varepsilon(ht(r))$, and the antipode of a Hopf algebroid is an algebra antihomomorphism (see [1, Proposition 4.4] respectively [9, Definition 4.1]

for the two different notions). Hence (2.3) defines a right action of *H* on *R*. Finally,

one has S 0 *t* = s (see [1, Definition 4.1 (iii)] respectively [9, Definition 4.1.2.]), so

$$rt(q) = \varepsilon(S(t(q))s(r)) = \varepsilon(s(q)s(r)) = qr \text{ for all } q, r \in R$$

3. A LIE–RINEHART ALGEBRA WITHOUT FLAT RIGHT CONNECTION ON R

We now prove Theorem 1.1, 1. and 2. We begin by considering more generally

Lie–Rinehart algebras(*R,L)* whose R-module structure on *L* is given by a character

X: R → K.

Lemma 3.1. *Let(R,.) be a commutative K-algebra,(L,-,-]*$_L$) *be a Lie algebra, and* ρ:L → Der$_K$(*R) be a Lie algebra map. Define an R-module structure on* L *by*

$r \cdot \xi := \chi(r)\xi$ *where x*:R → *K is a character on R. Then* (R,L)*is a Lie–Rinehart algebra if and only if* _ *is* R-*linear and* $\rho(\xi)(r) \in \ker \chi$ *for all* $r \in R, \xi \in L$

Proof. This follows as the Leibniz rule (2.1) takes the form

$$[\xi, \chi(r)\zeta]_L = \chi(r)[\xi, \zeta]_L + \chi(\rho(\xi)(r))\zeta$$

and hence by the K-linearity of the bracket becomes equivalent to $\rho(\xi)(r) \in \ker\chi$ □

Note that for these examples, $[-,-]_L$ is even R-linear, so L is a Lie algebra over R. However, in general we have $P \neq 0$.

Assume now that $(R\ L)$ is a Lie–Rinehart algebra as in the above lemma, and that multiplication in R can be extended to a right $V(R, L)$-module structure on R. Denote by $\partial(\xi) \in R$ the element obtained by acting with $\xi \in L$ on 1 ∈ R under this right action. This defines a K-linear map $\partial : L \rightarrow R$ and in $V(R, L)$ we have

$$\rho(\bar{\xi})(r) = [\bar{\xi}, \bar{r}] = \bar{\xi}\bar{r} - \bar{r}\bar{\xi} = \bar{\xi}\bar{r} - \overline{r \cdot \xi} = \bar{\xi}\bar{r} - \overline{\chi(r)}\bar{\xi},$$

so by acting with this element on 1 ∈ R, one sees that this map satisfies

$$\rho(\xi)(r) = \partial(\xi) \cdot (r - \chi(r)). \tag{3.1}$$

A K-linear map with this property defines a right $V(R\ L)$-module structure extending multiplication on R if and only if it satisfies the condition $\partial([\xi, \zeta]_L) = \rho(\xi)(\partial(\zeta)) - \rho(\zeta)(\partial(\xi))$. It also corresponds to a generator of the Gerstenhaber bracket on $\Lambda_R L$, see [4], but we shall not need these facts.

Proof of Theorem 1.1, 1. and 2. The first part is verified by explicit

computation; the Lie–Rinehart algebra is of the form as in Lemma 3.1 with _ given

by $\chi(x) = \chi(y) = 0$.

For 2., take $r = x$ and α in (3.1). One obtains $y = E(x) = \rho(\alpha)(x) = \partial(\alpha) \cdot x$ However, there is no element $z \in R$ such that $y = z.x$.

Carrying out Rinehart's construction explicitly yields a presentation of the

associative K-algebra $V(R,L)$ in terms of generators $x, y, \bar{\alpha}$ satisfying the relations

$$\bar{\alpha}x = y, \quad \bar{\alpha}y = x\bar{\alpha} = y\bar{\alpha} = x^2 = y^2 = xy = yx = 0$$

Hence *V(R,L)* has a *K*-linear basis given by $\{\bar{\alpha}^n, x, y\}_{n\in\mathbb{N}}$ The source and target maps are both the inclusion of *R* into *V (RL*). Hence one can also see directly that*V(R,L)* admits no antipode: *S* would satisfy S(*x)* = *x*, S(*y*) = *y* and one would have

$y = S(y) = S(\bar{\alpha}x) = S(x)S(\bar{\alpha}) = xS(\bar{\alpha})$, but there is no element z ∈ *V(R,L)* such that

y = *x*z.

ACKNOWLEDGMENT

A.R. thanks Jose Figueroa O'Farrill for his encouragement, and Gwyn Bellamy for remarks about differential operators.

FUNDING

U.K. acknowledges support by the EPSRC grant (No. EP/J012718/1) "Hopf algebroids and Operads" and the Polish Government Grants 2012/06/M/ST1/00169; A.R. is funded by an EPSRC DTA grant.

REFERENCES

1. Bohm, G. (2009). *Hopf Algebroids.* Handbook of Algebra, Vol. 6. Amsterdam: Elsevier/North-Holland, pp. 173–235
2. Bohm, G., Szlachanyi, K. (2004). Hopf algebroids with bijective antipodes: Axioms, integrals, and duals. *J. Algebra* 274(2):708–750
3. Evens, S., Lu, J.-H., Weinstein, A. (1999). Transverse measures, the modular class and a cohomology pairing for Lie algebroids. *Quart. J. Math. Oxford Ser. (2)* 50(200): 417–436
4. Huebschmann, J. (1998). Lie–Rinehart algebras, Gerstenhaber algebras and Batalin-Vilkovisky algebras. *Ann. Inst. Fourier (Grenoble)* 48(2):425–440
5. Huebschmann, J. (2008). *The Universal Hopf Algebra Associated with a Hopf- Lie-Rinehart Algebra.* Preprint arXiv:0802.3836
6. Kowalzig, N., Krahmer, U. (2010). Duality and products in algebraic (co)homology theories. *J. Algebra* 323(7):2063–2081
7. Kowalzig, N. (2009). Hopf algebroids and their cyclic theory. Ph.D. thesis, Universiteit Utrecht.
8. Kowalzig, N., Posthuma, H. (2011). The cyclic theory of Hopf algebroids. *J. Noncommut. Geom.* 5(3):423–476
9. Lu, J.-H. (1996). Hopf algebroids and quantum groupoids. *Internat. J. Math.* 7(1): 47–70

10. Mackenzie, K. (1987). Lie groupoids and Lie algebroids in differential geometry.*London Mathematical Society Lecture Note Series*, Vol. 124. Cambridge: Cambridge University Press.
11. Moerdijk, I., Mrˇcun, J. (2010). On the universal enveloping algebra of a Lie algebroid.*Proc. Amer. Math. Soc.* 138(9):3135–3145
12. Pradines, J. (1967). Theorie de Lie pour les groupoides differentiables. Calculdifferentiel dans la categorie des groupoides infinitesimaux. *C. R. Acad. Sci. Paris Sér. A-B*, 264:A245–A248.
13. Rinehart, G. S. (1963). Differential forms on general commutative algebras. *Trans. Amer. Math. Soc.* 108:195–222
14. Schauenburg, P. (2000). Duals and doubles of quantum groupoids (× R-Hopf algebras).*New trends in Hopf algebra theory (La Falda), Contemp. Math.* 267:273–299. Providence RI: Amer. Math. Soc., 2000
15. Takeuchi, M. (1977). Groups of algebras over A ⊗ A. *J. Math. Soc. Japan* 29(3): 459–492
16. Xu, P. (2001). Quantum groupoids. *Comm. Math. Phys.* 216(3):539–581

CHAPTER 8

Partial Ordering of Range Symmetric Matrices and M-Projectors with Respect to Minkowski Adjoint in Minkowski Space

D. Krishnaswamy, Mohd Saleem Lone

Department of Mathematics, Annamalai University, Chidambaram, India

ABSTRACT

In this paper, we obtain some new characterizations of the range symmetric matrices in the Minkowski Space M by using the Block representation of the matrices. These characterizations are used to establish some results on the partial ordering of the range symmetric matrices with respect to the Minkowski adjoint. Further, we establish some results regarding the partial ordering of m-projectors with respect to the Minkowski adjoint and manipulate them to characterize some sets of range symmetric elements in the Minkowski Space.M All the results obtained in this paper are an extension to the Minkowski space of those given by A. Hernandez, et al. in [The star partial order and the eigenprojection at 0 on EP matrices, Applied Mathematics and Computation, 218: 10669-10678, 2012].

Keywords: Partial Order, Minkowski Adjoint, Minkowski Inverse, Range Symmetric, M-Projectors`

1 INTRODUCTION AND PRELIMINARIES

Let us denote by $M_{(m,n)}(\mathbb{C})$ the set of $m\times n$ matrices and when $m=n$ we write $M_n(\mathbb{C})$ for $M_{(m,n)}(\mathbb{C})$. The symbols A^*, $A^\sim$, $A^\oplus$, $A^\dagger$, $R(A)$and $N(A)$ de- note the conjugate transpose, Minkowski adjoint, Minkowski inverse, Moore-Penrose inverse, range space and null space of a matrix $\boldsymbol{A}$ respectively. $\boldsymbol{I_n}$ denote the iden- tity matrix of order.$n\times n$ Further we denote by $\mathbb{C}_n^{mp}$ the set of all m-projections. i.e. $\mathbb{C}_n^{mp}=\left\{\boldsymbol{P}:\boldsymbol{P}^2=\boldsymbol{P}=\boldsymbol{P}^\sim\right\}$. Also we use the convection according to which $\boldsymbol{P}_w=\boldsymbol{W}\boldsymbol{W}^\oplus$ and $\tilde{\boldsymbol{P}}_w=\boldsymbol{I}_k-\boldsymbol{W}\boldsymbol{W}^\oplus$. Where I_k is the identity matrix of suitable order. r and s will denote the rank of the matrices $\boldsymbol{R}$ and $\boldsymbol{B}$.

Indefinite inner product is a scalar product defined by

$$[\boldsymbol{u},\boldsymbol{v}]=\langle\boldsymbol{u},\boldsymbol{M}\boldsymbol{v}\rangle=\boldsymbol{u}^*\boldsymbol{M}\boldsymbol{v}, \tag{1}$$

where $\langle\ ,\ \rangle$ denotes the conventional Hilbert Space inner product and $\boldsymbol{M}$ is a Hermitian matrix. This Hermitian matrix $\boldsymbol{M}$ is referred to as metric matrix. Min- kowski Space $\boldsymbol{M}$ is an indefinite inner product space in which the metric matrix is denoted by $\boldsymbol{G}$ and is defined as $\boldsymbol{G}=\begin{bmatrix}1 & 0\\ 0 & -\boldsymbol{I}_{n-1}\end{bmatrix}$ satisfying $\boldsymbol{G}^2=\boldsymbol{I}$ and $\boldsymbol{G}^*=\boldsymbol{G}$.

$\boldsymbol{G}$.is called the Minkowski metric matrix. In case $\boldsymbol{u}=(u_0,u_1,\cdots,u_{n-1})\in\mathbb{C}^n$, then G is called the Minkowski metric tensor and is defined as $\boldsymbol{Gu}=(u_0,-u_1,\cdots,-u_{n-1})$. For detailed study of indefinite linear algebra refer to [1].

The minkowski inverse of a matrix $\boldsymbol{R}\in M_{(m,n)}(\mathbb{C})$, introduced by Meenakshi in [2] , is the unique solution to the following four matrix equations:

[MI-1]: $\boldsymbol{RXR}=\boldsymbol{R}$.

[MI-2]: $\boldsymbol{XRX}=\boldsymbol{X}$.

[MI-3]: $(\boldsymbol{RX})^\sim=.\boldsymbol{RX}$

[MI-4]: $(\boldsymbol{XR})^{\sim} = \boldsymbol{XR}$

However unlike the Moore-Penrose inverse of a matrix, the Minkowski inverse of a matrix does not exist always. In [2] , Meenakshi showed that the Minkowski inverse of a matrix $\boldsymbol{R} \in M_{(m,n)}(\mathbb{C})$ exists if and only if $(\boldsymbol{RR}^{\sim}) = \text{rk}(\boldsymbol{R}^{\sim}\boldsymbol{R}) = \text{rk}(\boldsymbol{R})$, where $\boldsymbol{R}^{\sim} = \boldsymbol{G}_1\boldsymbol{R}^*\boldsymbol{G}_2$ is called the Minkowski adjoint of the matrix $\boldsymbol{R}$ and $\boldsymbol{G}_1$ and $\boldsymbol{G}_2$ are the Minkowski metric matrices of suitable order m and n. A matrix $\boldsymbol{R} \in \boldsymbol{M}_{\boldsymbol{n}}(\mathbb{C})$ is said to be m-symmetric if $\boldsymbol{R} = \boldsymbol{R}^{\sim}$ and is said to be G-unitary if and only if $(\boldsymbol{RR}^{\sim})$.=rk $(\boldsymbol{R})$, In [3] , Meenakshi introduced the concept of range symmetric matrices in Minkowski Space and developed the Minkowski inverse of the range symmetric matrices and some equivalent conditions for a matrix to be range symmetric. A matrix $R \in M_n(\mathbb{C})$ is said to be range symmetric if and only if $N(\boldsymbol{R}) = N(\boldsymbol{R}^{\sim})$. In [4] , the authors produced the necessary and sufficient conditions for the product of range symmetric matrices to be range symmetric and further showed that any block matrix in Minkowski space can be expressed as the product of range symmetric matrices. In [5] the authors studied the range symmetric matrices in relation with their Minkowski inverse and m-projectors. Summarizing the equivalent conditions for the definition of a range symmetric matrix form [3] [5] [6] the following equivalent con- ditions will be used in the forthcoming results:

[RS-1]: $\boldsymbol{R}$ is range symmetric.

[RS-2]: $N(\boldsymbol{R}) = N(\boldsymbol{R}^{\sim})$.

[RS-3]:.$\boldsymbol{RR}^{\oplus} = \boldsymbol{R}^{\oplus}\boldsymbol{R}$

[RS-4]: $R(\boldsymbol{R}) = R(\boldsymbol{R}^{\sim})$..

[RS-5]: their exist a G-unitary matrix $\boldsymbol{U}$ such that $R = U_R(D_R \oplus 0).U^{\sim}{}_R$.

Partial orders on matrices has remained the topic of interest for many authors in the area of matrix theory and generalized inverse. Almost all authors who have worked on partial ordering of matrices have formulated the definition involving different kinds of generalized inverses and in particular the Moore-Penrose Inverse. Results involving partial orders on matrices in relation with their generalized inverse are scattered in the literature of the matrix theory and generalized inverses for instance see [7] - [19]. Partial ordering on matrices has a wide range of applications in different fields which include electrical networks, statistics, generalized inverses etc. see [20] [21] [22] [23]. Different kinds of partial orders on matrices have been studied which include Star partial ordering $\leq^*$ introduced by Drazin [24] , minus partial order $\leq^-$ introduced by Hartwig [25] , Sharp partial $\leq^\#$ order introduced by Mitra [19] , followed by left star ordering $^*\leq$ and right star ordering$\leq^*$. In [26] , Punithavalli introduced the partial ordering on matrices in Minkowski space w.r.t the Minkowski adjoint. She studied the partial ordering, left partial ordering and right partial ordering w.r.t the Minkowski adjoint on Range symmetric matrices. She also established some equivalent conditions for the reverse order law to hold in relation to the partial ordering w.r.t Minkowski adjoint. Form ([26] , page 79), we have for any two matrices $\boldsymbol{R}, \boldsymbol{S} \in M_{(m,n)}$ (C), $\boldsymbol{R}$ is said to be below $\boldsymbol{S}$ under the partial order w.r.t Minkowski adjoint, denoted by $\boldsymbol{R} \leq^{\sim} \boldsymbol{S}$, if one of the following equivalent condition is satisfied:

[PO-1]: $\boldsymbol{R}^{\sim} \boldsymbol{R} = \boldsymbol{R}^{\sim}$ and $\boldsymbol{R}\boldsymbol{R}^{\sim}$.= $\boldsymbol{S}\boldsymbol{R}^{\sim}$.

[PO-2]: $\boldsymbol{R}^{\oplus}\boldsymbol{R} = \boldsymbol{R}^{\oplus}\boldsymbol{S}$ and. $\boldsymbol{R}\boldsymbol{R}^{\oplus}$.= $\boldsymbol{S}\boldsymbol{R}^{\oplus}$.

[PO-3]: $R(\boldsymbol{R}) \subset R(\boldsymbol{S})$ and.$R(\mathbf{R}^{\sim}) \subset \boldsymbol{R}(S^{\sim})$.

In any of the above cases we say $\boldsymbol{R}$ is predecessor of $\boldsymbol{S}$ or $\boldsymbol{S}$ is successor of $\boldsymbol{R}$. We will use the notation $M_n^k(\mathbb{C}) = M_n^k$ to denote the set of all the matrices of index k.

In this paper we obtain some characterizations of range symmetric matrices and utilize them to study the partial ordering of range symmetric matrices w.r.t the Min- kowski adjoint in Minkowski space and hence different characterizations of partial orders on range symmetric matrices are obtained. Finally we study the partial ordering on m-Projectors w.r.t the Minkowski adjoint. All the results obtained in this paper are an extension of those given in [27] to the Minkowski space M.

2. PROPERTIES OF RANGE SYMMETRIC MATRICES

In this section we develop some properties of Range Symmetric matrices by utilizing the representation obtained in corollary in [5]. Let $\boldsymbol{R}, \boldsymbol{S} \in M_{,n}$ (C) be non-zero range symmetric matrices of rank r and s respectively. Then $\boldsymbol{R}$ and$\boldsymbol{S}$, accord- ing to the above mentioned result, can be written as

$$\boldsymbol{R} = \boldsymbol{U}_R \begin{bmatrix} \boldsymbol{D}_R & 0 \\ 0 & 0 \end{bmatrix} \boldsymbol{U}_R^{\sim} \tag{2}$$

and

$$S = U_S \begin{bmatrix} D_S & 0 \\ 0 & 0 \end{bmatrix} U_S^{\sim}$$

(3)

where U_R and U_S are G-unitary and D_R and U_R are invertible matrices of order ***r.***

Theorem 1 Let ***R, S*** $\in M_{,n}$ (C) be such that ***R*** is range symmetric. Then the following statements are equivalent:

1. ***RS = SR***

2. If ***R*** is given by (2), then there exists ***J*** $\in M_{,r}$ (C) and $M \in M_{n-r}(\mathbb{C})$ such that $S = U_R \begin{bmatrix} J & 0 \\ 0 & M \end{bmatrix} U_R^{\sim}$ with $D_R J = J D_R$.

Proof. We consider the decomposition of the matrix S, according to the size of blocks of R, as:

$$S = U_R \begin{bmatrix} J & K \\ L & M \end{bmatrix} U_R^{\sim}.$$

From the statement (i) of the theorem, we get

$$\begin{bmatrix} D_R & 0 \\ 0 & 0 \end{bmatrix} \begin{bmatrix} J & K \\ L & M \end{bmatrix} = \begin{bmatrix} J & K \\ L & M \end{bmatrix} \begin{bmatrix} D_R & 0 \\ 0 & 0 \end{bmatrix}.$$

This gives $D_R J = J D_R$, $L = 0$ and $K = 0$ and hence the result follows.

If both the matrices ***R*** and ***S*** are range symmetric, then we have the following result for the commutativity.

Theorem 2 Let ***R, S*** $\in M_{,n}$ (C) be range symmetric matrices. If $U_R^{\sim} U_S = \begin{bmatrix} J & K \\ L & M \end{bmatrix}$ Then the following statements are equivalent:

1. ***RS = SR***.

2. $(J^{\sim} D_R)\, D_S = (D_S J^{\sim} D_R) J$.

3. $D_R (J D_S J^{\sim}) = (J D_S J^{\sim}) D_R$.

Proof. (i)⇔(ii) Consider the representations of $\boldsymbol{R}$ and $\boldsymbol{S}$ given by (2) and (3) res- pectively. With given $U_R^{\sim} U_S = \begin{bmatrix} J & K \\ L & M \end{bmatrix}$, we have

$$RS = U_R \begin{bmatrix} D_R & 0 \\ 0 & 0 \end{bmatrix} \begin{bmatrix} J & K \\ L & M \end{bmatrix} \begin{bmatrix} D_S & 0 \\ 0 & 0 \end{bmatrix} U_S^{\sim} = U_R \begin{bmatrix} D_R J D_S & 0 \\ 0 & 0 \end{bmatrix} U_S^{\sim} \tag{4}$$

Also

$$U_S^{\sim} U_R = \left(U_R^{\sim} U_S\right)^{\sim} = \begin{bmatrix} J^{\sim} & -L^{\sim} G_r \\ -G_r K^{\sim} & G_{n-r} M^{\sim} G_{n-r} \end{bmatrix}$$

Therefore

$$SR = U_S \begin{bmatrix} D_S & 0 \\ 0 & 0 \end{bmatrix} \begin{bmatrix} J^{\sim} & -L^{\sim} G_r \\ -G_r K^{\sim} & G_{n-r} M^{\sim} G_{n-r} \end{bmatrix} \begin{bmatrix} D_R & 0 \\ 0 & 0 \end{bmatrix} U_R^{\sim}$$

$$= U_S \begin{bmatrix} D_S J^{\sim} D_R & 0 \\ 0 & 0 \end{bmatrix} U_R^{\sim} \tag{5}$$

From Equations (4) and (5) we have

$$U_R \begin{bmatrix} D_R J D_S & 0 \\ 0 & 0 \end{bmatrix} U_S^{\sim} = U_S \begin{bmatrix} D_S J^{\sim} D_R & 0 \\ 0 & 0 \end{bmatrix} U_R^{\sim} \tag{6}$$

Pre multiplying and post multiplying (6) by $U_S^{\sim}$ and U_S respectively and sub- stituting the matrix representation of $U_S^{\sim} U_R$ and $U_R^{\sim} U_S$ we get

$$\begin{bmatrix} J^{\sim} D_R J D_S & 0 \\ -G_r K^{\sim} D_R J D_S & 0 \end{bmatrix} = \begin{bmatrix} D_S J^{\sim} D_R J & 0 \\ D_S J^{\sim} D_R K & 0 \end{bmatrix}$$

From this equality, on using the fact that $\boldsymbol{D_R}$ and $\boldsymbol{D_S}$ are nonsingular, we have$(J^{\sim} D_R J)\ D_S = D_s\ (J^{\sim} D_R J)$, $k^{\sim} D_R J = 0$, and $J^{\sim} D_R K = 0$ and hence the equivalence follows.

(i)⇔(iii) From $U_R^{\sim}U_S = \begin{bmatrix} J & K \\ L & M \end{bmatrix}$, using the fact that U_R is G-unitary, we have $U_S = U_R\begin{bmatrix} J & K \\ L & M \end{bmatrix}$ and hence $U_S^{\sim} = \begin{bmatrix} J^{\sim} & -L^{\sim}G_r \\ -G_rK^{\sim} & G_{n-r}M^{\sim}G_{n-r} \end{bmatrix}U_R^{\sim}$.

Substituting the re- presentations of U_S and $U_S^{\sim}$ in the block representation of S given by (3) we have

$$S = U_R\begin{bmatrix} J & K \\ L & M \end{bmatrix}\begin{bmatrix} D_S & 0 \\ 0 & 0 \end{bmatrix}\begin{bmatrix} J^{\sim} & -L^{\sim}G_r \\ -G_rK^{\sim} & G_{n-r}M^{\sim}G_{n-r} \end{bmatrix}U_R^{\sim}$$

$$= U_R\begin{bmatrix} JD_SJ^{\sim} & -JD_SL^{\sim}G_r \\ -LD_SJ^{\sim} & LD_SL^{\sim}G_r \end{bmatrix}U_R^{\sim}$$

Furthermore, doing some algebra we have,

$$RS = U_R\begin{bmatrix} D_RJD_SJ^{\sim} & -D_RJD_SL^{\sim}G_r \\ 0 & 0 \end{bmatrix}U_R^{\sim}$$ and

$$SR = U_R\begin{bmatrix} JD_SJ^{\sim}D_R & 0 \\ -LD_SJ^{\sim}D_R & 0 \end{bmatrix}U_R^{\sim}$$

Therefore the equality $RS = SR$, on using the fact that D_R, D_S and G_r are nonsingular, gives

$D_1(JD_SJ^{\sim}) = (JD_SJ^{\sim})D_R$, $LD_SJ^{\sim} = 0$ AND $JD_SL^{\sim} = 0$ Hence the equivalence follows.

Theorem 3 Let $R \in M_n$ (c) be such that $R^{\oplus}$ ⊕ exists. Then the following state- ments are equivalent:

1. R is range symmetric.
2. $R(\boldsymbol{R}) = R(\boldsymbol{R}^{\oplus})$
3. $N(\boldsymbol{R}) = N(\boldsymbol{R}^{\oplus})$.

Proof. (i)⇔(ii) Since $RR^{\oplus}$ and $R^{\oplus}R$ are m-symmetric idempotents, in fact m-projectors, on using [RS-3], we have $\boldsymbol{R}$ is range symmetric if and only if $R(R^{\oplus}) = R(R^{\oplus}R)$. Also from [MI-1] and [MI-2] we have $R(R) = R(RR^{\oplus})$ and $R(R^{\oplus}) = R(R^{\oplus}R)$. Therefore $R(R) = R(R^{\oplus})$. Hence the equivalence follows.

(i)⇔(iii) Similarly $(I_n - RR^{\oplus})$ and $(I_n - R^{\oplus}R)$ are idempotents such that $N(R^{\oplus}) = R(I_n - RR^{\oplus})$ and $N(R) = R(I_n - R^{\oplus}R)$. Again using [RS-3], the result fol- lows.

Theorem 4 Let $R \in M_n(\mathbb{C})$ be a non zero matrix. Then the following statements are equivalent:

1. **R** is range symmetric.

2. There exists an invertible matrix $M \in M_{n-r}(\mathbb{C})$ and $L \in M_{(n-r,r)}(\mathbb{C})$ such that $R = R^{\sim}F$ with. $F = U_R \begin{bmatrix} D_R(D_R^{\sim})^{-1} & K \\ 0 & M \end{bmatrix} U_R^{\sim}$

3. There exists an invertible matrix $M \in M_{n-r}(\mathbb{C})$ and $L \in M_{(n-r,r)}(\mathbb{C})$ such that $R = R^{\oplus}E$ with $E = U_R \begin{bmatrix} (D_R)^2 & 0 \\ L & M \end{bmatrix} U_R^{\sim}$

Proof. (i)⇔(ii) Using [RS-4], there exists an invertible matrix $E \in M_n(\mathbb{C})$ such that $R = R^{\sim} E$. We partition E according to the blocks of **R** such that

$$E = U_R \begin{bmatrix} J & K \\ L & M \end{bmatrix} U_R^{\sim}$$

Now $R = R^{\sim} E$, gives $E = U_R \begin{bmatrix} (D_R)^{-1} D_R & 0 \\ L & M \end{bmatrix} U_R$, using the fact that D_R is invertible and G_r is G-unitary.

(i)⇔(iii) From statement (ii) of the Theorem 3 and [RS-4], we have $R(R^{\oplus}) = R(R^{-})$, the equivalence follows on the same lines as above

Theorem 5 Let $R \in M_n(\mathbb{C})$ be a nonzero matrix. Then the following statements are equivalent:

1. $\boldsymbol{R}$ is range symmetric.

2. There exists an invertible matrix $\boldsymbol{M} \in M_{n-r}(\mathbb{C})$ and $\boldsymbol{K} \in M_{(r,n-r)}(\mathbb{C})$ such that $R = R^{\sim}F$ with. $\boldsymbol{F} = \boldsymbol{U}_R \begin{bmatrix} \boldsymbol{D}_R(\boldsymbol{D}_R^{\sim})^{-1} & \boldsymbol{K} \\ 0 & \boldsymbol{M} \end{bmatrix} \boldsymbol{U}_R^{\sim}$ S

3. There exists an invertible matrix $\boldsymbol{M} \in M_{n-r}(\mathbb{C})$ and $\boldsymbol{K} \in M_{(r,n-r)}(\mathbb{C})$ such that $\boldsymbol{R} = \boldsymbol{R}^{\oplus}\boldsymbol{F}$ with $\boldsymbol{F} = \boldsymbol{U}_R \begin{bmatrix} (\boldsymbol{D}_R)^2 & \boldsymbol{K} \\ 0 & \boldsymbol{M} \end{bmatrix} \boldsymbol{U}_R^{\sim}$.

Proof. The proof follows on the same lines as in the above theorem, using the fact that two matrices $\boldsymbol{R}$ and $\boldsymbol{S}$ are row equivalent if and only if $N(R) = N(S)$ and utilizing the statement (iii) of Theorem 3 and [RS-2].

3. PARTIAL ORDERING OF RANGE SYMMETRIC MATRICES W.R.T MINKOWSKI ADJOINT

In this section some characterizations of predecessors of range symmetric matrices under the partial ordering w.r.t Minkowski adjoint. Using the equivalences of the defi- nition of Partial ordering w.r.t Minkowski adjoint that is [PO-1] and, [PO-2], it can

be easily verified that $R^{\sim}S$, $SR^{\sim}$ $\boldsymbol{R}^{\oplus}\boldsymbol{S}$ and $\boldsymbol{S}\boldsymbol{R}^{\oplus}$ are m-symmetric.

Theorem 6 Let $\boldsymbol{R}, \boldsymbol{S} \in M_n(\mathbb{C})$ such that $\boldsymbol{S}$ is a nonzero range symmetric matrix. Then the following statements are equivalent:

1. $\boldsymbol{R} \leq^{\sim} \boldsymbol{S}$.

2. There exists $J \in M_r$ (C)such that

$\boldsymbol{R} = \boldsymbol{U}_S \begin{bmatrix} \boldsymbol{J} & 0 \\ 0 & 0 \end{bmatrix} \boldsymbol{U}_S^{\sim}$ with $\boldsymbol{J} \leq^{\sim} \boldsymbol{D}_S$ (7)

Proof. (i)⇔(ii) We consider the following block representation of $\boldsymbol{R}$ according to the block size of $\boldsymbol{S}$ as:

$$R = U_S \begin{bmatrix} J & K \\ L & M \end{bmatrix} U_S^{\sim}$$

Then

$$R^{\sim}R = U_S \begin{bmatrix} J^{\sim}J - L^{\sim}G_r L & J^{\sim}K - L^{\sim}G_r M \\ -G_r K^{\sim}J + G_{(n-r)}M^{\sim}G_{(n-r)}L & -G_r K^{\sim}K + G_{(n-r)}M^{\sim}G_{(n-r)}M \end{bmatrix} U_S^{\sim}$$

and

$$R^{\sim}S = U_S \begin{bmatrix} JD_S & 0 \\ LD_S & 0 \end{bmatrix} U_S^{\sim}$$

Therefore the equality $R^{\sim}R = R^{\sim}S$ gives $J^{\sim}J - L^{\sim}G_r L = JD_S$ and $-G_r K^{\sim}K + G_{(n-r)}M^{\sim}G_{(n-r)}M = 0 \Rightarrow K = 0$ and $M = 0$ Also computing $RR^{\sim}$ and $SR^{\sim}$ and using the equality $RR^{\sim} = SR^{\sim}$, we get $JJ^{\sim} = D_S J$ and $L = 0$. Thus $J^{\sim}J = JD_S$ and $JJ^{\sim} = D_S J$ i.e., $J \leq^{\sim} D_S$.

However if S is range symmetric and $R \leq^{\sim} S$, then R need not be range sym- metric e.g. consider the matrices

Example 1

$$R = \begin{bmatrix} 1 & i \\ 0 & 0 \end{bmatrix} \text{ and } S = \begin{bmatrix} 1 & i \\ i & -1 \end{bmatrix}$$

Remark 1 If both the matrices $R, S \in M_n(\mathbb{C})$ are range symmetric and $R \leq^{\sim} S$, then using the statements [PO-1], [PO-2] and [RS-3], it can be easily observed that $R^{\oplus}S = SR^{\oplus}$. Using the representations (3) and (7) of S and R respectively and Theorem 6.8.3. from [26] , we have another equivalent condition for the partial ordering of range symmetric matrices w.r.t minkowski adjoint given by $R \leq^{\sim} S \Leftrightarrow RR^{\oplus} = S^{\oplus}R$ and $R^{\oplus}R = RS^{\oplus}$. Furthermore, R is range symmetric, we have $S^{\oplus}R = RS^{\oplus}$.

The next result gives some equivalent conditions for a matrix R to be range symmetric when S is range symmetric and S is the successor of R.

Theorem 7 Let $R, S \in M_n(\mathbb{C})$ such that S is a nonzero range symmetric matrix and $R \leq^{\sim} S$, where S is given by (3) and R is given by (7). Then the following statements are equivalent:

1. R is range symmetric.

2. $JD_s = D_sJ$.

3. $J^{\oplus} D_S = D_S J^{\oplus}$.

4. $J(J^{\sim} - D_S) = D_S(J^{\sim} - J)$

5. $(J^{\sim} - D_S)J = (J^{\sim} - J)D_S$.

6. J is range symmetric.

Proof. (i)⇔(ii) From remark 1, we have $S^{\oplus}R = RS^{\oplus}$. Now using the facts that $D_S^{\oplus} = D_S^{-1}$; D_S ;being invertible and U_S is G-unitary and substituting the repre- sentations of S and R from (3) and (7) respectively in the above equality and doing some simple algebra leads to $JD_s = D_sJ$

(ii)⇔(iii) For, $R = U_S \begin{bmatrix} J & 0 \\ 0 & 0 \end{bmatrix} U_S^{\sim}, \; R^{\oplus} = U_S \begin{bmatrix} J^{\oplus} & 0 \\ 0 & 0 \end{bmatrix} U_S^{\sim}$. Again using Remark 1 and substituting the respective representations of $R^{\oplus}$ and S, the equivalence follows.

(ii)⇔(iv) Using [PO-1] and substituting the representations of R, $R^{\sim}$ and S, the equivalence follows after some computation.

On the same lines the equivalences (ii)⇔(v) and (iii)⇔(vi) follow by using the Remark 1 and statements [PO-1] and [PO-2].

The next result similar to Theorem 6 holds if we consider R to be range symmetric and decompose S in terms of representation for R

Theorem 8 Let $R, S \in M_n(\mathbb{C})$ such that R is a nonzero range symmetric matrix. Then the following statements are equivalent:

1. $R \leq^{\sim} S$.

2. There exists $M \in M_{n-r}(\mathbb{C})$ such that

$$S = U_R \begin{bmatrix} D_R & 0 \\ 0 & M \end{bmatrix} U_R^{\sim} \tag{8}$$

Proof. The proof follows on the same line as in Theorem 6

We again note that if $R \leq^{\sim} S$ and R is range symmetric, then S need not be range symmetric. Consider Example 1. In the following result we establish some equivalent conditions for S when R is range symmetric and $R \leq^{\sim} S$.

Theorem 9 Let $R, S \in M_n(\mathbb{C})$ be given by (2) and (8) respectively such that R is a nonzero range symmetric matrix and $R \leq^{\sim} S$. Then the following statements are equivalent:

1. S is range symmetric.
2. $M$is range symmetric.
3. $S(R^{\oplus} - S^{\oplus}) = (R^{\oplus} - S^{\oplus})S$.
4. $S^{\oplus}(R - S) = (R - S)S^{\oplus}$.

Proof. (i)⇔(ii) For $S = U_R \begin{bmatrix} D_R & 0 \\ 0 & M \end{bmatrix} U_R^{\sim}$, since D_R is nonsingular and U_R is G- unitary, direct verification shows that $S^{\oplus} = U_R \begin{bmatrix} D_R^{\oplus} & 0 \\ 0 & M^{\oplus} \end{bmatrix} U_R^{\sim}$. Therefore $SS^{\oplus} = U_R \begin{bmatrix} I_r & 0 \\ 0 & MM^{\oplus} \end{bmatrix} U_R^{\sim}$ and. $S^{\oplus}S = U_R \begin{bmatrix} I_r & 0 \\ 0 & M^{\oplus}M \end{bmatrix} U_R^{\sim}$. S being range sym- metric, by [RS-3] we have $SS^{\oplus} = S^{\oplus}S$. This gives $MM^{\oplus} = M^{\oplus}M$ and the equi- valence follows.

(i)⇒(iii) Since $R \leq^{\sim} S$ and R and S are range symmetric, using the observation mentioned in Remark 1 i.e., $R^{\oplus}S = SR^{\oplus}$, we have $SR^{\oplus} - SS^{\oplus} = R^{\oplus}S - S^{\oplus}S$, the equivalence follows.

(iii)⇒(i) Since $R \leq^{\sim} S$ and R is range symmetric, again by the same fact that $R^{\oplus}$ and S commute, using (iii) i.e., $S(R^{\oplus} - S^{\oplus}) = (R^{\oplus} - S^{\oplus})S$, we get S is range symmetric.

(i)⇔(iv) From Remark 1, we have $RS^{\oplus} = S^{\oplus}R$. This gives $RS^{\oplus} - SS^{\oplus} = S^{\oplus} - SS^{\oplus}R$. Now using the fact that S is range symmetric the equivalence follows.

In the above results we have used the commutativity of R and $S^{\oplus}$ and $R^{\oplus}$ and S. However if we assume the conditions given in the above theorem with an additional assumption that $RS = SR$, then the conditions obtained by interchanging R and S are also equivalent.

Theorem 10 Let $R, S \in M_n(\mathbb{C})$ be range symmetric such that R is a non zero matrix. Then the following statements are equivalent:

1. $R \leq^{\sim} S$.

2. There exists a G-unitary matrix, $U \in M_n(\mathbb{C})$, $D \in M_r(\mathbb{C})$ and $M \in M_{r-s}(\mathbb{C})$ such that $R = U\begin{bmatrix} D & 0 & 0 \\ 0 & 0 & 0 \\ 0 & 0 & 0 \end{bmatrix}U^{\sim}$ and $S = U\begin{bmatrix} D & 0 & 0 \\ 0 & M & 0 \\ 0 & 0 & 0 \end{bmatrix}U$

Proof. (i)⇒(ii) Consider the decomposition of S given by (3) i.e., $S = U_S\begin{bmatrix} D_S & 0 \\ 0 & 0 \end{bmatrix}U_S^{\sim}$. Since S is range symmetric, therefore by Theorem 6, there exists $J \in M_r(\mathbb{C})$ such that $R = U_S\begin{bmatrix} J & 0 \\ 0 & 0 \end{bmatrix}U_S^{\sim}$ with $J \leq^{\sim} D_S$. Using Theorem 7, we have J is range symmetric. We consider the following block representation of J as $J = U_J\begin{bmatrix} D_J & 0 \\ 0 & 0 \end{bmatrix}U_J^{\sim}$, where U_J is G-unitary and $D \in M_r(\mathbb{C})$ is invertible. Since $J \leq^{\sim} D_S$, by Theorem 8, we can find $M \in M_{r-s}(\mathbb{C})$ such that $D_S = U_J\begin{bmatrix} D_J & 0 \\ 0 & M \end{bmatrix}U_J^{\sim}$. Thus M is nonsingular when. $M \neq 0$ Taking D_J = D, we have $S = U\begin{bmatrix} D & 0 & 0 \\ 0 & M & 0 \\ 0 & 0 & 0 \end{bmatrix}U^{\sim}$,

where $U = U_S\begin{bmatrix} D_J & 0 \\ 0 & I \end{bmatrix}$ is G-unitary.

Follows at once by direct verification.

4. PARTIAL ORDERING OF M-PROJECTORS

In this section we obtain some results on partial ordering of m-projectors w.r.t Minkowski adjoint. The following result from [5] , with two more obvious conditions, will be used extensively in the forthcoming results.

Lemma 1 Let $R \in M$ be range symmetric, then

1. $R^{\oplus} = \left(R + \tilde{P}_R\right)^{-1}\left(I_n - \tilde{P}_R\right)$.

2. $\tilde{P}_R$ is idempotent.

3. $R\tilde{P}_R = \tilde{P}_R R = R^{\oplus}\tilde{P}_R = \tilde{P}_R R^{\oplus} = 0$.

4. $\tilde{P}_R = 0$ if and only if R is nonsingular.

5. $R = P(D \oplus 0)P^{\sim}$ then $\tilde{P}_R = P(0 \oplus I_{n-r})P^{\sim}$.

6.rank $\left(\tilde{P}_R\right) = n -$ rank (R)

7. $\tilde{P}_R$ is invertible then $R = 0$ and if $R = 0$, then $\tilde{P}_R = I_n$.

8. $\tilde{P}_R$ has index atmost one.

Lemma 2 Let $R, S \in M_n^0 \cup M_n^1$. Then

1. If $R \leq^{\sim} S$, then $\text{rk}(R) \leq \text{rk}(S)$.

2. $R \leq^{\sim} \tilde{P}_R \Leftrightarrow R = 0$.

3. $\tilde{\boldsymbol{P}}_R \leq^{\sim} \boldsymbol{R} \Leftrightarrow \boldsymbol{R}$ is invertible.

4. If $\boldsymbol{R}$ is nonzero singular matrix then $\boldsymbol{R}$ and $\tilde{\boldsymbol{P}}_R$ are incomparable under the partial ordering w.r.t Minkowski adjoint.

5. $\tilde{\boldsymbol{P}}_S \leq^{\sim} \tilde{\boldsymbol{P}}_R$, then $\operatorname{rk}(\boldsymbol{R}) \leq \operatorname{rk}(\boldsymbol{S})$.

6. $\boldsymbol{R} \leq^{\sim} \boldsymbol{R} + \tilde{\boldsymbol{P}}_R$.

Proof. (i) Since $R(\boldsymbol{R}) = R(\boldsymbol{R}\boldsymbol{R}^{\oplus}\boldsymbol{R}) \subseteq R(\boldsymbol{R}\boldsymbol{R}^{\oplus}) = R(\boldsymbol{S}\boldsymbol{R}^{\oplus}) \subseteq R(\boldsymbol{S})$. This gives $(\boldsymbol{R}) \leq \operatorname{rk}(\boldsymbol{S})$.

(ii) $\boldsymbol{R} \leq^{\sim} \tilde{\boldsymbol{P}}_R$, then $\boldsymbol{R}\boldsymbol{R}^{\oplus}\boldsymbol{R} = \boldsymbol{R}^{\oplus}\tilde{\boldsymbol{P}}_R = 0 \Rightarrow \boldsymbol{R} = 0$. Conversely if R = 0, $\boldsymbol{R}^{\oplus} = 0$ and $\tilde{\boldsymbol{P}}_R = \boldsymbol{I}_n$ and hence $\boldsymbol{R} \leq^{\sim} \tilde{\boldsymbol{P}}_R$.

(iii) From statement (ii) of Lemma 1 and the fact that $\left(\tilde{\boldsymbol{P}}_R\right)^{\oplus} = \tilde{\boldsymbol{P}}_R$, if $\tilde{\boldsymbol{P}}_R \leq^{\sim} \boldsymbol{R}$, then $\tilde{\boldsymbol{P}}_R = \tilde{\boldsymbol{P}}_R\boldsymbol{R} = 0$ and hence by point (iv) of Lemma 1 $\boldsymbol{R}$ is invertible. Again by the same argument i.e, point (iv) of Lemma 1 converse holds.

(iv) It is obvious from (ii) and (iii).

(v) Follows at once by using point (i) of the Lemma 2 and point (vi) of Lemma 1.

(vi) The statement follows at once on using the fact that $\boldsymbol{R}^{\oplus}\tilde{\boldsymbol{P}}_R = 0$.

Lemma 3 Let $\boldsymbol{R} \in M_n^0 \cup M_n^1$. Then

1. If $\boldsymbol{R}$ is range symmetric, then $\tilde{\boldsymbol{P}}_R$ is m-symmetric and hence range symmetric

2. If $\tilde{\boldsymbol{P}}_R$ is range symmetric, then so is $\boldsymbol{R}$

Proof. (i) The statement follows at once on using the [RS-3], [MI-3] and [MI-4].

(ii) If $\tilde{P}_R = 0$, the the result is trivial. Let $\tilde{P}_R \neq 0$ such that $\text{rk}(\tilde{P}_R) = r$, then by point (v) of Lemma 1 we have $\tilde{P}_R = U\begin{bmatrix} D & 0 \\ 0 & 0 \end{bmatrix}U$. Also using point (ii) of Lemma 1 we get $D^2 = D$ i.e., $D = I_r$ Thus we have $RR^{\oplus} = U\begin{bmatrix} 0 & 0 \\ 0 & I_{n-r} \end{bmatrix}U$. Consider the block re- presentation of R, where the partition is done according to the blocks of $\tilde{P}_R$ such that $R = U\begin{bmatrix} J & K \\ L & M \end{bmatrix}U$. Using [MI-1] and [MI-2], we get J, K and $L = 0$ Therefore $RR^{\oplus} = U\begin{bmatrix} 0 & 0 \\ 0 & MM^{\oplus} \end{bmatrix}U$. This shows that M is nonsingular and the result follows.

Remark 2 Since $\tilde{P}_R$ is a m-projector [5] , we have $\left(\tilde{P}_R\right)^{\oplus} = \tilde{P}_R$. If we write $\tilde{P}_R = \tilde{P}_1(R)$ i.e., we take $\tilde{P}$ as a function of R, then

$$\tilde{P}_2(R) = \tilde{P}\left(\tilde{P}_1(R)\right) = I_n - \tilde{P}_R\left(\tilde{P}_R\right)^{\oplus} = RR^{\oplus} = 0 \Leftrightarrow R = 0$$

Thus $\tilde{P}_2(0) = 0$. However $\tilde{P}_2(R) \neq R$ in general. Consider the decomposition $R = U\begin{bmatrix} D_R & 0 \\ 0 & 0 \end{bmatrix}U$, we have

$\tilde{P}_2(R) = R \Leftrightarrow D_R = I_r$ i.e, $R = U\begin{bmatrix} I_r & 0 \\ 0 & 0 \end{bmatrix}U$, which is the fundamental represen- tation of a m-projector. Hence we conclude that $\tilde{P}_2(R) = R$ if and only if R is a m- projector.

We generalize the function $\tilde{P}(R)$ by defining it as:

$$\tilde{P}_k(R) = \left(\tilde{P}_{k-1}\tilde{P}_1\right)(R) = \tilde{P}_{k-1}\left(\tilde{P}_1(R)\right) \tag{9}$$

Thus we have the following equations

$$\tilde{P}_k(R) = \begin{cases} \tilde{P}_1(R) & \text{if } k \text{ is odd} \\ RR^{\oplus} & \text{if } k \text{ is even} \end{cases} \tag{10}$$

and hence if R = 0, we get

$$\tilde{P}_k(R) = \begin{cases} I_n & \text{if } k \text{ is odd} \\ 0 & \text{if } k \text{ is even} \end{cases} \tag{11}$$

Let us consider some sets with following notations:

$$\Gamma = \{R \in \mathcal{M} : R \text{ is range symmetric}\} \tag{12}$$

$$\Delta = \left\{R \in M_n^0 \cup M_n^1 : \tilde{P}_1(R) \text{ is range symmetric}\right\} \tag{13}$$

and

$$\Lambda = \left\{R \in M_n^0 \cup M_n^1 : \tilde{P}_1(R) \text{ is range symmetiric and } \tilde{P}_1(R) \neq 0\right\} \tag{14}$$

Theorem 11 Let Γ, Δ and Λ be the sets defined in (12), (13) and (14) respectively. Then, $\Gamma = \Delta$ and $\Gamma \cap M_n^1 = \Lambda$.

Proof. The proof follows easily by utilizing Lemmas 1 and 3.

From the statement (i) of Lemma 3, it is obvious that $\tilde{P}_1(\Gamma) \subseteq \Gamma$. However the reverse inclusion does not hold in general. Consider the matrix $R = \begin{bmatrix} 4 & 0 \\ 0 & 0 \end{bmatrix}$. If possible suppose there exist a matrix $J \in M_2(\mathbb{C}) \cap \Gamma$ such that $R = \tilde{P}_1(J)$, then by Lemma 3, we have,

$J = U\begin{bmatrix} 0 & 0 \\ 0 & 1 \end{bmatrix}U^{\sim}$ which is absurd, since $4 \in \sigma(R)$ but $4 \notin \sigma(J)$. Therefore $\Gamma \not\subset \tilde{P}_1(\Gamma)$.

Remark 3 Let $R \in M_n^1 - \{0\}$, then by using Lemma 1 $\tilde{P}_1(R) \subseteq M_n^1 - \{0\}$. If $R \in \Lambda - \{0\}$, then by Remark 2, we have $\tilde{P}_1(R) \in \mathbb{C}_n^{mp} \cap M_n^1 - \{0\} = (\mathbb{C}_n^{mp} \cap \Lambda) - \{0\}$ and if $R \in \Lambda$, then by Theorem 11, we have $\tilde{P}_1(R) \in \left[(\mathbb{C}_n^{mp} \cap \Lambda) - \{0\}\right] \cup \{I_n\}$.

The next result provides a characterization of the set

$$\Psi = \left\{S \in M_n^0 \cup M_n^1 : 0 \leq^{\sim} S \leq^{\sim} \tilde{P}_1(S)\right\}$$

Theorem 12 Let $R \in M_n(\mathbb{C})$ be range symmetric given by (2), then

$$\Psi = \left\{U_R(0 \oplus M)U_R^{\sim} : M \in \mathbb{C}_{n-r}^{mp}\right\} \subseteq \mathbb{C}_n^{mp}$$

Proof. Let $R = U\begin{bmatrix} D_R & 0 \\ 0 & 0 \end{bmatrix}U^{\sim}$. Then $\tilde{P}_1(R) = U_R\begin{bmatrix} 0 & 0 \\ 0 & I_{n-r} \end{bmatrix}U_R^{\sim}$. Therefore for $S \leq^{\sim} R$, we have $S = U_R\begin{bmatrix} 0 & 0 \\ 0 & M \end{bmatrix}U_R^{\sim}$ such that $M \leq^{\sim} I_{n-r}$ i.e., $M^2 = M = M^{\sim}$ and the result follows.

The next result shows that the function $\tilde{P}_1(R)$, when restricted to the set $\boldsymbol{\Gamma}$ is mo- notonically decreasing w.r.t the partial ordering w.r.t Minkowski adjoint.

Theorem 13 Let $R, S \in \Gamma$, such that $R \leq^{\sim} S$. Then $\tilde{P}_1(R) \leq^{\sim} \tilde{P}_1(S)$

Proof. Let $R, S \in \Gamma$, such that $R \leq^{\sim} S$. Since R is range symmetric we have

$$R = RR^{\oplus}R = R^2R^{\oplus} = R^{\oplus}R^2$$

and therefore

$$RS = R^2R^{\oplus}S = R^2R^{\oplus}R = R^2 = RR^{\oplus}R^2 = SR^{\oplus}R^2 = SR \quad (15)$$

Also

$$RR^{\oplus}SS^{\oplus} = RSR^{\oplus}S^{\oplus} = SRR^{\oplus}S^{\oplus} = SR^{\oplus}RS^{\oplus} = SS^{\oplus}RS^{\oplus} = SS^{\oplus}RR^{\oplus} \quad (16)$$

Finally using (15) and (16) we get $\tilde{P}_1(S) = \tilde{P}_1(R)\tilde{P}_1(S)$ and $\tilde{P}_1(S) = \tilde{P}_1(S)\tilde{P}_1(R)$. Hence $\tilde{P}_1(R) \leq^{\sim} \tilde{P}_1(S)$.

However for the range symmetric matrices $R = \begin{bmatrix} 1 & 0 \\ 0 & 0 \end{bmatrix}$ and $S = \begin{bmatrix} 2 & 0 \\ 0 & 0 \end{bmatrix}$, we have $\tilde{P}_1(R) \leq^{\sim} \tilde{P}_1(S)$ but $S \,{}^{\sim}\, R$. Thus we have the following result.

Theorem 14 Let $R, S \in \mathbb{C}_n^{mp}$, such that $\tilde{P}_1(S) \leq^{\sim} \tilde{P}_1(R)$. Then $R \leq^{\sim} S$.

Proof. The proof follows at once by using Theorem 13 and Remark 2.

Theorem 15 Let R be range symmetric and $S \in M_n^0 \cup M_n^1$ be such that $R \leq^{\sim} S$. Then $\tilde{P}_1(S) \leq^{\sim} \tilde{P}_1(R) \Leftrightarrow \tilde{P}_1(S) \in \mathbb{C}_n^{mp}$.

Proof. Consider the decomposition of R as given in (2). Then from Theorem 8 we get $S = U\begin{bmatrix} D_R & 0 \\ 0 & M \end{bmatrix}$ and hence $\tilde{P}_1(S) = \begin{bmatrix} 0 & 0 \\ 0 & \tilde{P}_1(M) \end{bmatrix}$. Thus if we assume $\tilde{P}_1(S) \leq^{\sim} \tilde{P}_1(R)$, then from Theorem 6, we get $\tilde{P}_1(S) = \begin{bmatrix} 0 & 0 \\ 0 & \tilde{P}_1(M) \end{bmatrix}$ with $J \leq^{\sim} I_{n-r}$. There- fore $J = \tilde{P}_1(M)$. Hence $\tilde{P}_1(S) \in \mathbb{C}_n^{mp}$, Conversely if we assume that $\tilde{P}_1(S) \in \mathbb{C}_n^{mp}$, then S is range symmetric $\tilde{P}_1(S) = \begin{bmatrix} 0 & 0 \\ 0 & J \end{bmatrix}$ and finally from Theorem 13, the result follows.

Theorem 16 Let $R \neq 0 \in M_n^0 \cup M_n^1$ has the representation as given in point (*V*) of Lemma 1 and $S \in M_n^0 \cup M_n^1$. Then $\tilde{P}_1(R) \leq^- \tilde{P}_1(S)$ if and only if there exists $E \in M_n^0 \cup M_n^1$ such that $S = P\begin{bmatrix} E & 0 \\ 0 & 0 \end{bmatrix}P$

Proof. Assume that $\tilde{P}_1(R) \leq^- \tilde{P}_1(S)$. Then from Theorem 13 we have $\tilde{P}_1(S)\tilde{P}_1(R) = \tilde{P}_1(R) = \tilde{P}_1(R)\tilde{P}_1(S)$. Let $\tilde{P}_1(S) = P\begin{bmatrix} J & K \\ L & M \end{bmatrix}P$, where the blocks are partitioned according to the blocks of ***R***. Using the above mentioned equality we have *K* = 0 *L* = 0 and *M* = I_{n-r} and hence $\tilde{P}_1(S) = P\begin{bmatrix} J & 0 \\ 0 & I_{n-r} \end{bmatrix}P$ Also $SS^{\oplus} = S^{\oplus}S = I_n - \tilde{P}_1(S) = P\begin{bmatrix} I_r - J & 0 \\ 0 & 0 \end{bmatrix}P$. Taking $S = P\begin{bmatrix} E & F \\ G & H \end{bmatrix}P$, where the decomposition is done according to the blocks of $\tilde{P}_1(S)$. Then from the equation $S = (SS^{\oplus})S = S(S^{\oplus}S)$ we get *G* = 0, *H* = 0, *F* = 0, *JE* = 0 = *EJ* and *JF* = 0 and therefore $S = P\begin{bmatrix} E & 0 \\ 0 & 0 \end{bmatrix}P$ with *JE* = *EJ*. Clearly $E \in M_r(\mathbb{C})$ has index one. The converse is obvious.

ACKNOWLEDGEMENTS

The second author was supported by UGC-BSR through grant No. F25-1/2014-15(BSR)/ 7-254/2009(BSR) (20.01.2015). This support is greatly appreciated.

REFERENCES

1. Gohberg, I., Lancaster, P. and Rodman, L. (2005) Indefinite Linear Algebra and Applications. Birkhäuser, Verlag, Basel, Boston, Berlin.

2. Meenakshi, A.R. (2000) Generalized Inverse of Matrices in Minkowski Space. Proceedings of National Seminar on Algebra and Its Applications , 1, 1-14.
3. Meenakshi, A.R. (2000) Range Symmetric Matrices in Minkowski Space. Bulletin of the Malaysian Mathematical Sciences Society , 23, 45-52.
4. Meenakshi, A.R. and Krishnaswamy, D. (2006) Product of Range Symmetric Block Matrices in Minkowski Space. Bulletin of the Malaysian Mathematical Sciences Society , 29, 59-68.
5. Lone, M.S. and Krishnaswamy, D. (2016) m-Projections Involving Minkowski Inverse and Range Symmetric Property in Minkowski Space. Journal of Linear and Topological Algebra .
6. Krishnaswamy, D. (2005) Contributions to the Study on Range Symmetric Matrices in Minkowski Space. Ph.D. Dissertation, Annamalai University, India.
7. Ben-Isreal, A. and Greville, T. (2003) Generalized Inverse: Theory and Applications. 2nd Edition, Springer Verlag, New York.
8. Campbell, S.L. and Meyer Jr., C.D. (1991) Generalized Inverse of Linear Transformations. 2nd Edition, Dover, New York.
9. Prasolov, V.V. (1994) Problems and Theorems in Linear Algebra. American Mathematical Society, Providence.
10. Meyer, C.D. (2000) Matrix Analysis and Applied Linear Algebra. SIAM, Philadelphia. https://doi.org/10.1137/1.9780898719512
11. Mitra, S.K., Bhimasankaram, P. and Malik, S.B. (2010) Matrix Partial Orders, Shorted Operators and Applications. World Scientific Publishing Company, Singapore.
12. Rao, C.R. and Mitra, S.K. (1971) Generalized Inverse of Matrices and Its Applications. John Wiley & Sons, New York.
13. Tosic, M. and Cvetkovic-Ilic, D.S. (2012) Invertibility of a Linear Combination of Two Matrices and Partial Orderings. Applied Mathematics and Computation, 218, 4651-4657. https://doi.org/10.1016/j.amc.2011.10.052
14. Malik, S.B. (2013) Some More Properties of Core Partial Order. Applied Mathematics and D. Krishnaswamy, M. S. Lone 145 Computation, 221, 192-201. https://doi.org/10.1016/j.amc.2013.06.012
15. Malik, S.B., Ruedab, L. and Thome, N. (2014) Further Properties on the Core Partial Order and Other Matrix Partial Orders. Linear Multilinear Algebra, 62, 1629-1648. https://doi.org/10.1080/03081087.2013.839676
16. Baksalary, J.K. and Mitra, S.K. (1991) Left-Star and Right-Star Partial Orderings. Linear Algebra and Its Applications , 149, 73-89. https://doi.org/10.1016/0024-3795(91)90326-R
17. Deng, C.Y. and Wang, S.Q. (2012) On Some Characterizations of the Partial Orderings for Bounded Operators. Mathematical Inequalities & Applications , 15, 619-630. https://doi.org/10.7153/mia-15-54
18. Liu, F.X. and Yang, H. (2011) Some Results on the Partial Orderings of Block Matrices. Journal of Inequalities and Applications , 2011, 1-7. https://doi.org/10.1186/1029-242x-2011-54
19. Mitra, S.K. (1987) On Group Inverses and the Sharp Order. Linear Algebra and Its Applications , 92, 17-37. https://doi.org/10.1016/0024-3795(87)90248-5
20. Baksalary, J.K., Hauke, J. and Styan, G.P.H. (1994) On Some Distributional Properties of Quadratic Forms in Normal Variables and on Some Associated Matrix Partial Orderings. Multivariate Analysis and Its Applications , 24, 111-121.
21. Baksalary, J.K. and Puntanen, S. (1990) Characterizations of the Best Linear Unbiased Estimator in the General Gauss Markov Model with the Use of Matrix Partial Orderings. Linear Algebra and Its Applications , 127, 363-370. https://doi.org/10.1016/0024-3795(90)90349-H
22. Puntanen, S. and Styan, G.P.H. (2015) Best Linear Unbiased Estimation in Linear Models (Version 8). StatProb: The Encyclopedia Sponsored by Statistics and Probability Societies.

23. Stepniak, C. (1987) Ordering of Nonnegative Definite Matrices with Application to Comparison of Linear Models. Linear Algebra and Its Applications , 70, 67-71. https://doi.org/10.1016/0024-3795(85)90043-6
24. Drazin, M.P. (1978) Natural Structures on Semi Groups with Involution. Bulletin American Mathematical Society , 84, 139-141. https://doi.org/10.1090/S0002-9904-1978-14442-5
25. Hartwig, R.E. (1980) How to Partially Order Regular Elements? Japanese Journal of Mathematics , 25, 1-13.
26. Punithavalli, G. (2014) Contributions to the Study on Various Solutions of the Matrix Equation AXB=C in Minkowski Space M. PhD Dissertation, Annamalai University, Annamalai Nagar.
27. Hernnandez, A., Lattanzi, M., Thome, N. and Urquiza, F. (2012) The Star Partial Order and the Eigenprojection at 0 on EP Matrices. Applied Mathematics and Computation, 218, 10669-10678. https://doi.org/10.1016/j.amc.2012.04.034

CHAPTER 9

Superpotentials, Calabi–Yaualgebras, and PBW Deformations

J.karmazyn

ABSTRACT

The paper [9]by Bocklandt, Schedler and Wemyss considers path algebras with relations given by the higher derivations of a superpotential, giving a condition for such an algebra to be Calabi–Yau. In particular they show that the algebra $\mathbb{C}[V] \rtimes G$ for *V* a finite dimensional $\mathbb{C}$ vector space and *G* a finite subgroup of GL(*V*), is Morita equivalent to a path algebra with relations given by a superpotential, and is Calabi–Yau for G <SL(V). In this paper we extend these results, giving a condition for a PBW deformation of a Calabi–Yau, Koszul path algebra with relations given by a superpotential to have relations given by a superpotential, and proving these are Calabi–Yau in certain cases.

We apply our methods to symplectic reflection algebras, where we show that every symplectic reflection algebra is Morita equivalent to a path algebra whose relations are given by the higher derivations of an inhomogeneous superpotential. In particular we show these are Calabi–Yau regardless of the deformation parameter.

Also, for *G* a finite subgroup of $GL_2(\mathbb{C})$not contained in $SL_2(\mathbb{C})$, we consider PBW deformations of a path algebra with relations which is Morita equivalent to C[*x*, *y*] $\rtimes$ G. We show there are no non-trivial PBW deformations when *G* is a small subgroup.

Keywords: Path algebra, Superpotential, Calabi–Yau algebras, Koszul algebras, PBW deformations

1. INTRODUCTION

1.1. Introduction

In this paper we consider path algebras of quivers with certain relations, in particular studying relations produced from a superpotential. Given a quiver Q, a homogeneous superpotential of degree n is an element, $\Phi_n = \sum c_{a_1 \ldots a_n} a_1 \ldots a_n$,in the path algebra of Q satisfying the n *superpotential condition*: caq=$(-1)^{n-1}$$_{cqa}$for all arrows a and paths q. From such a superpotential Φ_n and a non-negative integer k we construct an algebra $\mathcal{D}(\Phi_n, k) := \frac{\mathbb{C}Q}{R}$as a path algebra with relations R. These relations are constructed by the process of *differentiation*, where we define the left derivative of a path p by a path q, denoted δ_qp, to be t if $p = qt$ and 0 otherwise, and the relations are given by R=_{$\delta_p\Phi_n$:|p| =k}

Algebras of this form are considered in [9], where they are related to Calabi–Yau (CY), N-Koszul algebras. In [9]a complex, W•, is defined which depends only on the superpotential, and a path algebra with relations is N-Koszul and CY if and only if it is of the form $D(\Phi_n, k)$for a superpotential $\Phi_n$and $W^\bullet$is a resolution.

Skew group algebras, $\mathbb{C}[V] \rtimes G$, for G a finite subgroup of GL(V), are Morita equivalent to path algebras of this form. These are 2-Koszul, and CY when $G < SL(V)$, and hence their relations can be given by a superpotential. An explicit way to calculate this superpotential is given in [9, Theorem 3.2].

We prove two results concerning the PBW deformations of $(n-k)$-Koszul, $(k+2)$-CY algebras of the form $D(\Phi_n, k)$. We define an inhomogeneous superpotential of degree n to be an element of the path algebra Φ:=$\Phi_n+\varphi_{n-1}+\ldots+\varphi k$, such that each $\varphi_j := \sum c_p p$is a sum of elements of the path algebra of length j, and each φj satisfies the n superpotential condition. Such a superpotential defines relations P=_{$\delta_p\Phi$:|p| =k},

and we define $\mathcal{D}(\Phi', k) := \frac{\mathbb{C}Q}{P}$. Theorem3.1.1classifies which PBW deformations are of this form. This is known for the case of PBW deformations of $D(\Phi_n, 1)$[7, Theorem 3.1, 3.2], but here we extend this to include higher differentials, k>1.

We then prove that certain classes of PBW deformations of a 2-Koszul, n-CY D(Φ_n,n−2)are n-CY in Theorem3.2.1. This is already known in the 3-CY case due to [7]which proves that any PBW deformation of an N-Koszul, 3-CY D(Φ_n, 1)given by inhomogeneoussuperpotential is 3-CY[7, Theorem 3.6], and in the case of a one vertex quiver there is a result[17, Theorem 3.1], which finds a necessary and sufficient condition for a PBW deformation of a Noetherian, 2-Koszul, n-CY algebra to be n-CY.

Next we give an application of these results to symplectic reflection algebras. Sym-plectic reflection algebras are defined in [12] as PBW deformations of certain skew group algebras $\mathbb{C}[V] \rtimes G$, and hence we consider the Morita equivalent path algebra with relations. Applying the previous results, and a result of [12], we deduce that these path algebras with relations are of the form $D(\Phi_{2n}+\varphi 2_{n-2}, 2n-2)$and are $2n$-CY.

We go on to consider PBW deformations of the path algebras with relations Morita equivalent to $\mathbb{C}[\mathbb{C}^2] \rtimes G$when G is a finite subgroup of $GL_2(\mathbb{C})$not contained is $SL_2(\mathbb{C})$. We show that if G does not contain pseudo-reflections there are no PBW deformations.

1.2. Main results

1) A classification of the PBW deformations of a $(k+2)$-CY, $(n-k)$-Koszul, super-potential algebra $D(\Phi_n, k)$, whose relations are given by inhomogeneous superpotentials. These are proved to be the PBW deformations satisfying one additional property, which we call the zero PBW condition, Definition3.0.6, and the corresponding superpotentials are shown to be k-coherent superpotentials, Definition3.0.5.

Theorem *(Theorem3.1.1). Let* $A = D(\Phi_n, k)$, *for* Φ_n *a homogeneous superpotential of degree* n, *be* $(n-k)$*-Koszul and* $(k+2)$*-CY. Then the zero PBW deformations of* A *correspond exactly to the algebras* $D(\Phi\text{?}, k)$*defined by k-coherent inhomogeneous super-potentials of the form* $\Phi\text{?}=\Phi_n+\varphi_{n-1}+...+\varphi_k$.

2) A proof that these PBW deformations are CY in certain cases. We consider $A = D(\Phi_n, n-2)$which is n-CY and 2-Koszul, and a zero PBW deformation, A, which by the previous results is of the form $A := D(\Phi\text{?}, n-2)$for $\Phi\text{?}=\Phi_n+\varphi_{n-1}+\varphi_{n-2}$.

Theorem *(Theorem3.2.1). Suppose* $\varphi n-1=0$, *then* A *is n-CY.*

3) An application to symplectic reflection algebras. Let H be a path algebra with rela-tions Morita equivalent to an undeformed symplectic reflection algebra. So H is 2-Koszul and $2n$-CY and $H=D(\Phi_{2n}, 2n-2)$for some homogeneous degree $2n$ superpotentialΦ_{2n}.

Theorem *(Theorem4.1.1). Any PBW deformation of H is a zero PBW deformation and of the form* $D(\Phi\text{?}, n-2)$*for* $\Phi\text{?}=\Phi_{2n}+\varphi_{2n-2}$*an inhomogeneous superpotential. Hence any PBW deformation of H is 2n-CY, and all symplectic reflection algebras are Morita equivalent to 2n-CY algebras of the form* $D(\Phi\square$, $2n-2)$.

1.3. Contents

We outline the structure of the paper.

Section 2: We discuss preliminaries, listing definitions and results fundamental to the rest of the paper. These concern quivers, superpotentials, CY algebras, *N*-Koszul algebras, and PBW deformations. In particular we recall a Theorem classifying PBW deformations from [5], and several Theorems from [9] concerning path algebras with relations defined by superpotentials and CY algebras.

Section 3: The main results of the paper are stated and proved here. We begin by making two definitions required to state our theorems, k-*coherent* superpotentials and *zero PBW*PBW deformations. We then prove Theorem3.1.1classifying which PBW de-formations of an *N*-Koszul, CY, superpotential algebra $D(\Phi_n, k)$are of the form $D(\Phi\square, k)$for an inhomogeneous superpotential $\Phi\square=\Phi_n+\varphi_{n-1}+...+\varphi_k$. We then specialise to the case of 2-Koszul, *n*-CY, $D(\Phi_n, n-2)$, and consider PBW deformations of the form $D(\Phi_n+\varphi n-2, n-2)$, proving they are n-CY in Theorem3.2.1. We note known related results.

Section 4: We apply our results to symplectic reflection algebras. We recall the definition of symplectic reflection algebras, and their classification as PBW deformations by [12]. We consider the Morita equivalent path algebras with relations, which are of the form $D(\Phi_{2n}+\varphi 2_{n-2}, 2n-2)$and 2*n*-CY by the previous results. We calculate several examples, including the case of preprojective algebras.

Section 5: Finally we give an analysis of PBW deformations of algebras $D(\Phi_2, 0)$Morita equivalent to $\mathbb{C}[\mathbb{C}^2] \rtimes G$for *G* a finite subgroup of $GL_2(\mathbb{C})$. In particular we show there are no nontrivial PBW deformations when *G* is a small subgroup not contained in $SL_2(\mathbb{C})$.

2. PRELIMINARIES

In this section we set up the definitions and give a summary of results which we use later.

2.1. Quivers and superpotentials

Here we give the definition of a quiver, its path algebra, and its path algebra with relations. We define certain elements of the path algebra to be superpotentials, and give a construction to create relations on the quiver from the superpotential. We are following the set up of [9].

Definitions 2.1.1. A *quiver* is a directed multigraph. We will denote a quiver Q by $Q = (Q_0, Q1)$, with $Q_0$the set of vertices and $Q_1$the set of arrows. The set of arrows is equipped with head and tail maps $h, t : Q_1 \to Q_0$which take an arrow to the vertices that are its head and tail respectively.

A *non-trivial path* in the quiver is defined to be a sequence of arrows

$$p = a_r \ldots a_2 a_1 \text{ with } a_i \in Q_1 \text{ satisfying } h(a_i) = t(a_{i+1}) \text{ for } 1 \leq i \leq r-1$$

.

$$\bullet \xrightarrow{a_1} \bullet \xrightarrow{a_2} \cdots \xrightarrow{a_r} \bullet$$

We will reuse the notation of the head and tail maps for the head and tail of a path, defining $h(p) = h(a_r)$and $t(p) = t(a1)$when $p = a_r \ldots a_2 a_1$. We also define a *trivial path* e_v for each vertex $v \in Q_0$, which has both head and tail equal to v. *A* path p is called *closed* if $h(p) = t(p)$. The *pathlength* of a nontrivial path $p = a_r \ldots a_1$, where each a_i is an arrow, is defined to be r. *A* trivial path is defined to have pathlength 0. We will denote the pathlength of a path p by $|p|$.

Definition 2.1.2. Let $\mathbb{k}$be a field. We define the *path algebra* of the quiver Q, $\mathbb{k}Q$, as follows: $\mathbb{k}Q$ as a vector $\mathbb{k}$-vector space has a basis given by the paths in the quiver; an associative multiplication is defined by concatenation of paths.

$$p.q = \begin{cases} pq & \text{if } h(q) = t(p) \\ 0 & \text{otherwise} \end{cases}$$

We define S to be the subalgebra of this generated by the trivial paths, and V to be the $\mathbb{k}$ vector subspace

$\mathbb{k}Q$ spanned by the arrows, $a \in Q1$. Then S is a semisimple algebra, with one simple module for each vertex.

For an arrow a, we have $a = e_{h(a)}.a.e_{t(a)}$, and so V has the structure of a left $S^e := S \otimes_{\mathbb{k}} S^{op}$ module and

$\mathbb{k}Q$ can be identified with the tensor algebra $T_S(V) = S \oplus V \oplus (V \otimes_S V) \oplus \ldots$, equating the path $a_1 \ldots a_r$ with $a_1 \otimes_S \ldots \otimes_S a_r$.

The algebra $T_S(V) = \mathbb{k}Q$ is equipped with a grading and filtration by pathlength, the graded part in degree n is $T_S(V)^n := V^{\otimes_S n}$, and the filtered part is $F^n(T_S(V)) := V^{\otimes_S n} \oplus \ldots \oplus V \oplus S$.

Given $R \subset TS(V)$ we define $I(R)$ to be the two sided ideal in $T_S(V)$ generated by R. We then define

$$\frac{\mathbb{k}Q}{R} := \frac{\mathbb{k}Q}{I(R)}$$

and refer to it as the path algebra with *relations* R.

We define $\langle R \rangle$ to be the S^e submodule of $T_S(V)$ generated by R.

Superpotentials We define superpotentials and twisted superpotentials in the homogeneous and inhomogeneous cases. In the majority of this text we will be working with non-twisted superpotentials, but in Section 5 we will consider the twisted case.

Definition 2.1.3. Let $\Phi_n \in T\,S(V)^n$, written $\Phi_n = \Sigma_p c_p p$, where the sum is taken over all paths p of pathlength n with coefficients c_p in $\mathbb{k}$. Our convention will be that $c_0 = 0$. For example if $t(a) = h(b)$ then $ab = 0$ and hence $c_{ab} = c_0 = 0$.

We will say Φ_n satisfies the *n superpotential condition* if $c_{\sigma aq} = (-1)^{n-1} c_{qa}$ for all $a \in Q1$ and paths q, and if Φ_n satisfies the n superpotential condition we will call it a *homogeneous superpotential of degree n*. In particular this requires $c_p = 0$ if p is not a closed path.

Let $\sigma \in \mathrm{Aut}_{\mathbb{K}}(\mathbb{k}Q)$ be a graded automorphism, so that $\sigma(TS(V)^n) = TS(V)^n$ for all $n \geq 0$, and assume that $\sigma(h(p)) = h(\sigma(p))$, $\sigma(t(p)) = t(\sigma(p))$ for any path p.

We will say Φ_n satisfies the *twisted n superpotential condition* if $c\sigma(a)q = (-1)^{n-1} c_{qa}$ *for* $a \in Q1$ and paths q. We will say Φ_n is a σ-*twisted homogeneous superpotential of degree n* if it satisfies the *twisted n superpotential condition.*

Definition 2.1.4. Consider an element $\Phi \in F^n(T_S(V))$written $\Phi = \Sigma_{m \leq n} \Sigma_{|p|=m} c_p p$ as above. We will say Φ is an *inhomogeneous superpotential of degree n* if it satisfies the conditions $c_{aq} = (-1)^{n-1} c_{qa}$ for any path q and arrow a, and for some p of length n the coefficient c_p is non-zero. An inhomogeneous superpotential Φ can be written in homogeneous parts $\Phi = \varphi_n + \varphi_{n-1} + ... + \varphi_0$where each of the φ_m satisfies the n-superpotential condition and φ_n is non-zero. Note that if n is even, m odd, and char $\Bbbk \neq 2$then $\varphi_m = 0$, and also that any $\varphi_0 \in S$ satisfies the n superpotential condition.

We will say this is a *σ-twisted inhomogeneous superpotential of degree n* if it satisfies $c_{\sigma(a)q} = (-1)^{n-1} c_{qa}$for any path q and arrow a. *A* twisted inhomogeneous superpotential Φ can be written in homogeneous parts $\Phi = \varphi_n + \varphi_{n-1} + ... + \varphi_0$where each of the φ_m satisfies the n-twisted superpotential condition.

Differentiation We define differentiation by paths.

Definition 2.1.5. Let p be a path in $\Bbbk Q$, and a $\in Q1$. We then define left/right derivative of p by a as;

$$\delta_a p = \begin{cases} q, & \text{if } p = aq \\ 0, & \text{otherwise} \end{cases} \qquad p\delta'_a = \begin{cases} q, & \text{if } p = qa \\ 0, & \text{otherwise} \end{cases}$$

We extend this so that if $q = a_1 ... a_n$ is a path of length n then $\delta q = \delta_{a_n} ... \delta_{a_1}$, and $\delta'_q = \delta_{a_n} ... \delta'_{a_1}$. We will call the operation δ'_q right differentiation by the path $_q$

Note that for Φ a degree n inhomogeneous superpotential, b, c $\in Q_1$, and p any path

$$\delta_b \Phi = (-1)^{n-1} \Phi \delta'_b,$$

$$\delta_b \Phi \delta'_c = (-1)^{n-1} \delta_b \delta_c \Phi = (-1)^{n-1} \delta_{cb} \Phi,$$

and

$$\delta_p \Phi = \sum_{a \in Q_1} a \delta_a \delta_p \Phi = \sum_{a \in Q_1} \delta_p \Phi \delta'_a a.$$

Definition 2.1.6. Given an inhomogeneous superpotential Φ of degree n we can consider the S^e-modules $W_{n-k} \subset \Bbbk Q$ given by:

$$\mathcal{W}_{n-k} = \langle \delta_p \Phi : |p| = k \rangle$$

We then define the *algebra D(Φ, k)*to be

$$\mathcal{D}(\Phi, k) := \frac{\Bbbk Q}{I(\mathcal{W}_{n-k})}$$

We will call this the *superpotential algebra* $D(\Phi, k)$.

Example 2.1.7. Consider the quiver, Q, with one vertex, •, and two arrows x and y. Then $\Bbbk Q$ is the free algebra on two elements, $\Bbbk\langle x, y\rangle$

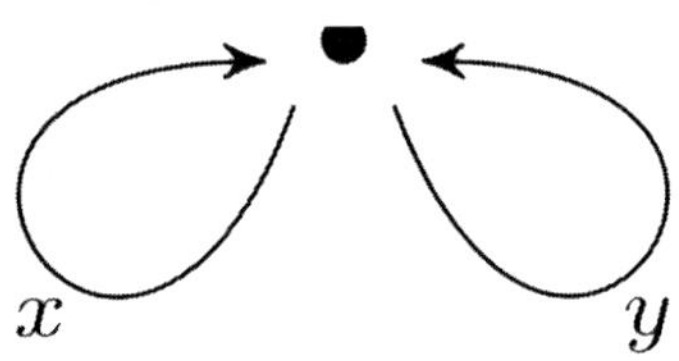

We consider some superpotentials on Q.

1)Φ_2=$xy- yx$ is a degree 2 homogeneous superpotential.

2)Φ =$xy- yx$ −e**.** is an inhomogeneous superpotential of degree 2.

3)Φ_3=$xyx + xxy + yxx$ is a degree 3 homogeneous superpotential.

These give us the algebras

1)$D(\Phi_2, 0)$ = $\Bbbk$ $[x, y]$, as W_2=$\{xy- yx\}$gives relations R=$\{xy -yx =0\}$

2)$D(\Phi, 0)$ =A_1($\Bbbk$), the first Wey l algebra, is the quiver with relations given by R=$\{xy-yx =e.\}$

3)$D(\Phi_3,1)$is the quiver with relations given by R=$\{yx +xy=0, xx =0\}$

2.2. Calabi–Yau algebras

We define Calabi–Yau algebras following the definitions of [13]and [1]. We recall the definition of self-dual used in [9], and note that the existence of a self-dual finite projective A^e-module resolution of length n implies an algebra is n-Calabi–Yau.

Let A be an associative $\Bbbk$algebra, set $A^e = A \otimes_{\Bbbk} A^{op}$and $D(A^e)$to be the unbounded derived category of left A^e modules. We note that there are two A^e-module structures on $A \otimes_{\Bbbk} A$, the *inner structure*given by $(a \otimes b)(x \otimes y) = (xb \otimes ay)$, and the *outer structure* given by $(a \otimes b)(x \otimes y) = (ax \otimes yb)$. By considering A^e with the outer structure for any A^e-module, M, we can make $\mathrm{Hom}_{A^e}(M, A^e)$an A^e-module using the inner structure.

Definition 2.2.1. Let n ≥2. Then A is n *bimodule Calabi–Yau (n-CY)*if A has a finite length resolution by finitely generated projective A^e-modules, and

$$\mathrm{RHom}_{A^e}\left(A, A^e\right)[n] \cong A \quad \text{in } D\left(A^e\right)$$

In the paper [9]self-duality of a complex is used. We give the definition here. Denote $(-)^{\vee} := \mathrm{Hom}_{A^e}(-, A^e) : A^e\text{-Mod} \to A^e\text{-Mod}$.

Definition 2.2.2. Define a complex of A^e modules, $C^\bullet$, of length n to be self-dual, written

$$\mathrm{Hom}_{A^e}\left(C^\bullet, A^e\right) \cong C^{n-\bullet},$$

if there exist A^e-module isomorphisms α_i such that the following diagram commutes:

$$\begin{array}{ccccccccc}
C_n & \xrightarrow{d_n} & C_{n-1} & \xrightarrow{d_{n-1}} & \cdots & \xrightarrow{d_2} & C_1 & \xrightarrow{d_1} & C_0 \\
\downarrow{\scriptstyle \alpha_n} & & \downarrow{\scriptstyle \alpha_{n-1}} & & & & \downarrow{\scriptstyle \alpha_1} & & \downarrow{\scriptstyle \alpha_0} \\
C_0^{\vee} & \xrightarrow{-d_1^{\vee}} & C_1^{\vee} & \xrightarrow{-d_2^{\vee}} & \cdots & \xrightarrow{-d_{n-1}^{\vee}} & C_{n-1}^{\vee} & \xrightarrow{-d_n^{\vee}} & C_n^{\vee}
\end{array}$$

By construction the existence of a length n self-dual projective A^e-module resolution of A implies that A is n-CY, and we will use this later in Section3to show algebras are n-CY.

We note some properties of Calabi–Yau algebras

Lemma 2.2.3. *Let Abe n-CY* $\Bbbk$*-algebra, where* $\Bbbk$ *is algebraically closed. Then*

1.*If there exists a non-zero finite dimensional A-module, then A has global dimension n.*

2.*For X, Y*∈D(*A*)*with finite dimensional total homology*

$$\mathrm{Hom}_{D(A)}(X,Y) \cong \mathrm{Hom}_{D(A)}(Y,X[n])^*$$

where **denotes the standard* $\Bbbk$ *dual*

Proof. These are some of the standard properties of CY algebras, see for example [1,Proposition2.4]and[7, Section 2]for proofs. □

2.3. Koszul algebras

Here we give the definitions of *N*-Koszul algebras.

The concept of an *N*-Koszul algebra was introduced by Berger, and is defined and studied in the papers [3–6,14]. It generalises the concept of a Koszul algebra, which is defined here to be a 2-Koszul algebra.

Definition 2.3.1. Let *S* be a semisimple ring, *V* a left S^e-module, and $T_S(V)$the tensor algebra. An algebra is *N-homogeneous*if it is given in the form A= $\frac{T_S(V)}{I(R)}$ for *R* an S^e-submodule of $V\otimes_S{}^N$.

We now define the *Koszul* N-*complex.* For *N*-homogeneous *A* = $\frac{T_S(V)}{I(R)}$ define

$$K_i = \bigcap_{j+N+k=i} (V^{\otimes j} \otimes_S R \otimes_S V^{\otimes k}) \quad i \geq N$$

$$K_i = V^{\otimes s i} \quad 0 < i < N$$

$$K_0 = S$$

and let

$$K^i(A) = A \otimes_S K_i \otimes_S A$$

Now we can define an *N*-complex (In an *N*-complex d^N=0rather than d^2=0.)

$$\ldots \xrightarrow{d} K^n(A) \xrightarrow{d} \ldots \xrightarrow{d} K^0(A) \xrightarrow{\mu} A \to 0$$

To define the differentials let μ be multiplication, and d_l, d_r:$A \otimes SV^{\otimes Si} \otimes S^A \to A \otimes SV^{\otimes S(i-1)} \otimes S^A$ be defined by

As $R \subset V \otimes_S^N$ we see $K_i \subset V \otimes^{Si}$ and hence can consider the restriction of d_l, d_r to $K^i(A)$. Now choose $q \in \mathbb{k}$ a primitive Nth root of unity, which we need to assume exists in $\mathbb{k}$, and define d :$K^i(A) \to K^{i-1}(A)$as $d = d_l|_{K^i(A)} - (q)^{i-1} d_r|K_i(A)$.

As d_l and d_r commute with $d^N{}_l=0$and $d^N{}_r=0$,d defines an N differential.

To such an N-complex we can associate a complex by contracting several terms together. This is done by splitting the N-complex into sections of N consecutive differentials from the right. Each one of these is collapsed to a term with two differentials by keeping the right most differential, and composing the other $N-1$. This defines a complex of the form

$$\ldots \to K^{N+1}(A) \xrightarrow{d} K^N \xrightarrow{d^{N-1}} K^0(A) \xrightarrow{\mu} A \to 0$$

which exists regardless of the field $\mathbb{k}$, with $d = (d_l - d_r)|K^j(A)$and $d^{N-1} = (d^{N-1}{}_l + d^{N-2}{}_l d_r + \ldots + d^{N-2}{}_r)|K^j(A)$.

We say that A is N-*Koszul* if this complex is exact, and hence gives an A^e-module resolution of A.

We call 2-Koszul algebras *Koszul*. In this case the N-complex is in fact a complex.

2.4. Superpotentials and higher order derivations

We summarise some results from the paper [9]. Later we will be particularly interested the applications to skew group algebras so we recall their definition and the relevant results from [9].

Superpotentials and higher order derivations Here we state several results from the paper [9]concerning superpotential algebras.

Let Q be a quiver, with path algebra $\mathbb{C}Q$. Let Φ_n be a homogeneous superpotential of degree n, and $A = D(\Phi_n, n-2)$. For i =0,...,n we have the Se-modules W_i as in Definition2.1.6 above, and we can define a complex $W^\bullet$ by

$$0 \to A \otimes_S W_n \otimes_S A \xrightarrow{d_n} A \otimes_S W_{n-1} \otimes_S A \xrightarrow{d_{n-1}} \ldots \xrightarrow{d_1} A \otimes_S W_0 \otimes_S A \to 0$$

. (1)

To define the differential first define $d_i^l, d_i^r : A \otimes_S W_i \otimes_S A \to A \otimes_S W_{i-1} \otimes_S A$

by

$$d_i^l(\alpha \otimes \delta_p \Phi_n \otimes \beta) = \sum_{a \in Q_1} \alpha a \otimes \delta_a \delta_p \Phi_n \otimes \beta$$

$$d_i^r(\alpha \otimes \delta_p \Phi_n \otimes \beta) = \sum_{a \in Q_1} \alpha \otimes \delta_p \Phi_n \delta'_a \otimes a\beta$$

for *p* any path of length *n* –*i*.

The differential d_i is defined as

$$d_i = \epsilon_i(d_i^l + (-1)^i d_i^r) \quad \text{where } \epsilon_i := \begin{cases} (-1)^{i(n-i)} & \text{if } i < (n+1)/2 \\ 1 & \text{otherwise} \end{cases}$$

This is a complex as d_l, d^r commute and have square 0. We show this for d^l, the other case being similar. Writing Φ_n= _|*t*|=n$c_t{}^t$, we have

$$\begin{aligned} d_{j-1}^l \circ d_j^l(1 \otimes \delta_p \Phi_n \otimes 1) &= \sum_{a,b} ab \otimes \delta_{ab} \delta_p \Phi_n \otimes 1 \\ &= \sum_{a,b,q} ab \otimes c_{pabq} q \otimes 1 \\ &= \sum_{a,b,q} (-1)^{(n-1)(j-2)} c_{qpab} ab \otimes q \otimes 1 \\ &= \sum_{q} (-1)^{(n-1)j} \delta_{qp} \Phi_n \otimes q \otimes 1 = 0 \end{aligned}$$

Where *a*, *b* ∈*Q* 1, *p* is a path of length (*n* –*j*), *q a* path of length *j*–2, and the sums are taken over all such *a*, *b*and *q*.

Theorem 2.4.1. *(See [9, Section 6].)Let A =D(*Φ_n*,* n –2)*, then W• is a self-dual complex of projective* A^e*-modules.*

Theorem 2.4.2. *(See [9, Theorem 6.2].)Let A =* $\frac{\mathbb{C}Q}{R}$ *be a path algebra with relations. Then A is 2-Koszul and* n-*CY if and only if A is of the form D(*Φ_n*,* n *−2)for some homogeneous superpotential* Φ_n *of degree n and the attached complex* W• *is a resolution of A. In this case the resolution equals the Koszul complex of Definition2.3.1with each* $W_i = K_i$*.*

The complex (1) is the relevant complex for *D(*Φ_n*,* n −2), where the relations are obtained by differentiation by paths of length n −2. More generally, differentiating by paths of length *k*, we need another complex. Let *N*=n −*k*, and in this case we define an *N*-complex $\widetilde{\mathcal{W}}^{\bullet}$, again making use of the S^e-modules W_j.

$$0 \to A \otimes_S \mathcal{W}_n \otimes_S A \xrightarrow{d_n} A \otimes_S \mathcal{W}_{n-1} \otimes_S A \xrightarrow{d_{n-1}} \dots \xrightarrow{d_1} A \otimes_S \mathcal{W}_0 \otimes_S A \to 0 \tag{2}$$

with differential $d_i : A \otimes_S \mathcal{W}_i \otimes_S A \to A \otimes_S \mathcal{W}_{i-1} \otimes_S A$ defined by $d_i = d_i^l + (q)^i d_i^r$,for *q* a primitive *N* throot of unity, which we assume exists in $\Bbbk$. This can be contracted into a 2-complex, $\widehat{\mathcal{W}}^{\bullet}$

$$0 \to A \otimes_S \mathcal{W}_{mN+1} \otimes_S A \xrightarrow{d} A \otimes_S \mathcal{W}_{mN} \otimes_S A \xrightarrow{d^{N-1}} A \otimes_S \mathcal{W}_{(m-1)N+1} \otimes_S A \xrightarrow{d} \dots$$
$$\xrightarrow{d} A \otimes_S \mathcal{W}_0 \otimes_S A \to 0$$

by composing the differentials in $\widehat{\mathcal{W}}^{\bullet}$, as in Definition2.3.1, where *m* is the largest integer such that m ≤n/*N*.

Theorem 2.4.3. *(See [9, Theorem 6.8].)Let* A = $\frac{\mathbb{C}Q}{R}$ *be a path algebra with relations. Then A is* (*n* −*k*)*-Koszul and* (*k*+2)*-Calabi–Yauif and only if it is of the form D(Φn, k) where* Φ_n *is ahomogeneous superpotential of degree n and* $\widehat{\mathcal{W}}^{\bullet}$ *is a resolution of A. In this case* $\widehat{\mathcal{W}}^{\bullet}$*equals the Koszul* (*n* −*k*)*-complex as in Definition2.3.*

Skew group algebras Let *G* be a finite subgroup of *GL(V)*. We can form the smash product $\Bbbk[V] \rtimes G$ which is the semi direct product with the action of *G* given by that of GL(*V*). The multiplication of (f_1, g_1), (f_2, g_2) $\in \Bbbk[V] \rtimes G$ is given by

$$(f_1, g_1).(f_2, g_2) = (f_1 f_2^{g_1}, g_1 g_2)$$

This is a graded algebra with G in degree 0 and V^* in degree 1, where $\Bbbk[V] = Sym(V^*)$.

For such a group G and representation V we can construct the McKay quiver. We now assume our field has characteristic not dividing the order of G and is algebraically closed. The McKay quiver for (G, V) has a vertex, i, for each irreducible representation, V_i, of G over $\Bbbk$, and $\dim_{\Bbbk} \mathrm{Hom}\,(V_i \otimes \Bbbk V, V_j)$ arrows from I to j.

Theorem 2.4.4. *(See [9, Theorem 3.2, Lemma 6.1].)Let* V *be an n-dimensional* $\mathbb{C}$*-vector space and G be a finite subgroup of* GL(V)*. Then* $\mathbb{C}[V] \rtimes G$ *is Morita equivalent to* $D(\Phi_n, n-2)$ *for some homogeneous superpotential* Φ_n *of degree* n *attached to the McKay quiver* (G, V)*. Moreover* $D(\Phi n, n-2)$ *is* n*-CY and Koszul for* $G \leq SL(V)$.

There is a recipe to compute a superpotential Φ_n such that $\mathbb{C}[V] \rtimes G$ is Morita equivalent to the superpotential algebra $D(\Phi_n, n-2)$ attached to the McKay quiver (G, V). The recipe is given in [9, Theorem 3.2]. When G is abelian the superpotential algebra $D(\Phi_n, n-2)$ is in fact isomorphic to $\mathbb{C}[V] \rtimes G$.

2.5. PBW deformations

In this subsection we will recall the definition of the PBW deformations of a graded algebra A. We will be considering PBW deformations of A an N-Koszul algebra relative to S a semisimple algebra, and we will make use of the setup and results of [5].

Let S be a semisimple algebra, V a left S^e-module, and $T_S(V)$the tensor algebra of V over S. Consider the $(Z\geq 0)$ grading on this with degree n part $T_S(V)_n = V^{\otimes_S n}$, and filtered parts $F^n = V^{\otimes_S n} \oplus \ldots \oplus V \oplus S$. For Ran S^e-sub module of $V^{\otimes_S N}$ define $I(R)_n = I(R)$to be the two sided ideal in $T_S(V)$ generated by R, which is graded by $I(R)_n = I(R) \cap V^{\otimes_S n}$.

Define $A := \frac{T_S(V)}{R} := \frac{T_S(V)}{I(R)}$, and as R is homogeneous this is a graded algebra with degree n part $A_n := \frac{V^{\otimes_S n}}{I(R)_n}$.

Now define the projection map $\pi_N : F^N \rightarrow V^{\otimes_S N}$, and let P be an S^e-submodule of F^N such that $\pi(P) = R$. Let $I(P)$ be the 2 sided ideal in $T_S(V)$generated by P.

This is not graded but is filtered by $I(P)_n = I(P) \cap F^n$, and hence $A = \frac{T_S(V)}{I(P)}$

is also a filtered algebra with $A_n = \frac{F^n}{I(P)^n}$

We can construct the associated graded algebra *gr* (*A*), which is graded with $gr(A)_n = \frac{A^n}{A^{n-1}}$. Identify *gr*(*A*)*n* with the S_e-module $\frac{F^n}{I(P)^n + F^{n-1}} \to gr(A)_n$ by noting $I(P)^n \cap F^{n-1} = I(P)^{n-1}$ and consider the maps $\varphi_n : V^{\otimes n} \to gr(A)_n$ defined as the composition $V^{\otimes n} \hookrightarrow F^n \to \frac{F^n}{I(P)^n + F^{n-1}}$. This allows us to define a surjective algebra morphism

$$\phi = \bigoplus_{n \geq 0} \phi_n : T_S(V) \to gr(A). \text{ Now } \phi_N(R) = 0 \text{ as } P + F^{N-1} = R \oplus F^{N-1}$$

, and hence this defines a surjective morphism of Z≥0-graded S^e-modules

Definition 2.5.1. We say that $A = \frac{T_S(V)}{P}$ is a *PBW deformation* of A $= \frac{T_S(V)}{R}$ if *p* is an isomorphism. We say that *P* is of *PBW type* if *A* is a PBW deformation of *A*.

The papers [7] and [5] prove a collection of conditions on *P* equivalent to it being PBW type which we state here.

First note that for $P \subset F^N$ to be of PBW type it must be the case that $P \cap F^{N-1} = \{0\}$. Hence any PBW type Pcan be given as $P = \{r - \theta(r) : r \in R\}$ for an S^e-module map $\theta : R \to F^{N-1}$. Such a map can be written in homogeneous components as $\theta = \theta_{N-1} + \ldots + \theta 0$, with $\theta_j : R \to V \otimes_{Sj}$.

Consider the map $\psi(\theta_j)$ defined for each θ_j as

where $(V \otimes_S R) \cap (R \otimes_S V) \subset V^{\otimes S(N+1)}$

Theorem 2.5.2. *(See [5, Section 3].)The* S^e*-module P is of PBW type if and only if the following conditions are satisfied*

PBW1) $P \cap F^{N-1} = \{0\}$

PBW2) Im $\psi(\theta_{N-1}) \subset R$

PBW3) $\mathrm{Im}(\theta_j(\psi(\theta_{N-1}))+\psi(\theta_{j-1}))=\{0\}$ *for all* $1 \leq j \leq N-1$

PBW4) $\mathrm{Im}\theta_0(\psi(\theta_{N-1}))=\{0\}$

Example 2.5.3. Let $A=D(\Phi_n, k)$ be an N-Koszul $(k+2)$-CY superpotential algebra, where $N=n-k$. Then **Theorem2.4.3 tells** us that the Koszul N-complex equals the complex $\widehat{\mathcal{W}^{\bullet}}$, and $W_i=K_i$, of Definition2.3.1. Hence $(V\otimes_S R)\cap(R\otimes_S V)=W_{N+1}$, and $R=W_N$. We can also use the expression $\delta_p\Phi_n=\sum_{a\in Q_1} a\otimes\delta_a\delta_p\Phi_n=\sum_{a\in Q_1}\delta_p\Phi_n\delta'_a\otimes a$ to give an explicit form for $\psi(\theta_j)$ with $j=0, \ldots, N-1$:

$$\psi(\theta_j): \mathcal{W}_{N+1}\to V^{\otimes_S(j+1)}$$

$$\delta_p\Phi \mapsto \sum_{a\in Q_1} a\otimes\theta_j(\delta_a\delta_p\Phi)-\theta_j\left(\delta_p\Phi\delta'_a\right)\otimes a \quad \text{where } |p|=k-1$$

We note that in the case of S being a field, and A being 2-Koszul, Theorem2.5.2 was proved by Braverman and Gaitsgory [10]. They also show any PBW deformation gives a graded deformation. With the notation as above we define a graded deformation of A, A_t, to be a graded $\mathbb{k}[t]$ algebra with t in degree 1. That is A_t is free as a module over $\mathbb{k}[t]$ and there is an isomorphism $A_t/tA_t\to A$. It is shown that for a PBW deformation A there is a graded deformation of A with fibre at $t=1$canonically isomorphic to A [10, Theorem4.1].

In the case with $D(\Phi_n, n-2)$ Koszul and S a field there is also the following theorem of Wu and Zhu[17], which classifies when PBW deformations of a Noetherian n-CY algebra are also n-CY. In our language this applies when there is only a single vertex in the quiver.

Theorem 2.5.4. *Let Φ_n be a homogeneous superpotential on a single vertex quiver. Let $A=D(\Phi_n, n-2)$. If A is n-CY Noetherian and Koszul and* A *is a PBW deformation given by* θ_0, θ_1 *then* A *is* n-*CY if and only if*

$$\sum_{i=0}^{n-2}(-1)^i id^{\otimes i}\otimes\theta_1\otimes id^{\otimes(n-2-i)}(\Phi_n)=0$$

Proof. See [17, Theorem 3.1]. This proves such a PBW deformation is n-CY if and only if

$$\sum_{i=0}^{n-2}(-1)^i id^{\otimes i} \otimes \theta_1 \otimes id^{\otimes(n-2-i)} : \bigcap_i V^{\otimes i} \otimes R \otimes V^{\otimes(n-2-i)} \to V^{\otimes n-1}$$

is 0. Note that $\bigcap_i V^{\otimes i} \otimes R \otimes V^{\otimes(n-2-i)} = \mathcal{W}_n = \langle \Phi_n \rangle$ in this case, giving the result.

3. MAIN RESULTS

We state and prove our main results concerning the PBW deformations of $a(k+2)$-CY, $(n-k)$-Koszul, superpotential algebra $A := D(\Phi n, k)$, where Φ_n is a homogeneous superpotential. We will prove a condition for a PBW deformation to be of the form $D(\Phi', k)$for an inhomogeneous superpotential $\Phi' = \Phi_n + \varphi_{n-1} + ... + \varphi_k$, and prove that inhomogeneous superpotentials with degree n part Φn define PBW deformations of A. In the Koszul case we will show that certain PBW deformations of such an n-CY algebra are also n-CY.

Throughout this section consider $\Phi_n \in \mathbb{C}Q$ to be a homogeneous superpotential of degree n on some quiver Q. This will have in homogeneous superpotentials associated to it, denoted $\Phi' = \Phi_n + \varphi_{n-1} + ... + \varphi_k$.

Throughout this section we work over in $\mathbb{C}$ in order to apply the results of [9], in particular Theorem 2.4.3. We note that this is the only way in which the assumption $\Bbbk = \mathbb{C}$ is used in the proofs of Theorems3.1.1 and 3.2.1, and we could instead assume the conclusions of Theorem 2.4.3as additional conditions on A; that $A = D(\Phi n, k)$ and that the Koszul resolution of A equals the resolution $\widehat{\mathcal{W}^\bullet}$.

We introduce two new definitions we make use of in the proof

Definition 3.0.5. We will call Φ', an inhomogeneous superpotential, k-*coherent* if for any $\lambda_p \in \mathbb{C}$

$$\sum_{|p|=k} \lambda_p \delta_p \Phi_n = 0 \in \mathbb{C}Q \quad \Rightarrow \quad \sum_{|p|=k} \lambda_p \delta_p \Phi' = 0 \in \mathbb{C}Q$$

Definition 3.0.6. Let $A = \frac{T_S(V)}{P}$ and $A = \frac{T_S(V)}{R}$ be as in Definition 2.5.1. We will say that A is a *zero PBW deformation* of A if it is a PBW deformation which also satisfies the *zero PBW condition*

$$\mathrm{Im}\big(\psi(\theta_{N-1})\big) = \{0\},$$

where ψ and θ are defined as in Section 2.5.

We will say P is of *zero PBW type* if A is a zero PBW deformation of A.

Lemma 3.0.7. *An* S^e*module* P *is of zero PBW type if and only if*

$$\begin{aligned}&\mathrm{PBW1)}\ P \cap F^{N-1} = \{0\}\\ &\mathrm{ZPBW)}\ \mathrm{Im}(\psi(\theta_j)) = \{0\}\ for\ j = 0 \ldots N-1\end{aligned}.$$

Proof. By definition P is of zero PBW type if and only if it is of PBW type and also satisfies the zero PBW condition. It is of PBW type if and only if it satisfies conditions PBW1, 2, 3, 4) of Theorem 2.5.2. Satisfying the zero PBW condition is equivalent to reducing the conditions PBW 2, 3, 4) to the condition ZPBW). □

3.1. Deformations of superpotential algebras

We state and prove our results relating superpotentials and PBW deformations.

Theorem 3.1.1. *Let* $A = \frac{\mathbb{C}Q}{R}$ *be a path algebra with relations which is* $(n-k)$*-Koszul and* $(k+2)$*-CY. Then* A $= D(\Phi n, \mathrm{k})$*, for* Φn *a homogeneous superpotential of degree n, and the zero PBW deformations of* A *correspond exactly to the algebras* $D(\Phi', \mathrm{k})$ *defined by k-coherent inhomogeneous superpotentials of the form* $\Phi' = \Phi n + \varphi_{n-1} + \ldots + \varphi_k$.

Proof. As A is $(k+2)$-CY and $(n-k)$-Koszul by Theorem 2.4.3 $A = D(\Phi_n, k)$ for Φ_n a degree n homogeneous superpotential.

We define inverse maps between P of zero PBW type and coherent superpotentials Φ' which will give us the correspondence. We note that zero PBW deformations exactly correspond to P of zero PBW type.

We first define a map, F, taking P of zero PBW type to k-coherent superpotentials. We define $F(P)$ as an element of $\mathbb{C}Q$, then show it is k-coherent and a superpotential. Any zero PBW deformation defined by P is given by a map $\theta = \theta_{n-k-1} + \ldots + \theta_0$ with $\theta j: R \to V^{\otimes Sj}$ such that P and θ satisfy the conditions PBW1) and ZPBW) of **Lemma3.0.7**. We construct *F(P)* :=Φ_n+$\varphi_{n-1}(\theta_{n-k-1})$ +...+$\varphi_k(\theta_0)$ by defining

$$\phi_{n-j}(\theta_{n-k-j}) := -\sum_{|p|=k} p\theta_{n-k-j}(\delta_p \Phi_n) =: \sum_{|s|=n-j} c_s s$$

and show that this is a *k*-coherent superpotential. We note that by this definition $\delta_p \phi_{n-j}(\theta_{n-k-j}) = -\theta_{n-k-j}(\delta_p \Phi_n)$ for any $|p| = k$.

Firstly *F(P)* is not k-coherent precisely when there exist coefficients $\lambda_p \in \mathbb{C}$ such that $\sum_{|p|=k} \lambda_p \delta_p \Phi_n = 0$ with ω=$\sum_{|p|=k} \lambda_p \delta_p \Phi'(\theta) \neq 0$. But this only occurs when there exists ω ∈ P∩ F^{n-1} which is non-zero. But as *P* is of zero PBW type PBW1) holds, and P∩Fn−1={0}. Hence F(P)is *k*-coherent.

Now we show that *F*(*P*) is a superpotential. As *P* is of zero PBW type by Lemma3.0.7 Im ψ (θ*j*) ={0} for *j*=0, ..., *n* −*k*−1. We evaluate ψ (θ*j*) on elements of the form $\delta_q \Phi_n$ where $|q| = k-1$, as in Example 2.5.3, and use $\theta_j(\delta_p \Phi_n) = -\delta_p \phi_{k+j}(\theta_j)$ to deduce that *F*(*P*)is a superpotential.

$$\begin{aligned} 0 = \psi(\theta_j)(\delta_q \Phi_n) &= \sum_a \left(a\theta_j(\delta_{qa}\Phi_n) + (-1)^n \theta_j(\delta_{aq}\Phi_n)a\right) \\ &= -\sum_a \left(a\delta_{qa}\phi_{k+j}(\theta_j) + (-1)^n \delta_{aq}\phi_{k+j}(\theta_j)a\right) \\ &= -\sum_{a,p} \left(c_{qap}ap + (-1)^n c_{aqp}pa\right) \\ &= -\sum_{a,p} \left(\left(c_{qap_1\ldots p_j} + (-1)^n c_{p_j qap_1\ldots p_{j-1}}\right)ap\right) \end{aligned}$$

with the sums over all $a \in Q1$and p =p_1...p*j* of length *j*.

Hence considering the coefficient of a path *ap* we see φ_{k+j}= $\sum_S C_S S$ satisfies the *n* superpotential condition

$$c_{qap_1\ldots p_j} = (-1)^{n-1} c_{p_j qap_1\ldots p_{j-1}}$$

Hence each φ_{j+k} satisfies the n superpotential condition, and F(P)is an inhomogeneous superpotential of degree n.

We note that the zero PBW deformation defined by P is, by construction, $D(F(P), k)$ as $P = \langle\{r - \theta(r) : r \in R\}\rangle = \langle\{\delta_p \mathcal{F}(P) : |p| = k\}\rangle$

Now we define a map, G, sending k-coherent superpotentials $\Phi'=\Phi_n+\varphi_{n-1}+...+\varphi_{n-k}$ to P of zero PBW type. We define $G(\Phi')$ as an S^e-module, show that $G(\Phi')$ satisfies the PBW1) condition, and then define a map $\theta^{\Phi'}$_as in **Lemma3.0.7**. We then show that $G(\Phi')$is of zero PBW type.

The map G is defined by $G(\Phi')$:=_{$\delta_p\Phi'$:|p| =k}]. We first check that $G(\Phi')$satisfies PBW1). Indeed as Φ' is k-coherent there are no ω= $\sum_{|p|=k} \lambda_p \delta_p \Phi' \neq 0$, with $\lambda_p \in \mathbb{C}$, satisfying $\sum_{|p|=k} \lambda_p \delta_p \Phi_n = 0$.

This implies $G(\Phi'_n)$ ∩Fn−1={0}, hence $G(\Phi'_n)$ satisfies PBW1).

Then, as $G(\Phi'n)$satisfies PBW1), G$(\Phi') = \langle\{r - \theta^{\Phi'}(r) : r \in R\}\rangle$

where θΦ'are the maps defined by $\theta^{\Phi'} := \sum_{j=0}^{n-k-1} \theta_j^{\phi_{j+k}}$ with each $\theta_j^{\phi_{j+k}}$ defined on elements $\delta_p \Phi_n$ by

$$\theta_j^{\phi_{j+k}} : R \to V^{\otimes_S j}$$

$$\delta_p \Phi_n \mapsto -\delta_p \phi_{j+k}$$

We next show that $G(\Phi')$is of zero PBW type. By Lemma3.0.7, to show that $G(\Phi')$is of zero PBW type it is enough to show that $G(\Phi')$and $\theta^{\Phi'}$ satisfy the conditions PBW1) and ZPBW). We have already shown PBW1) is satisfied, so need only check ZPBW); $\operatorname{Im} \psi(\theta_j^{\phi_{j+k}}) = \{0\}$ for $j = 0, \ldots, n-k-1$. As in Example 2.5.3($V \otimes_S R$) ∩ ($R \otimes_S V$) = $\mathcal{W}_{n-k+1} = \langle \delta_p \Phi_n : |p| = k-1 \rangle$, hence we calculate

$\psi(\theta_j^{\phi_{j+k}})(\delta_q\Phi_n)$, for $j=0,\ldots,n-k-1$, where $|q|=k-1$.

$$\psi(\theta_j^{\phi_{j+k}})(\delta_q\Phi_n)=\sum_a\left(a\theta_j^{\phi_{j+k}}(\delta_{qa}\Phi_n)+(-1)^n\theta_j^{\phi_{j+k}}(\delta_{aq}\Phi_n)a\right)$$
$$=-\sum_a\left(a\delta_{qa}\phi_{j+k}+(-1)^n\delta_{aq}\phi_{j+k}a\right)$$
$$=-\sum_{a,p}\left(c_{qap}ap+(-1)^n c_{aqp}pa\right)$$
$$=-\sum_{a,p}\left(\left(c_{qap_1\ldots p_j}+(-1)^n c_{p_jqap_1\ldots p_{j-1}}\right)ap\right)$$

Where we write $\varphi_j=\sum_{|t|=j}c_t t$, and the sums are over all $a\in Q1$ and $p=p_1\ldots p_j$ of length j. Hence as φ_j satisfies the n superpotential condition

$$c_{qap_1\ldots p_r}=(-1)^{n-1}c_{p_rqap_1\ldots p_{r-1}}$$

then $\psi(\theta_j^{\phi_{j+k}})(\delta_q\Phi_n)=0$

for all q and j. Therefore $\operatorname{Im}\psi(\theta_j^{\varphi_{k+j}})=\{0\}$ for all j, and $G(\Phi')$ is of zero PBW type.

We note that as $G(\Phi')=\langle\{\delta_p\Phi' : |p|=k\}\rangle$ by construction the zero PBW deformation defined by $G(\Phi')$ is indeed $D(\Phi', k)$.

Now the maps F and G are inverses by construction, and provide a bijection between zero PBW deformations and k-coherent superpotentials of the form $\Phi_n+\varphi_{n-1}+\ldots+\varphi_k$. Moreover the zero PBW deformation defined by P equals $D(F(P), k)$, and the zero PBW deformation defined by $G(\Phi')$ equals $D(\Phi', k)$. Hence the theorem is proved. □

Remark 3.1.2. Let $A=D(\Phi_n, n-2)$ be a 2-Koszul n-CY algebra, and $A=D(\Phi', n-2)$, as in Theorem3.1.1, be a zero PBW deformation of defined by θ_1,0. Then if n is even $\theta_1=0$.

Proof .Note that if n is even then any φ_{n-1} satisfying the n superpotential condition is 0, as observed after Definition 2.1.4. In particular a zero PBW deformation defined by θ_0, $\theta_1$corresponds to a superpotential $\Phi_n+\varphi_{n-1}+\varphi_{n-2}$ by the bijection of Theorem3.1.1, and superpotentials with $\varphi_{n-1}=0$correspond to zero PBW deformations with $\theta_1=0$. Hence for even *n* we have $\varphi_{n-1}=0$and so $\theta_1=0$.

3.2. CY property of deformations

We now consider when zero PBW deformations are CY. We consider the case where $A=D(\Phi_n, n-2)$ is *n*-CY and 2-Koszul. We currently prove a result only in the case of a zero PBW deformation with $\theta_1=0$, which— by Remark3.1.2— covers all even dimensional cases.

We then briefly mention two results concerning PBW deformations of CY algebras. Our results are weaker, but in a more general setting. One is the result of [17, Theo-rem3.1], which is quoted as Theorem2.5.4above, giving a complete characterisation of CY PBW deformations of a Noetherian CY algebra over a field. The other is the paper [7]which proves that the zero PBW deformations of a 3-CY superpotential algebra are 3-CY PBW deformations.

Theorem 3.2.1. *Let A be a zeroPBW deformation of A with* $\theta_1=0$. *Then A is n-CY.*

Proof. By Theorem3.1.1such a deformation is given by a superpotential $\Phi'=\Phi_n+\varphi_{n-2}=\sum_t c_t t$ and $A=D(\Phi', n-2)$. We construct a resolution for *A* and show it is self-dual. We recall the complex $W^\bullet$, defined as (1) in Section2.4, and, by Theorem2.4.1, that $W^\bullet$ is an A^e-module resolution of *A*. We then define a complex $W^\bullet A$

The S^e-modules W_k, and differential *d*, are defined as in Section2.4, i.e. $W_j=\langle\{\delta_p\Phi_n : |p|=n-j\}\rangle$,and $d_i=\epsilon_i(d_i^l+(-1)^i d_i^r$ where

$$\epsilon_i := \begin{cases} (-1)^{i(n-i)} & \text{if } i<(n+1)/2 \\ 1 & \text{otherwise} \end{cases}$$

and $d_i^l, d_i^r : A\otimes_S W_i\otimes_S A \to A\otimes_S W_{i-1}\otimes_S A$ are defined by

$$d_i^l(\alpha \otimes \delta_p \Phi_n \otimes \beta) = \sum_{a \in Q_1} \alpha a \otimes \delta_a \delta_p \Phi_n \otimes \beta$$

$$d_i^r(\alpha \otimes \delta_p \Phi_n \otimes \beta) = \sum_{a \in Q_1} \alpha \otimes \delta_p \Phi_n \delta'_a \otimes a\beta$$

for p any path of length $n-i$. We will show this is a self-dual resolution of A.

We first check this is a complex, checking $d_{j-1} \circ d_j = 0$ for $j=2, \ldots, n$ and that $\mu \circ d_1 = 0$. In the calculations $|p| = n-j$, and the sums are taken over all a, b $\in Q1$and paths q of length $j-2$.

$$\begin{aligned}
d_{j-1} \circ d_j(1 \otimes \delta_p \Phi_n \otimes 1) &= \epsilon_j \epsilon_{j-1} \sum_{a,b} (ab \otimes \delta_{pab} \Phi_n \otimes 1 - 1 \otimes \delta_p \Phi_n \delta'_{ab} \otimes ab) \\
&= \epsilon_j \epsilon_{j-1} \sum_{a,b,q} (c_{pabq} ab \otimes q \otimes 1 - 1 \otimes q \otimes c_{pqab} ab) \\
&= \epsilon_j \epsilon_{j-1} \sum_{a,b,q} ((-1)^{(n-1)j} c_{qpab} ab \otimes q \otimes 1 - 1 \otimes q \otimes c_{pqab} ab) \\
&= \epsilon_j \epsilon_{j-1} \sum_{q} ((-1)^{(n-1)j} \delta_{qp} \Phi_n \otimes q \otimes 1 - 1 \otimes q \otimes \delta_{pq} \Phi_n) \\
&= -\epsilon_j \epsilon_{j-1} \sum_{q} ((-1)^{(n-1)j} \delta_{qp} \phi_{n-2} \otimes q \otimes 1 - 1 \otimes q \otimes \delta_{pq} \phi_{n-2}) \\
&= -\epsilon_j \epsilon_{j-1} \sum_{q} ((-1)^{(n-1)j} c_{qp} 1 \otimes q \otimes 1 - 1 \otimes q \otimes 1 c_{pq}) \\
&= -\epsilon_j \epsilon_{j-1} \sum_{q} (c_{pq} 1 \otimes q \otimes 1 - 1 \otimes q \otimes 1 c_{pq}) = 0
\end{aligned}$$

$$\begin{aligned}
\mu \circ d_1(1 \otimes \delta_p \Phi_n \otimes 1) &= \epsilon_1 \mu \left(\sum_a a \otimes \delta_{pa} \Phi_n \otimes 1 - 1 \otimes \delta_p \Phi_n \delta'_a \otimes a \right) \\
&= \epsilon_1 \sum_a (c_{pa} a - (-1)^{n-1} c_{ap} a) = \epsilon_1 \sum_a (c_{pa} a - c_{pa} a) = 0
\end{aligned}$$

Since A is a PBW deformation of A we have $gr\ A \cong A$ and so $gr\ W\bullet_A \sim= W\bullet$, where we use the product filtration, $F_n(A \otimes_S W_j \otimes S_A) := \sum_{i+j+k \leq n} F^i(A) \otimes_S W_j \otimes_S F_k(A)$. Since $W^\bullet$ is exact, so is $_{W\bullet A}$ [16, Lemma 1], and since $W^\bullet$ consists of finitely generated projectives so does $W^\bullet_A$ [15, Lemmas 6.11, 6.16]. Hence $W^\bullet_A$ gives a resolution of A by finitely generated projectives. Now we need only check that this is still self-dual, implying A is n-CY.

We now construct isomorphisms α_k between the complex $W^\bullet_A$ and its dual

$$\begin{array}{ccccccccc}
\mathcal{A}\otimes_S \mathcal{W}_n \otimes_S \mathcal{A} & \xrightarrow{d_n} & \mathcal{A}\otimes_S \mathcal{W}_{n-1} \otimes_S \mathcal{A} & \xrightarrow{d_{n-1}} \cdots \xrightarrow{d_2} & \mathcal{A}\otimes_S \mathcal{W}_1 \otimes_S \mathcal{A} & \xrightarrow{d_1} & \mathcal{A}\otimes_S \mathcal{W}_0 \otimes_S \mathcal{A} \\
\downarrow{\scriptstyle \alpha_n} & & \downarrow{\scriptstyle \alpha_{n-1}} & & \downarrow{\scriptstyle \alpha_1} & & \downarrow{\scriptstyle \alpha_0} \\
(\mathcal{A}\otimes_S \mathcal{W}_0 \otimes_S \mathcal{A})^\vee & \xrightarrow{-d_1^\vee} & (\mathcal{A}\otimes_S \mathcal{W}_1 \otimes_S \mathcal{A})^\vee & \xrightarrow{-d_2^\vee} \cdots \xrightarrow{-d_{n-1}^\vee} & (\mathcal{A}\otimes_S \mathcal{W}_{n-1} \otimes_S \mathcal{A})^\vee & \xrightarrow{-d_n^\vee} & (\mathcal{A}\otimes_S \mathcal{W}_n \otimes_S \mathcal{A})^\vee
\end{array}$$

such that all the squares commute.

We will use the notation and result of [8, Section 4], working over $\mathbb{C}$. For T a finite dimensional S^e-module let F_T be the A^e-module A $\otimes_{ST} \otimes_{SA}$, and let T^* denote the $\mathbb{C}$ dual of T. Then

$$F_{T^*} \to F_T^\vee$$

$$(1 \otimes \phi \otimes 1) \mapsto \left((1 \otimes p \otimes 1) \mapsto \sum_{e,f \in Q_0} \phi(epf) e \otimes_{\mathbb{C}} f \right)$$

gives an isomorphism of A^e-modules. Moreover, as S is semisimple, tensoring $A \otimes_S(-) \otimes_{SA}$ is flat, hence constructing an isomorphism of S^e-modules $W_j \to W^*_{n-j}$ will give us an isomorphism of A^e-modules $FW_j \to FW^*_{n-j}$, and composing with the above isomorphism a will give us an isomorphism $F_{\mathcal{W}_j} \to F^\vee_{\mathcal{W}_{n-j}}$.

For $|p| = n - j$ define $\partial_p \in W^*_{n-j}$ by $\partial_p(\delta_q \Phi_n) = c_{qp}$, where $\Phi_n = \sum_{|t|=n} c_t t$ and $|q| = j$. Then there are S^e-module homomorphisms

$$\eta_j : \mathcal{W}_j \to \mathcal{W}^*_{n-j}$$

$$\delta_p \Phi_n \mapsto \gamma_j \partial_p$$

for γj arbitrary nonzero constants.

To see η_j is injective suppose $\eta_j(\sum \lambda_p \delta_p \Phi_n) = 0$. Then $\sum_p \lambda_p \partial_p(\delta_q \Phi_n) = \sum_p \lambda_p c_{qp} = 0$ for all q. Hence $\sum_p \lambda_p \delta_p \Phi_n = \sum_{p,q} \lambda_p c_{pq} q = \sum_q (\sum_p \lambda_p c_{pq}) q = 0$, so the map is injective.

There are $\mathbb{C}$ vector space isomorphisms between e$W_j f$ and $(fW_{n-j}e)^*$ or e, f $\in Q_0$ arising from the pairings,

$$\begin{aligned} eW_jf \otimes_{\mathbb{C}} fW_{n-j}e &\to \mathbb{C} \\ e\delta_p\Phi_nf \otimes f\delta_q\Phi_ne &\mapsto \gamma_jc_{qp} = \eta_j(e\delta_p\Phi_nf)(f\delta_q\Phi_ne) \\ &= (-1)^{(n-1)|q|}\frac{\gamma_j}{\gamma_{n-j}}\eta_{n-j}(f\delta_q\Phi_ne)(e\delta_p\Phi_nf) \end{aligned}$$

where we note that $\partial_p(\delta_q\Phi_n) = c_{qp} = (-1)^{(n-1)|q|}c_{pq} = (-1)^{(n-1)|q|}\partial_q(\delta_p\Phi_n)$, and to be non-zero we require $h(p) = t(q)$ and $t(p) = h(q)$.

As eW_j fand $fW_{n-j}e$ are finite dimensional $\mathbb{C}$ vector spaces the injectivity of η_j and η_{n-j} implies this pairing is perfect, thus η_j is an isomorphism.

Now we use this to define isomorphisms α_j

$$\begin{aligned} \alpha_j : \mathcal{A} \otimes_S W_j \otimes_S \mathcal{A} &\to (\mathcal{A} \otimes_S W_{n-j} \otimes_S \mathcal{A})^\vee \\ (a_1 \otimes \delta_p\Phi_n \otimes a_2) &\mapsto ((b_1 \otimes \delta_q\Phi_n \otimes b_2) \mapsto (b_1a_2 \otimes \eta_j(\delta_p\Phi_n)(\delta_q\Phi_n) \otimes a_1b_2)) \\ &= ((b_1 \otimes \delta_q\Phi_n \otimes b_2) \mapsto (\gamma_jc_{qp}b_1h(p)a_2 \otimes_{\mathbb{C}} a_1t(p)b_2)) \end{aligned}$$

where we note that for this to be non-zero $h(p) = t(q) = h(a_2) = t(b_1)$ and $h(q) = t(p) = h(b_2) = t(a_1)$.

It remains to check $\alpha_{j-1} \circ d_j = -d^\vee_{n-j+1} \circ \alpha_j$, i.e. the following diagram commutes;

$$\begin{array}{ccc} \mathcal{A} \otimes_S W_j \otimes_S \mathcal{A} & \xrightarrow{d_j} & \mathcal{A} \otimes_S W_{j-1} \otimes_S \mathcal{A} \\ \downarrow{\scriptstyle \alpha_j} & & \downarrow{\scriptstyle \alpha_{j-1}} \\ (\mathcal{A} \otimes_S W_{n-j} \otimes_S \mathcal{A})^\vee & \xrightarrow{-d^\vee_{n-j+1}} & (\mathcal{A} \otimes_S W_{n-j+1} \otimes_S \mathcal{A})^\vee \end{array}$$

If $|p| = n-j$ and $|q| = j-1$ then

$$(\alpha_{j-1} \circ d_j(1 \otimes \delta_p \Phi \otimes 1))(1 \otimes \delta_q \Phi \otimes 1)$$
$$= \alpha_{j-1}\left(\epsilon_j \sum_a a \otimes \delta_{pa}\Phi \otimes 1 + (-1)^{j+(n-1)} 1 \otimes \delta_{ap}\Phi \otimes a\right)(1 \otimes \delta_q \Phi \otimes 1)$$
$$= \gamma_{j-1}\epsilon_j \sum_a c_{qpa} t(q) \otimes a + \gamma_{j-1}\epsilon_j(-1)^{j+n-1} \sum_a c_{qap} a \otimes h(q)$$

whilst

$$(-d^{\vee}_{n-j+1} \circ \alpha_j(1 \otimes \delta_p \Phi \otimes 1))(1 \otimes \delta_q \Phi \otimes 1)$$
$$= \alpha_j(1 \otimes \delta_p \Phi \otimes 1)\left(-\epsilon_{n-j+1} \sum_a a \otimes \delta_{qa}\Phi \otimes 1 + (-1)^j 1 \otimes \delta_{aq}\Phi \otimes a\right)$$
$$= -\gamma_j \epsilon_{n-j+1} \sum_a c_{qap} a \otimes t(p) - \gamma_j \epsilon'_{n-j+1}(-1)^j \sum_a c_{aqp} h(p) \otimes a$$

where the sums are taken over all $a \in Q_1$. Thus for these to be equal we require

$$-\gamma_j \epsilon_{n-j+1} c_{qap} = \gamma_{j-1}\epsilon_j(-1)^{j+n-1} c_{qap}$$

and

$$-\gamma_j \epsilon_{n-j+1}(-1)^j c_{aqp} = \gamma_{j-1}\epsilon_j c_{qpa} = (-1)^{n-1}\gamma_{j-1}\epsilon_j c_{aqp}$$

for all a $\in Q_1$. These follow if $(-1)^n \gamma_{j-1}\epsilon_j = \gamma_j \epsilon_{n-j+1}(-1)^j$ for j=1, ..., n. As the γj were arbitrary non-zero scalars, we can choose the γj so that this is satisfied, and the proof is completed. □

Let Q be a single vertex quiver and A $=D(\Phi_n, n-2)$be a Noetherian, Koszul, n-CY algebra. We consider a PBW deformation, A, given by θ_1, θ_0, and set φ_{n-1}= $\sum_p p\theta_1(\delta_p \Phi_n) = \sum c_q q$. If the deformation is a zeroPBW deformation φ_{n-1} has the n-superpotential property.

Theorem 3.2.2. *Keeping the above notation and assumptions*

1. *Any zer oPBW deformation of A is n-CY*

2.*Let n =*2*or* 3. *Then the zero PBW deformations of* A*are exactly the n-CY PBW deformations of A, and moreover any superpotential* $\Phi' = \Phi_n + \varphi_{n-1} + \varphi_{n-2}$ *is* (n −2)*-coherent.*

Proof.1. Let A be a PBW deformation of A, defined by a map θ. Then define $\Phi'(\theta) = \Phi_n + \varphi_{n-1} + \varphi_{n-2}$by

$$\phi_{n-2+j} := -\sum_{|p|=n-2} p\theta_j(\delta_p \Phi_n)$$

.

Write φ$_{n-1}$= $\sum_t c_t t$. We will show that *A* is *n*-CY if and only if

$$0 = \sum_{i=0}^{n-2} (-1)^{in} c_{q_{i+2} \cdots q_{n-1} q_1 \cdots q_{i+1}}$$

for any *q*=$q_1 \ldots q_{n-1}$, with *qj* ∈Q_1. In particular this shows any zeroPBW deformation is *n*-CY, as then *Φ* '*θ*) is a superpotential and when *n* is even φ$_{n-1}$= ∑c_tt =0 and when *n* is odd the terms cancel in pairs by the superpotential property.

Referring to Theorem2.5.4A is *n*-CY if and only if

$$0 = \sum_{i=0}^{n-2} (-1)^i id^{\otimes i} \otimes \theta_1 \otimes id^{\otimes(n-2-i)}(\Phi_n)$$

$$= \sum_i (-1)^i \sum_{p=p_1 \ldots p_{n-2}} p_1 \ldots p_i \theta_1 \left(\delta_{p_1 \ldots p_i} \Phi_n \delta'_{p_{i+1} \ldots p_{n-2}}\right) p_{i+1} \cdots p_{n-2}$$

$$= \sum_i (-1)^{i+(n-1)(n-2-i)} \sum_p p_1 \ldots p_i \theta_1 (\delta_{p_{i+1} \cdots p_{n-2} p_1 \cdots p_i} \Phi_n) p_{i+1} \cdots p_{n-2}$$

$$\left(\text{using } \theta_1(\delta_q \Phi_n) = \delta_q \phi_{n-1} \text{ and writing } \phi_{n-1} = \sum_{|r|=n-1} c_r r\right)$$

$$= \sum_i (-1)^{in} \sum_{p,a} p_1 \ldots p_i c_{p_{i+1} \cdots p_{n-2} p_1 \cdots p_i a} a p_{i+1} \cdots p_{n-2}$$

$$= \sum_i (-1)^{in} \sum_q c_{q_{i+2} \cdots q_{n-1} q_1 \cdots q_i q_{i+1}} q_1 \cdots q_i q_{i+1} q_{i+2} \cdots q_{n-1}$$

Considering the coefficients of the paths, which are linearly independent, and calculating the coefficient of $q_1 \ldots q_{n-1}$we find the condition on φ$_{n-1}$to be

$$0 = \sum_{i=0}^{n-2} (-1)^{in} c_{q_{i+2} \cdots q_{n-1} q_1 \cdots q_{i+1}}$$

For all $q_1 \ldots q_{n-1}$.

2. By part1/Theorem2.5.4we have a condition for A to be n-CY. In the n =3case the condition gives that Im $id\otimes\theta_1-\theta_1\otimes id=\psi(\theta_1)=\{0\}$ which is the zero PBW condition. When n =2it gives the condition θ_1=0which is the zero PBW condition for even n.

Moreover when n =3any superpotential, Φ ', is 1-coherent. In particular the fact that A is 3-CY gives a duality in $W^{\bullet}$ between the 1st and 2nd terms— the arrows a $\in Q_1$ and the relations $\delta a\Phi 3$. Hence as the arrows are linearly independent so are the relations, and $\sum_{a\in Q_1}\lambda_a\delta_a\Phi_3=0\Rightarrow\lambda_a=0\Rightarrow\sum\lambda_a\delta_a$ Φ '=0, so Φ 'is 1-coherent. When n =2the superpotential is only differentiated by paths of length 0, so it is clearly 0-coherent. □

A similar result in the 3-CY case is already known in the context of a general quiver due to the following result of Berger and Taillefer. We note that their result is for potentials rather than superpotentials, where a potential is defined to be an element $W\in\mathbb{C}Q/[\mathbb{C}Q,\mathbb{C}Q]$, and there is a map into $\mathbb{C}Q$ defined by c :$a_n...a_1\_\rightarrow\sum_i a_i...a_1a_n...a_{i+1}$, see [7, Section 2].

Theorem 3.2.3. *(See [7, Section 3].)Let* A =*D*(c(W_{N+1}), 1)*be* N *Koszul and* 3-*CY, with* W_{N+1} *a potential on some quiver Q. Then a PBW deformation* A *of A is* 3-*CY if it is a zero PBW deformation. Moreover the zero PBW deformations correspond to* *D*(c(*W*'), 1)*for* W'=W_{N+1}+W_N+...+W_1*an inhomogeneous potential with each* W_i *in grade i.*

We also note that Theorem3.2.3makes use of the results of [8], in particular using the bimodule resolution of a graded 3-CY algebra given in [8, Section 4.2], whereas our results use the bimodule resolution of a Koszul n-CY algebra given in [9, Theorem 6.8].

4. APPLICATION: SYMPLECTIC REFLECTION ALGEBRAS

In this section we recall the definition of symplectic reflection algebras and deduce they are Morita equivalent to CY superpotential algebras by applying Theorems2.4.4, 3.1.1, and 3.2.1. We go on to calculate some examples, and consider the interpretation of the parameters of a symplectic reflection algebra in the superpotential setting.

Let V be a 2n dimensional space, equipped with symplectic form ω and G a symplectic reflection group which acts faithfully on V preserving the symplectic form. We say (G, V)is indecomposable if there is no G-stable splitting $V=V_1 \oplus V_2$with $\omega(V_1, V_2)=0$. We consider the skew group algebra $\mathbb{C}[V] \rtimes G$and the symplectic reflection algebras are defined to be the PBW deformations of this relative to $\mathbb{C}G$, and were classified by Eting of and Ginzburg [12].

Theorem 4.0.4. *(See [12, Theorem 1.3].)Any PBW deformation of such an indecomposable* $\mathbb{C}[V] \rtimes G$ *is of the form*

$$H_{t,\mathbf{c}} = \frac{T_{\mathbb{C}G}(V^*)}{\langle [x,y] - \kappa_{t,\mathbf{c}}(x,y) : x,y \in V^* \rangle}$$

where $\kappa_{t,\mathbf{c}}(x,y) = t\omega_{V^*}(x,y) - \Sigma_s \omega_s(x,y)\mathbf{c}(s)s$ *with the sum taken over the symplectic reflections* s, *and* c *a complex valued class function on symplectic reflections. The sym-plectic form* ω_{V^*}*on* V^**is induced from* ω *on V, and* ω_s *is defined as* ωV^* *restricted to* $(id-s)(V^*)$.

In particular they fall into the even dimensional case considered in Section3above, with the PBW parameter $\theta_1$equal to zero, and they are PBW deformations relative to$\mathbb{C}G$, which under Morita equivalence with a quiver of relations correspond to PBW deformations relative to *S*.

Two infinite classes of symplectic reflection algebras are the rational Cherednik and wreath product algebras.

•**Rational Cherednik algebras**.Let $\mathfrak{h}$ be a finite dimensional vector space and *G* a finite subgroup of GL($\mathfrak{h}$). Then V=h $\oplus$ $\mathfrak{h}^*$as the natural symplectic form $\omega((x, f), (y, g)) = f(y) - g(x)$, and an action of *G* as a symplectic reflection group. Then the symplectic reflection algebras given by PBW deformations of $\mathbb{C}[V] \rtimes G$are the Rational Cherednik Algebras.

•**Wreath product algebras**.Let *K* be a finite subgroup of $SL_2(\mathbb{C})$, and S_n the sym-metric group of order *n*. Then the wreath product group $G = S_n$ K is a symplectic reflection group acting on $V=(\mathbb{C}^2)^n$.

4.1. Symplectic reflection algebras as superpotential algebras

Here we show that symplectic reflection algebras are Morita equivalent to superpotential algebras.

Theorem 4.1.1. *Let* $H_{t,c}$ *be as in Theorem4.0.4.*

1.$H_{0,0}$ *is Morita equivalent to* $A = D(\Phi_{2n}, 2n-2)$*for some homogeneous superpotential* Φ_{2n}, *and is* 2_n*-CY and Koszul.*

2.*Any* $H_{t,c}$ *is Morita equivalent to* $A = D(\Phi', 2n-2)$ *for* $\Phi' = \Phi_{2n} + \varphi 2_{n-2}$ *an inhomogeneous superpotential which is* (2n −2)*-coherent*

3.*Any* (2n −2)*-coherent superpotential of the form* $\Phi' = \Phi_{2n} + \varphi_{2n-2}$*, gives an algebra* $A = D(\Phi', 2n-2)$ *that is Morita equivalent to a symplectic reflection algebra* $H_{t,c}$.

4.*All* $H_{t,c}$ *are* 2n-*CY algebras.*

Proof.1: $H_{0,0}$ is just $\mathbb{C}[V] \rtimes G$. Hence Theorem 2.4.4applies, and $H_{0,0}$ is Morita equivalent to a $2n$-CY, Koszul algebra $A := D(\Phi_{2n}, 2n-2)$for the McKay quiver for (G, V).

2 and 3: The Morita equivalence between $\mathbb{C}[V] \rtimes G$and A switches$\mathbb{C}G$ with S, respects the gradings, and respects the Koszul resolutions. Hence any zeroPBW deformation of $\mathbb{C}[V] \rtimes G$ corresponds to *a* zero PBW deformation of A, and 2 and 3 follow from Theorem3.1.1once we note by Theorem4.0.4all PBW deformations of $\mathbb{C}[V] \rtimes G$are zero PBW.

4: From 2 and 3 we know any symplectic reflection algebra is Morita equivalent to some $A := D(\Phi_{2n} + \varphi_{2n-2}, 2n-2)$. By Theorem3.2.1$A$ is $2n$-CY: A has a finite length resolution by finitely generated projective modules and $\mathrm{RHom}(A, A^e)[n] \cong A$. Then the Morita equivalent symplectic reflection algebra, $H_{t,c}$, also has a finite length resolution by finitely generated projective modules, as these properties are preserved under Morita equivalence. The isomorphism in the derived category $\mathrm{RHom}(A, A^e)[n] \cong A$ transfers to the isomorphism $\mathrm{RHom}(H_{t,c}, H^e_{t,c})[n] \cong H_{t,c}$ under the Morita equivalence, hence $H_{t,c}$ is $2n$-CY.

4.1.1. Examples

We give three examples.

Example 4.1.2. Here we consider the symplectic reflection algebra corresponding to the group S_3acting on $\mathfrak{h} * \mathfrak{h}^*$ in the manner of a rational Cherednik algebra, where the representation his given by

$$S_3 = \left\langle g = \begin{pmatrix} \varepsilon_3 & 0 \\ 0 & \varepsilon_3^2 \end{pmatrix}, h = \begin{pmatrix} 0 & 1 \\ 1 & 0 \end{pmatrix} \right\rangle$$

with ε_3a primitive third root of unity.

The McKay quiver is;

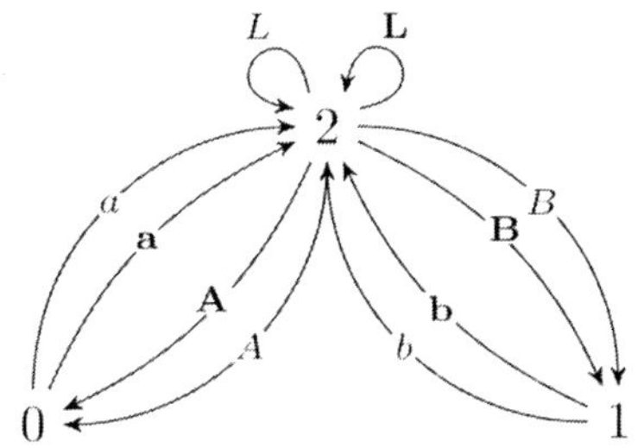

Now following the calculation in BSW [9, Theorem 3.2] we can calculate a superpotential to accompany this quiver, Φ_4

$$\begin{aligned}\Phi_4 = & -A\mathbf{a}A\mathbf{a} + 2\mathbf{A}aA\mathbf{a} + 4A\mathbf{L}L\mathbf{a} - 4\mathbf{A}LL\mathbf{a} - 4Ab\mathbf{Ba} + 2A\mathbf{b}B\mathbf{a} + 2\mathbf{A}bB\mathbf{a} \\ & - \mathbf{A}a\mathbf{A}a - 4A\mathbf{LL}a + 4\mathbf{A}L\mathbf{L}a + 2A\mathbf{bB}a + 2\mathbf{A}b\mathbf{B}a - 4\mathbf{Ab}Ba - 8LL\mathbf{LL} \\ & + 4bB\mathbf{LL} - 4b\mathbf{B}L\mathbf{L} + 4L\mathbf{b}B\mathbf{L} + 4\mathbf{bB}LL - 2bB\mathbf{bB} + b\mathbf{B}b\mathbf{B} + \mathbf{b}B\mathbf{b}B \\ & + \text{cyclic permutations}\end{aligned}$$

We now wish to look at the zero PBW deformations, which by Theorem4.1.1correspond to 2-coherent superpotentials $\Phi'=\Phi_4+\varphi_2$. Writing $\varphi_2=\sum c_{xy}xy$ the PBW deformations are parametrised by the c_{xy} such that $\Phi'$is a 2-coherent superpotential. We see that $\Phi'$is a superpotential if $c_{xy}=-c_{yx}$for all arrows $_{x,}$ $_{y}$. A superpotential $\Phi'$is 2-coherent if the $\delta_{xy}\Phi'=\delta_{xy}\Phi_4+c_{xy}e_h(x)e_t(y)$ satisfy the same linear relations as the $\delta_{xy}\Phi_4$. For instance as $\delta_{\mathrm{Aa}}\Phi$ =0and $\delta_{Aa}\Phi$ =$-\delta_{Aa}$we require that c_{Aa}=0and $c_{Aa}=-c_{Aa}$.

Making these calculations in this example we see the only non-zero c, and dependency relations among them, are:

$$c_{\mathbf{a}A} = -c_{A\mathbf{a}} = c_{\mathbf{A}a} = -c_{a\mathbf{A}}$$

$$c_{\mathbf{b}B} = -c_{B\mathbf{b}} = c_{\mathbf{B}b} = -c_{b\mathbf{B}}$$

$$c_{\mathbf{a}A} + c_{\mathbf{b}B} = c_{\mathbf{L}L} = -c_{L\mathbf{L}}$$

Hence we have a 2-coherent superpotential for any

$$\varphi_2 = c_{\mathbf{a}A}(\mathbf{a}A - A\mathbf{a} + \mathbf{A}a - a\mathbf{A}) + c_{\mathbf{b}B}(\mathbf{b}B - B\mathbf{b} + \mathbf{B}b - b\mathbf{B}) + (c_{\mathbf{a}A} + c_{\mathbf{b}B})(\mathbf{L}L - L\mathbf{L})$$

So we have 2 degrees of freedom in our parameters, exactly as we do for the t, c in $H_{t,c}$ for S_3acting on $\mathfrak{h}\oplus\mathfrak{h}^*$.

Example 4.1.3. Here we consider the symplectic reflection algebra corresponding to the dihedral group of order 8, D_8, acting on$\mathfrak{h}\oplus\mathfrak{h}^*$in the manner of a rational Cherednik algebra. The representation $\mathfrak{h}$ is given as

$$D_8 = \left\langle \sigma = \begin{pmatrix} \varepsilon_4 & 0 \\ 0 & \varepsilon_4^3 \end{pmatrix}, \tau = \begin{pmatrix} 0 & 1 \\ 1 & 0 \end{pmatrix} \right\rangle$$

where ε_4 is a primitive fourth root of unity.

This has McKay quiver

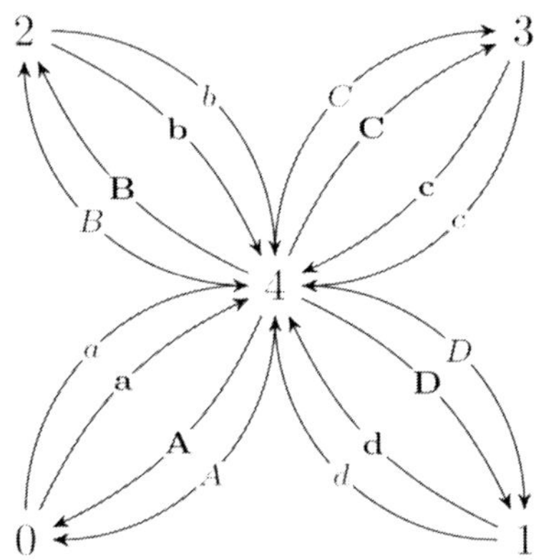

By choosing a *G*-equivariant basis we calculate the superpotential

$$\begin{aligned}\Phi = & -A\mathbf{a}A\mathbf{a} + 2\mathbf{A}aA\mathbf{a} - 4Ad\mathbf{Da} + 2A\mathbf{d}D\mathbf{a} + 2\mathbf{A}dD\mathbf{a} + 2A\mathbf{b}B\mathbf{a} - 2\mathbf{A}bB\mathbf{a} + 2A\mathbf{c}C\mathbf{a} \\ & - 2\mathbf{A}cC\mathbf{a} - \mathbf{A}a\mathbf{A}a - 4\mathbf{Ad}Da + 2\mathbf{A}d\mathbf{D}a + 2A\mathbf{dD}a + 2\mathbf{A}b\mathbf{B}a - 2A\mathbf{bB}a + 2\mathbf{A}c\mathbf{C}a \\ & - 2A\mathbf{cC}a - D\mathbf{d}D\mathbf{d} + 2\mathbf{D}dD\mathbf{d} + 2D\mathbf{b}B\mathbf{d} - 2\mathbf{D}bB\mathbf{d} + 2D\mathbf{c}C\mathbf{d} - 2\mathbf{D}cC\mathbf{d} - \mathbf{D}d\mathbf{D}d \\ & + 2\mathbf{D}b\mathbf{B}d - 2D\mathbf{bB}d + 2\mathbf{D}c\mathbf{C}d - 2D\mathbf{cC}d - B\mathbf{b}B\mathbf{b} + 2\mathbf{B}bB\mathbf{b} - 4Bc\mathbf{Cb} + 2B\mathbf{c}C\mathbf{b} \\ & + 2\mathbf{B}cC\mathbf{b} - 4\mathbf{Bc}Cb - \mathbf{B}b\mathbf{B}b + 2\mathbf{B}c\mathbf{C}b + 2B\mathbf{cC}b - C\mathbf{c}C\mathbf{c} + 2\mathbf{C}cC\mathbf{c} - \mathbf{C}c\mathbf{C}c \\ & + \text{cyclic permutations}\end{aligned}$$

We now calculate the zero PBW deformations of $A := D(\Phi_4, 2)$. We write $\varphi_2 = \sum c_{xy} xy$, and by Theorem4.1.1the zero PBW deformations of A are parameterised by the c_{xy} such that $\Phi' = \Phi_4 + \varphi_2$is a 2-coherent superpotential.

In particular we require $c_{xy} = -c_{yx}$ and for the c_{xy} to satisfy the same linear relations as the $\delta_{xy}\Phi_4$. Making these calculations in this example we see the only non-zero c_{xy}, and dependency relations among them, are;

$$c_{\mathbf{a}A} = -c_{A\mathbf{a}} = c_{\mathbf{A}a} = -c_{a\mathbf{A}}$$

$$c_{\mathbf{b}B} = -c_{B\mathbf{b}} = c_{\mathbf{B}b} = -c_{b\mathbf{B}}$$

$$c_{\mathbf{c}C} = -c_{C\mathbf{c}} = c_{\mathbf{C}c} = -c_{c\mathbf{C}}$$

$$c_{\mathbf{d}D} = -c_{D\mathbf{d}} = c_{\mathbf{D}d} = -c_{d\mathbf{D}}$$

$$c_{\mathbf{a}A} + c_{\mathbf{d}D} = -c_{b\mathbf{B}} - c_{c\mathbf{C}} = c_{\mathbf{b}B} + c_{\mathbf{c}C}$$

Hence to obtain a 2-coherent superpotential we require a $\varphi_2$of the form

So we see here there are 3 degrees of freedom in our parameters exactly as for the parameters *t* and *c* in the symplectic reflection algebra for D_8acting on $\mathfrak{h}\oplus\mathfrak{h}^*$.

Example 4.1.4. A special case of path algebras Morita equivalent to symplectic reflection algebras are the deformed preprojective algebras of [11]. These can be given as superpotential algebras in the *n* =2, differentiation by paths of length 0, case.

We consider a skew group algebras $\mathbb{C}[\mathbb{C}^2] \rtimes G$ for *G* is a finite subgroup of $SL_2(\mathbb{C})$. We construct the McKay quiver and label the arrows in a particular way; between any two vertices we choose a direction, label the arrows in this direction $a_1, \ldots, a_k$, and the arrows in the opposite direction $a^*_1,\ldots,a^*_k$. Then $\mathbb{C}[\mathbb{C}^2] \rtimes G$ is Morita equivalent to $A = D(\Phi_2, 0)$for the homogeneous superpotential, $\Phi_2 = \sum[a, a^*]$. This is the preprojective algebra

$$A = \mathcal{D}(\Phi_2, 0) = \frac{\mathbb{C}Q}{\sum[a, a^*]}$$

Now we consider the PBW deformations of *A*. We apply Theorem4.1.1, and deduce PBW deformations correspond to 0-coherent inhomogeneous superpotentials, $\Phi_2+\varphi_0$. We note that $\varphi_0:=-\sum_{i\in Q0}\lambda_i e_i \in S$can in fact be arbitrary as any element of *S* satisfies the superpotential property, and the 0-coherent property is always satisfied. Hence we recover that PBW deformations of the preprojective algebra are the deformed preprojective algebras

$$\mathcal{A} = \mathcal{D}(\Phi_2 + \phi_0, 0) = \frac{\mathbb{C}Q}{\sum[a, a^*] - \sum \lambda_i e_i},$$

which are parameterised by a scalar, λ_i, for each vertex, *i*. By Theorem4.1.1these are 2-CY.

5. APPLICATION: PBW DEFORMATIONS OF SKEW GROUP RINGS FOR GL2

So far we have been considering subgroups of SL(W)where W is a finite dimensional vector space. This corresponds to non-twisted superpotentials. Here we consider algebras Morita equivalent to $\mathbb{C}[V] \rtimes G$for G a finite subgroup of GL(W), and the existence of PBW deformations for$\mathbb{C}[V] \rtimes G$.

We recall

Theorem 5.0.5. *(See [9, 3.2, 6.1 and 6.8].)Let* W=$\mathbb{C}^n$, *and G be a finite subgroup of* GL(W). *Then is*$\mathbb{C}[V] \rtimes G$ *Morita equivalent to* $D(\Phi_n, n-2)$*for the McKay quiver* (G, W), Φ_n *a twisted homogeneous superpotential, and the twist automorphism given by* $(-) \otimes \mathbb{C}\det W$. *There is a recipe to construct the twisted superpotential.*

Working with homogeneous superpotentials, $\Phi_n=\sum c_p p$, we have $c_p=0$ for any p that is not a closed path. This is no longer the case for twisted homogeneous superpo-tentials, here we find $c_p=0$ unless $h(p)=\sigma(t(p))$. Since the twist for$\mathbb{C}[V] \rtimes G$is given by tensoring by det W, c_p is non-zero only for paths from W_i to det$W\otimes\mathbb{C}W_i$, where the Wiare the irreducible representations corresponding to vertices in the McKay quiver.

There are two different cases of finite subgroups of $GL_n(\mathbb{C})$we consider, those that contain pseudo-reflections, and those that do not. Those that do not are known as *small* subgroups.

As a particular case we will consider $GL_2(\mathbb{C})$, where differentiation is by paths of length 0, and so our relations are given by the superpotential, and any relations are a sum of paths with tail W_i and head det$W\otimes\mathbb{C}W_i$.

Theorem 5.0.6. *Let G be a small finite subgroup of* $GL_2(\mathbb{C})$, *which is not contained in* $SL_2(\mathbb{C})$. *Then* $\mathbb{C}[\mathbb{C}^2] \rtimes G$ *has no nontrivial (relative to*$\mathbb{C}$ G*) PBW deformations.*

Proof.The algebra$\mathbb{C}[\mathbb{C}^2] \rtimes G$ can be written as $\dfrac{T_{\mathbb{C}G}(\mathbb{C}^{2*}\otimes_{\mathbb{C}}\mathbb{C}G)}{\langle [x,y]\otimes 1\rangle}$ and is Morita equivalent to a path algebra with relations $\mathbb{C}Q/R = \frac{T_S(V)}{R} = \mathcal{D}(\Phi_2, 0)$for some twisted homogeneoussuper potential Φ_2, where we use notation as inSection2.1. In particular the Morita equivalence switches$\mathbb{C}$ G with S, and respects the gradings and Koszul resolutions. Hence considering PBW deformations as in Section 2.5 we see that under the Morita equivalence any PBW deformation of $\mathbb{C}[\mathbb{C}^2] \rtimes G$would give a PBW deformation of $\mathbb{C}Q/R$, noting that in one case considering PBW deformations relative to $\mathbb{C}G$, and in the other relative to *S*.

Hence it is enough to show that the Morita equivalent twisted superpotential algebra $A := D(\Phi_2, 0)$has no nontrivial PBW deformations.

There can only possibly exist PBW deformations if there exists some non-zero θ_0, $\theta_1$as in Section2.5. But $\theta_1 \in \mathrm{Hom}_{S^e}(R, V)$ and $\theta_0 \in \mathrm{Hom}_{S^e}(R, S)$, so if both these sets are {0}there are no nontrivial PBW deformations.

Define the *distance* between two vertices in the quiver to be the minimal length of a path from one to the other. It is shown in Appendix A, LemmaA.0.1, that the tail and head of any relation are vertices which are distance greater than one apart. Hence, as Semodule maps preserve heads and tails, the sets $\mathrm{Hom}_{S^e}(R, V)$and $\mathrm{Hom}_{S^e}(R, S)$are both {0}and there are no nontrivial PBW deformations. □

We look at examples of a small and non-small subgroup, using the calculations from [9]. We let ε_m denote a primitive mth root of unity.

Example 5.0.7. We first consider a small subgroup $\mathbb{D}_{5,2}$ with representation as

$$\left\langle \begin{pmatrix} \varepsilon_4 & 0 \\ 0 & \varepsilon_4^{-1} \end{pmatrix}, \begin{pmatrix} 0 & \varepsilon_4 \\ \varepsilon_4 & 0 \end{pmatrix}, \begin{pmatrix} 0 & \varepsilon_6 \\ \varepsilon_6 & 0 \end{pmatrix} \right\rangle$$

This is given as an example in [9, Example 5.4].

It has McKay quiver

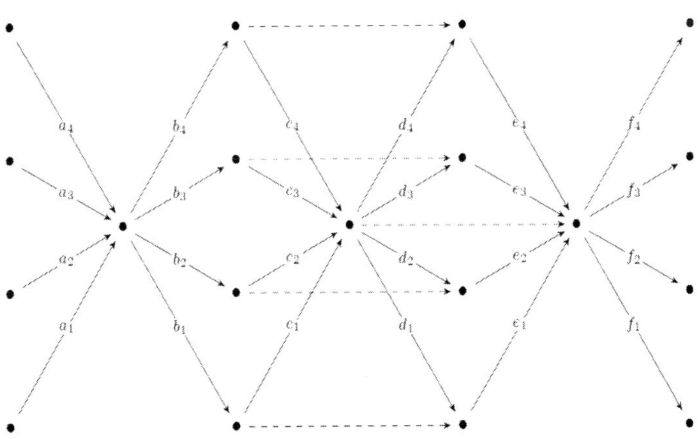

where the first column of vertices equals the final column of vertices. Tensoring by the determinant representation maps from a vertex to the next vertex in line to the right, as indicated by the dashed arrows, wrapping around from the right side of the diagram to the left.

The relations are

$$b_i a_i = 0, \qquad d_i c_i = 0, \qquad f_i e_i = 0 \quad \text{for } i = 1, 2, 3, 4$$

and

$$\sum_{j=1}^{4} c_j b_j = 0, \qquad \sum_{j=1}^{4} e_j d_j = 0, \qquad \sum_{j=1}^{4} a_j f_j = 0$$

Consider the head and tail of any element of R. These vertices are related by the twist, and we see the shortest path between the tail and head is always length 2. Then $\mathrm{Hom}_{S^e}(R, V)$ and $\mathrm{Hom}_{S^e}(R, S)$are both {0}, hence there can be no nontrivial PBW deformations.

Example 5.0.8. We now consider D_8 as the non-small subgroup with representation

$$\left\langle g = \begin{pmatrix} \varepsilon_4 & 0 \\ 0 & \varepsilon_4^3 \end{pmatrix}, h = \begin{pmatrix} 0 & 1 \\ 1 & 0 \end{pmatrix} \right\rangle$$

and McKay quiver and relations, Q and R.

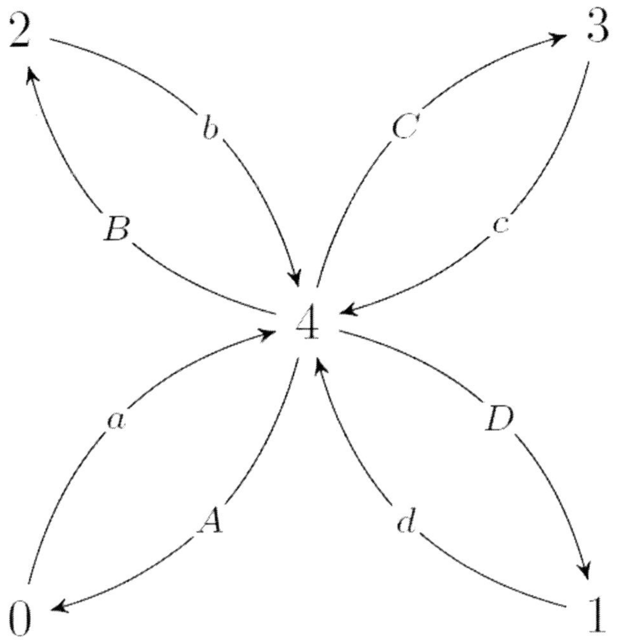

$$Da = 0 \qquad Cb = 0$$

$$Ad = 0 \qquad Bc = 0$$

$$aA + dD - bB - cC = 0$$

Then PBW deformations of $\mathbb{C}Q/R$ are classified by θ_1, θ_0:$R \rightarrow A$ which are S^e-module maps; $\theta_1 \in Hom_{S^e}(R, V)$ and $\theta_0 \in Hom_{S^e}(R, S)$.

As S^e-module maps preserve heads and tails we see that θ_1 must be zero, and θ_0 must be zero on all relations but the central one $aA + dD - bB - cC$. At this relation $\theta_0(aA + dD - bB - cC) = \lambda_{e4}$ for some $\lambda \in \mathbb{C}$. In this case $(R \otimes_S V) \cap (V \otimes_S R) = 0$, and so any such θ0 gives us a PBW deformation. Hence there is a one parameter collection of PBW deformations.

ACKNOWLEDGMENTS

The author is a student at the University of Edinburgh funded via a doctoral training grant from the Engineering and Physical Sciences Research Council [grant number EP/J500410/1], and this material will form part of his PhD thesis. The author would like to express his thanks to his supervisors, Prof. Iain Gordon and Dr. Michael Wemyss, for much guidance and patience, and also to the EPSRC.

APPENDIX A. MCKAY QUIVERS FOR FINITE SMALL SUBGROUPS OF $GL_2(\mathbb{C})$

We use the classification of McKay quivers for small finite subgroups of $GL_2(\mathbb{C})$[2], to prove the following:

Lemma A.0.1. *Let* $G < GL_2(\mathbb{C})$ *be a small finite subgroup, given by a representation* $W \cong \mathbb{C}^2$. *Let* W_i *be an irreducible representation of G. Then the shortest path from* W_i *to* $\det W \otimes W_i$ *has length* ≥ 2.

Proof. We outline this case by case by examining the McKay quivers, showing there are no length 0 or 1 paths between a vertex in the quiver and the vertex related by tensoring by the determinant.

We list the small finite subgroups of $GL_2(\mathbb{C})$ up to conjugacy, as in [2, Section 2].

Let $Z_n = \left\langle g = \begin{pmatrix} \varepsilon & 0 \\ 0 & \varepsilon \end{pmatrix} \right\rangle$, and ε be a primitive *n* throot of unity. Any finite small subgroup of $GL_2(\mathbb{C})$ is, up to conjugacy, one of the following:

1. Z a cyclic subgroup, $Z = \left\langle g = \begin{pmatrix} \varepsilon & 0 \\ 0 & \varepsilon^q \end{pmatrix} \right\rangle$ for $1 \leq q < n$.

2. $Z_n D = \{zd \mid z \in Z_n,\ d \in D\}$ for *D* a finite, non-cyclic, subgroup of $SL_2(\mathbb{C})$.

3. $\bar{H}$. To define $\bar{H}$ let $D < SL_2(\mathbb{C})$ be a binary dihedral group, with *A a* cyclic subgroup of index 2, and define $H < Z_{2n} \times D$ to be

$$H = \left\{(z, d) \in Z \times D \mid d + A = z + Z_m \text{ in } Z_2 = D/A = Z_{2n}/Z_n\right\}.$$

Then $\bar{H}$ is the image of H under the map $H \to GL_2(\mathbb{C})$.

4. $\bar{K}$. To define $\bar{K}$ let $D<SL_2(\mathbb{C})$be the binary tetrahedral group, with *A* a normal binary dihedral subgroup of index 3, let n ≥3, and define $K<Z_{3n}\times D$to be

$$K=\{(z,d)\in Z\times D \mid d+A=z+Z_n \text{ in } Z_3=D/A=Z_{3n}/Z_n\}.$$

Then $\bar{K}$ is the image of *K* under the map $K\rightarrow GL_2(\mathbb{C})$.

We note that if n =1, 2then $\bar{K}$ is the binary tetrahedral group with defining representation containing pseudo reflections, so is not small.

The McKay quivers for these groups are described in [2, Proposition7], and we look at the determinant representation in each case and show tensoring by it relates vertices at least distance two apart.

We first look at cyclic subgroups. Let *Z* be as above. Such a representation is in $SL_2(\mathbb{C})$only when $q+1=n$. We suppose $q+1=n$.

Such a group has *n* irreducible one dimensional representations, which we label $W_0,\ldots,W_{n-1}$, where W_i is given by $g\mapsto\varepsilon^i$. The defining representation is reducible as $\mathbb{C}2=W_1\oplus W_q$, and its determinant is the representation W_{1+q}. Hence the McKay quiver has *n* vertices corresponding to the W_i and at vertex *i* has two arrows to vertices *i* +1 and *i* +*q* modulo *n*. The relations on the McKay quiver have head and tail related by tensoring by the determinant, hence any relations with tail W_i have head W_{i+q+1}module *n*. We see that the two vertices W_i and $W_{i+q+1}$are distance 2 apart; they are not distance 0 as $i=i+1+q$ module *n*, and they are not distance 1 as the only arrows from *i* are to $i+1$ or $i+q$, and neither of these equals $i+1+q$ modulo *n*.

All the remaining groups are constructed by taking a subgroup of $Z_n\times D$ and then taking the image of this under the map to $GL_2(\mathbb{C})$. If we calculate the McKay quiver for the subgroup then the image in $GL_2(\mathbb{C})$ has McKay quiver which is a subquiver. Hence for our purposes it is enough to calculate the McKay quivers for the various subgroups of $Z_n\times D$.

We first do this for the case Z_nD. We note that this is contained in $SL_2(\mathbb{C})$for $n=1, 2$, hence we assume n >2. In this case we consider the McKay quiver of $Z_n\times D$. Let the irreducible representations of *D* be labeled $D_0, \ldots, D_{r-1}$where $D_0$is trivial, and D_1 is the given 2 dimensional representation. Let R_i, for i =0,...,n −1, be the none dimensional irreducible representations of Z_n with R_i given by $g\mapsto\varepsilon^i$. Then $Z_n\times D$ has *nr* irreducible representations given by $R_i\otimes D_j$ for $0\leq i<n$ and $0\leq j<r$. Then we consider the McKay quiver for the defining representation $R_1\otimes D_1$, as this corresponds to the defining

representation in $GL_2(\mathbb{C})$. In particular the McKay quiver has n groups of representations labeled by representation of Z_n, with group i corresponding to the set of representations $\{R_i \otimes D_j | j=0,...,r-1\}$. By definition any arrows in the quiver go from group i to group $i+1$ modulo n. As the defining representation is $R_1 \otimes D_1$ the determinant representation is $R_2 \otimes D_0$, and the determinant maps from group i to group $i+2$ modulo n. In particular, as $n>2$, any two vertices related by this are not distance zero or one apart.

For the cases $\bar{H}$ and $\bar{K}$ we take the McKay quiver for $Z_n \times D$, make some identifications to account for certain irreducible representations being identified for the subgroups H,K.

We first consider H. In this case we label the representations of the binary dihedral group as

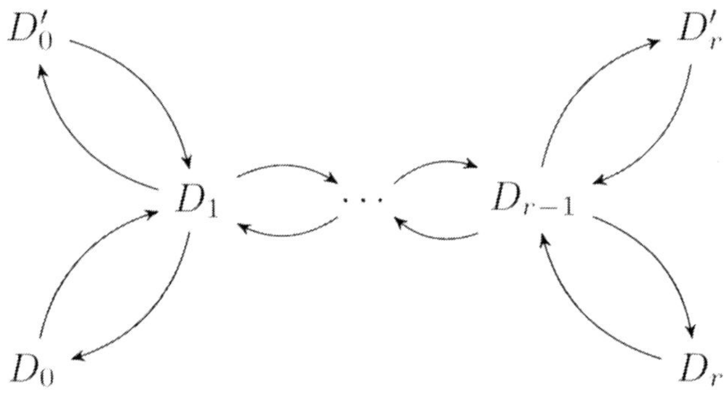

Where D_1 is the representation in $SL_2(\mathbb{C})$ and D_0 is the trivial representation.

Now we label the representations of Z_{2n} as Ri for i $=0 ... 2n-1$as above, and $Z_{2n} \times D$ has $2n(r+2)$ irreducible representations given by their tensor products. The defining representation in $GL_2(\mathbb{C})$is given as $R_1 \otimes D_1$, and hence the determinant representation is $R_2 \otimes D_0$. Now all representations of $Z_{2n} \times D$ are still irreducible for H but some are identified[2, Proposition 7(d)]. The representations which are identified are $Ri \otimes D_j$ with $R_{n+i} \otimes D_j$ for $j=1 ... r-1$, modulo $2n$, and $R_i \otimes D_j$ with $R_{n+i} \otimes D'j$ modulo $2n$ for $j=0$, r. The McKay quiver for H is the McKay quiver for $Z_{2n} \times D$ with these identifications made. In this case we group the irreducible representations into n groups labeled by the representation of Z_{2n} modulo n, so arrows go from group i to $i+1$ and determinant from group i to $i+2$module n. Hence, for $n>2$, the head and tail of two vertices related by the determinant are not distance zero or one apart. When $n=1$, 2 then $H=D$ and then the representation is contained in $SL_2(\mathbb{C})$.

The final case is to consider $K<Z_{3n} \times D$. Again we label the representations of Z_{3n} by R_i for $i=0 ... 3n-1$, and we label the representations of D by

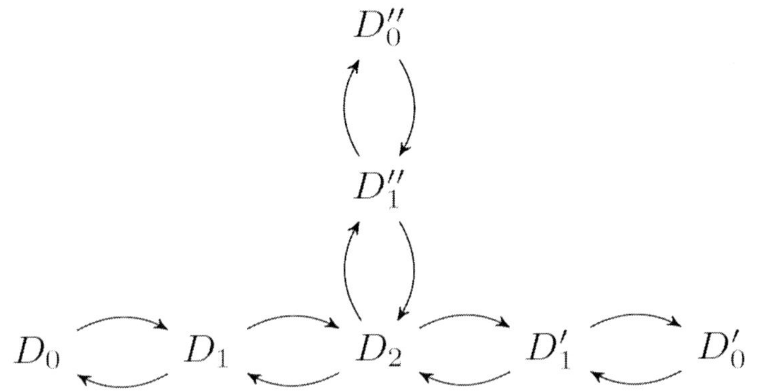

where D_0 is the trivial representation, and $D_1$the defining representation.

The representation defining the group in $GL_2(\mathbb{C})$is $R_1 \otimes D_1$, and the determinant representation is $R_2 \otimes D_0$.

Now all the irreducible representations for $Z_{3n} \times D$remain so for K, however some are identified [2, Proposition 7 *(f)*]. This time triples are identified: $R_i \otimes D_j \cong R_{i+n} \otimes D'_j \cong R_{i+2n} \otimes D''_j$ modulo 3_n as representation of k.

Once again we note that this splits the quiver into n groups, labeled by the representation of Z_{3n} module $_n$, with arrows from group i to i +1 and determinant from i to i +2modulo n. Hence, as $n > 2$, the determinant maps between vertices which are not distance 0 or 1 apart.

Hence for any small finite subgroup of $GL_2(\mathbb{C})$ not contained in $SL_2(\mathbb{C})$ the determinant in the McKay quiver maps between vertices which are distance greater than 1apart.

We also note that, while we do not know a complete proof without the use of the classification, given W_i=det $W \otimes W_i$ there is an elementary character theoretic proof of LemmaA.0.1.

Lemma A.0.2. *Let* G <$GL_2(\mathbb{C})$*be a small finite subgroup, given by a representation* $W \cong \mathbb{C}^2$. *Let* W_i *be an irreducible representation of G. Then, assuming* W_i=det $W \otimes W_i$, *there are no arrows from* W_i *to* det $W \otimes W_i$ *hence the shortest path from* W_i *to* det$W \otimes W_i$ *has length* ≥2.

Proof. We suppose that G is a small finite subgroup of $GL_2(\mathbb{C})$, and that W_i=det $W \otimes W_i$. Hence there are no length 0 paths from W_i to det$W \otimes W_i$. We show that there are no length one paths W_i to det $W \otimes W_i$

We let [–, –] denote the standard inner product for characters of G, and suppose that the determinant representation of W has order d. We note that for any two characters A, B

$$\langle \mathrm{ch}(W)A, B\rangle = \langle A, \mathrm{ch}(W)\,\mathrm{ch}(\det W)^{d-1}B\rangle$$

as if $g \in G$ has W(g)with eigenvalues ϵ_1, ϵ_2 then

Suppose there were an arrow from W_i to det $W \otimes W_i$, then

$$\mathrm{tr}\,W(g^{-1}) = (\epsilon_1^{-1} + \epsilon_2^{-1}) = (\epsilon_1\epsilon_2)^{-1}(\epsilon_1 + \epsilon_2) = \mathrm{tr}(\det W(g))^{-1}\,\mathrm{tr}(W(g))$$

By considering dimensions we see a =1, and $W \otimes W_i = W_i \oplus \det(W) \otimes W_i$.

Suppose $g \in G$, W(g)has eigenvalues ϵ_1, ϵ_2 , and $W_i(g)$ has eigenvalues $\mu_1,\ldots,\mu_r$. Then

$$\begin{aligned}(\epsilon_1 + \epsilon_2)(\mu_1 + \ldots + \mu_r) &= \mathrm{ch}(W \otimes W_i) = \mathrm{ch}(W_i \oplus \det(W) \otimes W_i)\\ &= (1 + \epsilon_1\epsilon_2)(\mu_1 + \ldots + \mu_r)\end{aligned}$$

As W_i is an irreducible representation, and G is nontrivial, we can choose $g \in G$ nontrivial such that $\mathrm{ch}W_i(g) = \mu_1+\ldots+\mu_r=0$. The eigenvalues of such a g satisfy $1 + \epsilon_1\epsilon_2 = \epsilon_1 + \epsilon_2$. Hence 1 is an eigenvalue of $W(g)$, so g is a pseudo reflection, which is a contradiction as G is small.

Hence there are no arrows from W_i to $W_i \otimes$det W.

REFERENCES

1. C. Amiot, O. Iyama, I. Reiten, Stable categories of Cohen–Macaulay modules and cluster categories, arXiv:1104.3658v1
2. M.Auslander, I. Reiten, McKay quivers and extended Dynkin diagrams, Trans. Amer. Math. Soc. 293(1) (1986) 293–301
3. R.Berger, Koszulity for nonquadratic algebras, J. Algebra 239(2) (2001) 705–734
4. R. Berger, M. Dubois-Violette, M. Wambst, Homogeneous algebras, J. Algebra 261(1) (2003) 172–185
5. R.Berger, V. Ginzburg, Higher symplectic reflection algebras and non-homogeneous N-Koszul prop-erty, J. Algebra 304(1) (2006) 577–601
6. R.Berger, N. Marconnet, Koszul and Gorenstein properties for homogeneous algebras, Algebr. Represent. Theory 9 (2006) 67–97, http://dx.doi.org 10.1007/s10468-005-9002-1
7. R.Berger, R. Taillefer, Poincaré–Birkhoff–Witt deformations of Calabi–Yau algebras, J. Noncom-mut. Geom. 1(2) (2007) 241–270
8. R.Bocklandt, Graded Calabi Yau algebras of dimension 3, J. Pure Appl. Algebra 212(1) (2008) 14–32.

9. R.Bocklandt, T. Schedler, M. Wemyss, Superpotentials and higher order derivations, J. Pure Appl. Algebra 214(9) (2010) 1501–1522
10. A.Braverman, D. Gaitsgory, Poincaré–Birkhoff–Witt theorem for quadratic algebras of Koszul type, J. Algebra 181(2) (1996) 315–328
11. W. Crawley-Boevey, M.P. Holland, Noncommutative deformations of Kleinian singularities, Duke Math. J. 92(3) (1998) 605–635
12. P.Etingof, V. Ginzburg, Symplectic reflection algebras, Calogero–Moser space, and deformed Harish Chandra homomorphism, Invent. Math. 147(2) (2002) 243–348
13. V.Ginzburg, Calabi–Yau algebras, arXiv:math/0612139v3
14. E.L. Green, E.N. Marcos, R. Martínez-Villa, P. Zhang, D-Koszul algebras, J. Pure Appl. Algebra 193(1–3) (2004) 141–162
15. J.C. McConnell, J.C. Robson, L.W. Small, Noncommutative Noetherian Rings, Grad. Stud. Math., American Mathematical Society, 2001
16. G. Sjödin, On filtered modules and their associated graded modules, Math. Scand. 33(1974) (1973) 229–249
17. Q. Wu, C. Zhu, Poincaré–Birkhoff–Witt deformation of Koszul Calabi–Yau algebras, Algebr. Rep-resent. Theory 16(2) (2013) 405–420, http://dx.doi.org/10.1007/s10468-011-9312-4

CHAPTER 10

Extensions of the Tensor Algebra and their Applications

Minoru Itoh

Faculty of Science, Department of Mathematics and Computer Science, Kagoshima University, Kagoshima, Japan

ABSTRACT

This article presents a natural extension of the tensor algebra. In addition to "left multiplications" by vectors, we can consider "derivations" by covectors as basic operators on this extended algebra. These two types of operators satisfy an analogue of the canonical commutation relations. This algebra and these operators have the following applications: (i) applications to invariant theory related to tensor products and (ii) applications to immanants. The latter includes a new method to study the quantum immanants in the universal enveloping algebras of the general linear Lie algebras and their Capelli type identities (the higher Capelli identities).

Keywords: Capelli identity, Central elements of universal enveloping algebras, Clifford algebra, Quantum immanants, Schur–Weyl duality, Symmetric group, Tensor algebra, Weyl algebra

INTRODUCTION

In this article, we introduce some extensions of the tensor algebra. The most basic one is constructed as a vector space as follows:

$$\overline{T}(V) = \bigoplus_{p \geq 0} V^{\otimes p} \otimes_{\mathbb{C}S_p} \mathbb{C}S_\infty$$

For this $\overline{T}(V)$, we can naturally define an associative algebra structure. The ordinary tensor algebra $T(V)$ can be regarded as a subalgebra of this algebra. This extended algebra $\overline{T}(V)$ is remarkable, because we can consider a natural "derivation" $L(v^*)$ determined from any covector $v^* \in V^*$ as an operator on $\overline{T}(V)$. An analogue of the canonical commutation relations holds between these derivations and the left multiplications $L(v)$ by vectors $v \in V$ (Theorem 2.3). It is also natural to call these multiplications and derivations "creation operators" and "annihilation operators," respectively (namely, we can regard this $\overline{T}(V)$ as an analogue of the Boson and Fermion Fock spaces). The algebra $\mathcal{L}(V)$ generated by these two types of operators is naturally isomorphic to $\bigoplus_{p,q \geq 0} V^{\otimes p} \otimes_{\mathbb{C}S_p} \mathbb{C}S_\infty \otimes_{\mathbb{C}S_q} V^{*\otimes q}$ as vector spaces, and we can regard this operator algebra $\mathcal{L}(V)$ as an analogue of the Weyl algebra and the Clifford algebra (actually this contains these algebras naturally as quotient algebras).

This framework has some applications to representation theory and invariant theory related to tensor products. For example, we can prove the Schur–Weyl duality and its generalization by a simple calculation in $\mathscr{L}(V)$ (Theorem 3.1). We also have an analogue of the $(GL_n(\mathbb{C}), GL_{n'}(\mathbb{C}))$ duality due to Howe [5], and this provides us with a natural correspondence between the center of the universal enveloping algebra $U(\mathfrak{gl}_n)$ and some invariant "differential operators" on $T(\mathbb{C}^n \otimes \mathbb{C}^{n'})$ (Theorem 3.7). We can describe this correspondence as a Capelli type identity on the tensor algebra $T(\mathbb{C}^n \otimes \mathbb{C}^{n'})$ (Theorem 4.1). Moreover, using this Capelli type identity, we can determine the $SL_n(\mathbb{C})$-invariants in $T(\mathbb{C}^n \otimes \mathbb{C}^{n'})$ (Theorem 4.4).

Our extensions of the tensor algebra are also useful to treat a matrix function called "immanant." This is parallel to the fact that the exterior algebras and the symmetric tensor algebras are useful to treat the determinant and the permanent. This method can be developed to study the "quantum immanants," a basis of the center of the universal enveloping algebra $U(\mathfrak{gl}_n)$ introduced by Okounkov [21]. The quantum immanants have been studied by using the R-matrix method, the fusion procedure, and representation theory of the Yangian $Y(\mathfrak{gl}_n)$ (see [16, 20-23]; Section 7.4 of [17] is helpful to look at the whole picture on this issue). Instead of these traditional approaches, we can use our extensions of the tensor algebra. Namely, making use of our algebras as formal variables, we can prove various fundamental relations for the quantum immanants by simple calculations. It is not to say that our approach is more powerful than established approaches. However, we can regard this approach as an advanced version of the exterior calculus used to study Capelli type identities in [4, 7-13, 26-30] (and also in Section 4 of this article), and we can manipulate the noncommutativity with similar skills. The author thinks that this approach is one of the best ways to study the quantum immanants and expects further developments.

Finally, combining several ideas in this article, we give Capelli type identities for the quantum immanants on $T(\mathbb{C}^n \otimes \mathbb{C}^{n'})$ as higher generalizations of the Capelli type identity given in Section 4 (Theorems 9.3 and 9.4).

This article is organized as follows. In Section 1, we introduce the algebra $\bar{\bar{T}}(\mathrm{v})$ as an extension of the tensor algebra $T(V)$. This is the base of all studies in this article. In Section 2, we define multiplication and derivation operators acting on $\bar{\bar{T}}(\mathrm{v})$, and give their quick applications. The algebra $\mathscr{L}(V)$ generated by these operators contains the Weyl algebra and the Clifford algebra naturally as quotient algebras. In Section 3, we use these operators to study a generalization of the Schur–Weyl duality and an analogue of Howe duality. In Section 4, we describe the action of the Capelli element on the tensor algebra $T(\mathbb{C}^n \otimes \mathbb{C}^{n'})$ using the multiplication and derivation operators. This description can be regarded as an analogue of the Capelli identity (we also give its "higher" generalization in Section 9). Moreover, we give the first fundamental theorem of invariant theory for the action of $SL_n(\mathbb{C})$ on $T(\mathbb{C}^n \otimes \mathbb{C}^{n'})$ as an application of this Capelli type identity. In Section 5, we give some variants of the algebras $\bar{\bar{T}}(\mathrm{v})$ and $\mathscr{L}(V)$. These variants work as a stage to study immanants and quantum immanants in later sections. In Section 6, we introduce some noncommutative immanants and see their fundamental properties. We also introduce the notion of "preimmanants." In Section 7, we express these noncommutative immanants using our algebras. In Section 8, we develop this method to treat the quantum immanants, and show their fundamental properties. Finally, in Section 9, we prove the higher Capelli identity and its analogue on $T(\mathbb{C}^n \otimes \mathbb{C}^{n'})$.

Recently, the author constructed q-analogues of $\bar{\bar{T}}(\mathrm{v})$ and $\mathscr{L}(V)$, which yield a simple proof of the q-Schur–Weyl duality between the quantum enveloping algebra $U_q(\mathfrak{gl}_n)$ and the Iwahori–Hecke algebra of type A. This result will be written somewhere else.

The author hopes that these extensions of the tensor algebra will be useful to study noncommutative invariant theory, various issues related to tensors, and furthermore supersymmetry theory.

1. DEFINITION OF THE ALGEBRA

First we define an algebra $\overline{T}$(v). This is the most fundamental algebra among the extensions of the tensor algebra which we discuss in this article.

1.1.

Let us consider an n dimensional ℂ-vector space V and its tensor algebra $T(V) = \bigoplus_{p\geq 0} T_p(V)$. The homogeneous part $T_p(V)$ of $T(V)$ is the p fold tensor product of V: $T_p(V) = V^{\otimes p}$. We consider the natural (right) action of the symmetric group S_p on this $V^{\otimes p}$. Namely, $\sigma \in S_p$ acts on $v_p \ldots v_1 \in V^{\otimes p}$ by $v_p \cdots v_1 \sigma = v_{\sigma(p)} \cdots v_{\sigma(1)}$.

Here, we omit the symbol " $\otimes$ " for elements of $T(V)$. Moreover, we often employ the numbering of vectors running from right to left, when considering a right action of the symmetric group. We regard $T_p(V) = V^{\otimes p}$ as a right $\mathbb{C}S_p$-module with this action. In addition, for $q \geq 0$, we regard $\mathbb{C}S_{p+q}$ as a $(\mathbb{C}S_p, \mathbb{C}S_{p+q})$-bimodule with left and right multiplications (we embed $\mathbb{C}S_p$ in $\mathbb{C}S_{p+q}$ according to the canonical inclusions $S_0 \subset S_1 \subset \ldots$ of symmetric groups). We consider the tensor product of these right module and bimodule (namely, an induced representation):

$$T_p^{(q)}(V) = V^{\otimes p} \otimes_{\mathbb{C}S_p} \mathbb{C}S_{p+q} \simeq \mathrm{Ind}_{\mathbb{C}S_p}^{\mathbb{C}S_{p+q}} V^{\otimes p}.$$

Since $T_p^{(0)}(V)$ is naturally isomorphic to $T_p(V) = V^{\otimes p}$, we have the inclusions

$$T_p(V) = T_p^{(0)}(V) \subset T_p^{(1)}(V) \subset \cdots. \tag{1.1}$$

Let us consider a much larger space $T_p^{(\infty)}(V) = V^{\otimes p} \otimes_{\mathbb{C}S_p} \mathbb{C}S_\infty \simeq \mathrm{Ind}_{\mathbb{C}S_p}^{\mathbb{C}S_\infty} V^{\otimes p}$. Here S_∞ denotes the infinite symmetric group, namely, the inductive limit of the sequence $S_0 \subset S_1 \subset \ldots$.

Let us denote this $T_p^{(\infty)}(V)$ by $\overline{T}_p(V)$ simply. We can regard $\overline{T}_p(V)$ as the inductive limit of the sequence (1.1).

Noting that $\overline{T}_0(V) \simeq \mathbb{C}S_\infty$, we consider the direct sum of $\overline{T}_0(V), \overline{T}_1(V), \ldots$:

$$\overline{T}(V) = \bigoplus_{p \geq 0} \overline{T}_p(V)$$
.

For this $\overline{T}(v)$, we can naturally define a structure of graded algebra. That is, for

$$\varphi = v_p \cdots v_1 \sigma \in \overline{T}_p(V), \qquad \varphi' = v'_{p'} \cdots v'_1 \sigma' \in \overline{T}_{p'}(V)$$
,

we define the product $\varphi\phi \in \overline{T}_{p+p'}(V)$, by Here α is the group endomorphism of S_∞ defined by

$$\varphi\varphi' = (v_p \cdots v_1 \sigma)(v'_{p'} \cdots v'_1 \sigma') = v_p \cdots v_1 v'_{p'} \cdots v'_1 \alpha^{p'}(\sigma)\sigma'$$

.

Here α is the group endomorphism of S∞ defined by

$$\alpha(\sigma) : k \mapsto \sigma(k-1)+1 \qquad \text{for } k \geq 2,$$

$$1 \mapsto 1,$$

Namely, we put $\alpha(s_i) = s_{i+1}$ for the adjacent transposition $s_i = (i\ i+1)$. We can easily see that this multiplication is well defined and moreover associative.

Let us see some relations for this multiplication. First, for $v, w \in V$, we have

$wv = vws_1$, $s_iv = vs_{i+1}$. (1.2)

Moreover, the defining relations of the symmetric group also hold in $\overline{T}(V)$:

$$s_i^2 = 1, \qquad s_is_{i+1}s_i = s_{i+1}s_is_{i+1}, \qquad s_is_j = s_js_i \qquad \text{for } |i-j| > 1. \tag{1.3}$$

Actually, these relations form the defining relations of the algebra as seen in Theorem 1.2 below.

It is easily seen that $T^{(q)}(V) = \bigoplus_{p \geq 0} T_p^{(q)}(V)$ is a subalgebra of $\overline{T}(V)$. In particular, $T(V) \approx T^{(0)}(V)$ is also a subalgebra, and the restriction of the multiplication of $T(V)$ is equal to the ordinary multiplication of the tensor algebra $T(V)$. Thus, we can regard $T^{(q)}(V)$ and $\overline{T}(V)$ as extensions of the ordinary tensor algebra.

1.2.

Let us consider a "canonical form" of elements in $T^{(q)}(V)$. We denote the set $\{1, \ldots, n\}$ by $[n]$ and consider a sequence $I = (i_1, \ldots, i_p) \in [n]^p$ (we assume that $I, J,$...mean the sequences $I = (i_1,..., i_p), J = (j_1,..., j_p),$...throughout this article). We put $I! = m_1!...m_n!$, where $m_1,..., m_n$ are the multiplicities of $1,..., n$ in the sequence $I = (i_1,..., i_p) \in [n]^p$, respectively. The symmetric group S_p naturally acts on $[n]^p$ by $\sigma(I) = (i_{\sigma(1)},..., i_{\sigma(p)})$. We denote by $(S_p)_I$ the stabilizer subgroup of the sequence $I \in [n]^p$. Namely, we put $(S_p)_I = \{\sigma \in S_p \mid \sigma(I) = I\}$. Then the order of this group $(S_p)_I$ is equal to $I!$, and the element

$$s_I = \frac{1}{I!} \sum_{\sigma \in (S_p)_I} \sigma$$

in $\mathbb{C}S_p$ is idempotent. Moreover, for a basis $e_1,..., e_n$ of V, we have the relation

$$e_{i_p} \cdots e_{i_1} = e_{i_p} \cdots e_{i_1} s_I .$$

From this, we see the following assertion.

Proposition 1.1 By fixing a basis $e_1,..., e_n$ of V, any element in $T_p^{(q)}(V)$ can be expressed uniquely as a sum of elements in the form $e_{i_p}...e_{i_1} t$. Here $I = (i_1,..., i_p)$ is a sequence in $\left(\binom{[n]}{p}\right)$, and t is an element of the left ideal $s_I \mathbb{C}S_{p+q}$ of $\mathbb{C}S_{p+q}$.

Here we put

$$\binom{[n]}{r} = \{(i_1, \ldots, i_r) \in [n]^r \mid i_1 < \cdots < i_r\},$$

$$\left(\binom{[n]}{r}\right) = \{(i_1, \ldots, i_r) \in [n]^r \mid i_1 \leq \cdots \leq i_r\}.$$

Proof of Proposition 1.1

It suffices to show the uniqueness. Indeed, we can change the order of p vectors freely by applying a suitable element of S_p from right, and we can assume that the factor in $\mathbb{C}S_{p+q}$ belongs $s_I \mathbb{C}S_{p+q}$ by (1.4).

The uniqueness is seen as follows. Let $I_1,..., I_r$ be distinct elements of $\left(\binom{[n]}{p}\right)$. Our task is to show $s_{I_i} t_i = 0$ on the assumption $\sum_{i=1}^{r} e_{I_i} t_i = \sum_{i=1}^{r} e_{I_i} s_{I_i} t_i = 0$ with $t_1,..., t_r \in \mathbb{C}S_{p+q}$. Here we denote $e_{i_p}...e_{i_1}$ simply by e_I. First of all, we see $e_{I_1} t_1 = ... = e_{I_r} t_r = 0$ from this assumption. Indeed the right action of S_p on $V^{\otimes p}$ only changes the order of p vectors. Thus it suffices to deduce $s_I t = 0$ from the equality $e_I t = e_I s_I t = 0$. This is done by considering the right cosets of $(S_p)_I$ in S_{p+q}. First, $e_{\sigma 1}(I),\ldots, e_{\sigma r}(I)$ are all equal if $\sigma_1,\ldots, \sigma_r \in S_{p+q}$ are in a same coset. Secondly, $e_{\sigma 1}(I),\ldots, e_{\sigma r}(I)$ are linearly independent if $\sigma_1,\ldots, \sigma_r$ are in different cosets. Thus, when $e_I t = e_I s_I t = 0$ with $t = \sum_{\sigma \in S_{p+q}} c_\sigma \sigma \in \mathbb{C}S_{p+q}$, we have $\sum_{\sigma \in C} c_\sigma = 0$ for each coset $C \in (S_p)_I \setminus S_{p+q}$, namely, $S_I t = 0$.

1.3.

We can also define the algebra $\overline{T}(V)$ in terms of generators and relations.

Theorem 1.2

The algebra $\mathcal{A}_n$ defined by the following generators and relations is isomorphic to $\overline{T}(V)$ by regarding $e_1,\ldots, e_n$ as a basis of V:

$$\textit{generators:}\ e_1, \ldots, e_n, s_1, s_2, \ldots,$$

$$\textit{relations:}\ e_b e_a = e_a e_b s_1, \qquad s_i e_a = e_a s_{i+1},$$

$$s_i^2 = 1, \quad s_i s_{i+1} s_i = s_{i+1} s_i s_{i+1}, \quad s_i s_j = s_j s_i \quad \textit{when } |i-j| > 1.$$

Proof

Noting (1.2) and (1.3), we can consider a natural homomorphism from $\mathcal{A}_n$ onto $\overline{T}(V)$. Moreover, any element in $\mathcal{A}_n$ can be expressed in the form as in Proposition 1.1 by using the first and second relations. Thus this homomorphism is injective.

2. MULTIPLICATIONS AND DERIVATIONS

Some interesting operators naturally act on $\overline{T}(V)$. In this section, we introduce the "multiplication" by $v \in V$ and the "derivation" by $v^* \in V^*$ (V^* is the linear dual of V) and discuss their fundamental properties and quick applications.

2.1.

We denote by $L(\varphi)$ the left multiplication by $\varphi \in$, $\overline{T}(V)$ i.e., we put $L(\varphi)\psi = \varphi\psi$ for φ and $\psi \in \overline{T}(V)$. The cases $\varphi = \sigma \in S_\infty \subset \mathbb{C}S_\infty = \overline{T}_0(V)$ and $\varphi =$

v ∈ V ⊂ $\overline{T}(V)$ are particularly fundamental, because the other cases are generated by these two cases.

In addition to these operators $L(\sigma)$ and $L(v)$, let us introduce the "derivation" $L(v^*)$ by a covector $v^* \in V^*$. Namely, for $v^* \in V^*$, we define the operator $L(v^*)$ on $T(V)$ by

$$L(v^*)v_p \cdots v_1 t = \sum_{k=1}^{p} \langle v^*, v_k \rangle v_p \cdots \hat{v}_k \cdots v_1 \cdot (p \; p-1 \; \cdots \; k+1 \; k)t$$

Here $v_p, \ldots, v_1$ are elements of V, and t is an element of $\mathbb{C}S_\infty$.

Moreover, means that we omit v_k. When $p = 0$, we put $L(v^*)t = 0$.

This definition seems natural, because each term in the right-hand side is obtained by moving v_k to the left end and taking the coupling with v^*. The permutation $(p \; p-1 \ldots k+1 \; k)$ appears following this movement of v_k.

By a direct calculation, we see that this operation is well defined. Note that this well-definedness is equivalent with the fact that $L(v^*)$ is commutative with the right multiplication by $\mathbb{C}S_\infty$.

The following relations are immediate from the definition of L.

Proposition 2.1

The operators $L(\sigma)$, $L(v)$, and $L(v^*)$ with $\sigma \in S_\infty$, $v \in V$, $v^* \in V^*$ commute with the right multiplication by $\mathbb{C}S_\infty$.

Proposition 2.2

For $v \in V$ and $v^* \in V^*$, the operators $L(v)$ and $L(v^*)$ map $T_p^{(q)}(V)$ as

$$L(v) : \; T_p^{(q)}(V) \to T_{p+1}^{(q-1)}(V) \quad \textit{for } q \geq 1,$$

$$T_p^{(0)}(V) \to T_{p+1}^{(0)}(V),$$

$$L(v^*) : \; T_p^{(q)}(V) \to T_{p-1}^{(q+1)}(V) \quad \textit{for } p \geq 1,$$

$$T_0^{(q)}(V) \to \{0\}.$$

Moreover, we have the following commutation relations. The proof is straightforward.

Theorem 2.3

For $v, w \in V$ and $v^*, w^* \in V^*$, we have

$$L(w)L(v) = L(v)L(w)L(s_1),$$

$$L(w^*)L(v^*) = L(s_1)L(v^*)L(w^*),$$

$$L(w^*)L(v) = L(v)L(s_1)L(w^*) + \langle w^*, v\rangle$$

and moreover

$$L(s_i)L(v) = L(v)L(s_{i+1}), \qquad L(v^*)L(s_i) = L(s_{i+1})L(v^*),$$

$$L(s_i)^2 = 1, \qquad L(s_i)L(s_{i+1})L(s_i) = L(s_{i+1})L(s_i)L(s_{i+1}),$$

$$L(s_i)L(s_j) = L(s_j)L(s_i) \qquad when\ |i-j| > 1.$$

We can regard the first three relations as an analogue of the canonical commutation relations (CCR) and the canonical anticommutation relations (CAR). The positions of $L(s_1)$ in the right-hand sides are interesting. We will discuss this analogue further in Section 2.3.

Moreover, we note the commutation relation with the natural action π of $GL(V)$ on $\overline{T}(V)$ as an algebra automorphism. We have the following relations for $v \in V$, $\sigma \in S_{p+q}$, $v^* \in V^*$ as seen from the definition of L:

$$\pi(g)L(v) = L(gv)\pi(g), \qquad \pi(g)L(\sigma) = L(\sigma)\pi(g), \qquad \pi(g)L(v^*) = L({}^t g^{-1}v^*)\pi(g) \, . \quad (2.1)$$

The following formula is also fundamental:

$$L(e^*_{i_1})\cdots L(e^*_{i_p})e_{i_p}\cdots e_{i_1} = I!s_I \, .$$

Moreover, when $I \in \left(\left(\begin{smallmatrix}[n]\\ r\end{smallmatrix}\right)\right)$ and $J \in \left(\left(\begin{smallmatrix}[n]\\ p\end{smallmatrix}\right)\right)$, we have

$$L(e^*_{i_1})\cdots L(e^*_{i_r})e_{j_p}\cdots e_{j_1} = \begin{cases} I!s_I, & I = J, \\ 0, & r \geq p \text{ and } I \neq J \end{cases} \, . \quad (2.2)$$

Indeed, when $r \geq p$ and $I \neq J$, there exists $k \in \{1,\ldots, p\}$ such that the multiplicity of e_k in I is greater than that in J.

Remark. For $\varphi \in T_k(V)$ and $\psi \in \overline{T}(V)$, we have

$$L(v^*)(\varphi\psi) = (L(v^*)\varphi)\psi + \varphi s_k s_{k-1} \cdots s_1 (L(v^*)\psi)$$

$$= (L(v^*)\varphi)\psi + \varphi \cdot (k+1\ k\ k-1\ \ldots\ 2\ 1)(L(v^*)\psi).$$

This can be regarded as an analogue of the Leibniz rule.

2.2.

Let $\mathcal{L}(V)$ be the subalgebra of $\mathrm{End}_{\mathbb{C}}(\overline{T}(V))$ generated by $L(v)$, $L(v^*)$, and $L(\sigma)$ with $v \in V$, $v^* \in V^*$, and $\sigma \in S_\infty$. This algebra $\mathcal{L}(V)$ can be identified with the algebra $\mathcal{B}_n$ defined by the following generators and relations:

$$\begin{aligned} &\textit{generators:}\ e_1, \ldots, e_n, e_1^*, \ldots, e_n^*, s_1, s_2, \ldots, \\ &\textit{relations:}\ e_b e_a = e_a e_b s_1, \quad e_b^* e_a^* = s_1 e_a^* e_b^*, \quad e_b^* e_a = e_a s_1 e_b^* + \delta_{ab}, \\ &s_i e_a = e_a s_{i+1}, \quad e_a^* s_i = s_{i+1} e_a^*, \\ &s_i^2 = 1, \quad s_i s_{i+1} s_i = s_{i+1} s_i s_{i+1}, \quad s_i s_j = s_j s_i \text{ when } |i-j| > 1. \end{aligned} \tag{2.3}$$

Indeed, regarding $e_1, \ldots, e_n$ and $e^*_1, \ldots, e^*_n$ as a basis of V and its dual basis of V^*, we have the following theorem.

Theorem 2.4

The following correspondence induces an isomorphism from $\mathcal{B}_n$ onto $\mathcal{L}(V)$:

This correspondence establishes a homomorphism f from $\mathcal{B}_n$ onto $\mathcal{L}(V)$ as seen from Theorem 2.3. Thus, to prove this theorem, it suffices to show that this f is injective. As a preparation for this, we note the following lemma.

Lemma 2.5

The algebra $\overline{T}(V)$ is isomorphic to the subalgebra $\mathcal{B}_n^\dagger$ of $\mathcal{B}_n$ of $\mathcal{B}_n$ generated by $e_1, \ldots, e_n$ and $s_1, s_2, \ldots$ through the natural correspondence $e_a \to e_a$, $s_i \to s_i$.

Indeed, Theorem 1.2 tells that this correspondence determines a homomorphism from $\overline{T}(V)$ onto $\mathcal{B}_n^\dagger$. This is also injective, because the inverse map is given by $\mathcal{B}_n^\dagger \to \overline{T}(V)$, $\psi \to f(\psi)1$.

Proof of Theorem 2.4

It suffices to show that $f : \mathcal{B}_n \to \mathcal{L}(V)$ is injective. Let us assume that there exists a nonzero $\Phi \in \mathrm{Ker}\, f$. By using the first five relations in (2.3), this can be expressed in the following form with $\psi_I \in \mathscr{B}_n^\dagger \simeq \overline{T}(V)$:

$$\Phi = \sum_{r=0}^{m} \sum_{I \in \left(\binom{[n]}{r}\right)} \psi_I e_{i_1}^* \cdots e_{i_r}^*.$$

Actually, we can take ψ_I to be an element of $\mathscr{B}_n^\dagger s_I$, because we have $e_{i_1}^* \cdots e_{i_r}^* = s_I e_{i_1}^* \cdots e_{i_r}^*$ (recall (1.4)). Let J be one of the shortest sequences among $\{I \mid \psi_I \neq 0\}$ (this set is not empty, because Φ is nonzero). Namely, we fix $J \in \left(\binom{[n]}{p}\right)$ such that $\psi_J \neq 0$ and $\psi_I = 0$ for any , $r < p$. Then, using (2.2), we can calculate $f(\Phi)e_{j_p} \ldots e_{j_1}$ as

$$f(\Phi)e_{j_p} \cdots e_{j_1} = \sum_{r=0}^{m} \sum_{I \in \left(\binom{[n]}{r}\right)} L(\psi_I)L(e_{i_1}^*) \cdots L(e_{i_r}^*)e_{j_p} \cdots e_{j_1} = J!\psi_J.$$

This ψ_J must be equal to zero. This is a contradiction, so that $\mathrm{Ker}\, f = \{0\}$.

We now see another realization of the algebra $\mathcal{B}_n \approx \mathcal{L}(V)$.

Corollary 2.6

The vector space

$$\bigoplus_{p,q\geq 0} V^{\otimes p} \otimes_{\mathbb{C}S_p} \mathbb{C}S_\infty \otimes_{\mathbb{C}S_q} V^{*\otimes q} = \bigoplus_{q\geq 0} \overline{T}(V) \otimes_{\mathbb{C}S_q} V^{*\otimes q}$$

is identified with $\mathcal{L}(V)$ as a vector space through the correspondence

$$v_p \cdots v_1 \sigma v_1^* \cdots v_q^* \mapsto L(v_p) \cdots L(v_1)L(\sigma)L(v_1^*) \cdots L(v_q^*).$$

Here v_i, v^*_i, σ and are elements of V, V^*, and S_∞, respectively.

Indeed, Theorem 2.3 teaches that this map is well-defined and surjective. Moreover, we can check that this is injective in the same way as the proof of Theorem 2.4.

2.3.

The first three relations of Theorem 2.3 are similar to the CCR and the CAR. Thus we can regard $\overline{T}(V)$ as an analogue of the Boson and Fermion Fock spaces. It is also natural to call $L(v)$ and $L(v^*)$ the "creation operator" by v and the "annihilation operator" by v^*, respectively.

More precisely, $\overline{T}(V)$ contains the Boson and Fermion Fock spaces as quotient algebras:

$$\overline{T}(V)/(\sigma - 1) \simeq S(V), \qquad \overline{T}(V)/(\sigma - \mathrm{sgn}(\sigma)) \simeq \Lambda(V).$$

Similarly, $\mathcal{L}(V)$ contains the Weyl algebra and the Clifford algebra as quotient algebras:

$$\mathcal{L}(V)/(\sigma - 1) \simeq \mathrm{Weyl}(V^* \oplus V), \qquad \mathcal{L}(V)/(\sigma - \mathrm{sgn}(\sigma)) \simeq \mathrm{Clifford}(V^* \oplus V)$$

.

Here $(\sigma - 1)$ and $(\sigma - \mathrm{sgn}(\sigma))$ are the two-sided ideals of $\overline{T}(V)$ or $\mathcal{L}(V)$ generated by $\{\sigma - 1 \mid \sigma \in S_\infty\}$ and $\{\sigma - \mathrm{sgn}(\sigma) \mid \sigma \in S_\infty\}$, respectively. The author wonders if the algebras $\overline{T}(V)$ and $\mathcal{L}(V)$ might be useful to deepen supersymmetry theory.

2.4.

These operators $L(v)$ and $L(v^*)$ are useful to express polarization operators on $T(V)$.

Let V be a vector space. The canonical action of $GL(V)$ on V is naturally extended to the actions on $T(V)$ and $\overline{T}(V)$. The infinitesimal action of $\mathfrak{gl}(V)$ on $\overline{T}(V)$ is expressed as

$$\pi(E_{ij}) = L(e_i)L(e_j^*). \tag{2.4}$$

Here E_{ij} $(1 \leq i, j \leq n)$ means the canonical basis of $\mathfrak{gl}(V)$, and $e_1, \dots, e_n$ and $e^*_1, \dots, e^*_n$ mean a basis of V and its dual basis of V^*, respectively. This relation (2.4) is easily seen by a direct calculation:

$$L(e_i)L(e_j^*)v_k \cdots v_1 = \sum_{a=1}^{k} \langle e_j, v_a \rangle v_k \cdots v_{a+1} e_i v_{a-1} \cdots v_1. \tag{2.5}$$

It is natural to regard this as the counterpart of the polarization operator $x_i \frac{\partial}{\partial x_j}$ on the polynomial space $\mathbb{C}[x_1,\ldots,x_n]$ [31]. As seen from Proposition 2.2, we can interpret both sides of the relation (2.4) as linear transformations on $T(V)$.

2.5.

We can also consider an analogue of the Euler operator:

$$A = \sum_{i=1}^{n} L(e_i)L(e_i^*) \qquad (2.6)$$

This satisfies $A\varphi = p\varphi$ for $\varphi \in \overline{T}_p(V)$ as seen from (2.5). We will use this operator (and its variants) to calculate some commutants in Section 3.

$$D = \sum_{i,j=1}^{n} L(e_i)L(e_j)L(e_i^*)L(e_j^*) = \sum_{i,j=1}^{n} L(e_j)L(e_i)L(s_1)L(e_i^*)L(e_j^*).$$

The following variant is also interesting:

This operator acts as the operation "exchange arbitrary two vectors and sum up all results." Namely, we have

$$De_{k_p}\cdots e_{k_1} = \sum_{\substack{1\le a,b\le p\\ a\neq b}} e_{k_p}\cdots \underbrace{e_{k_a}}_{b\text{th}} \cdots \underbrace{e_{k_b}}_{a\text{th}} \cdots e_{k_1}$$

(here we count vectors from right). Thus this operator can be rewritten as

$$D = \frac{1}{(p-2)!} R\left(\sum_{\sigma\in S_p} \sigma^{-1} s_1 \sigma\right).$$

Here $R(t)$ is the right multiplication by $t \in \mathbb{C}S_\infty$. More generally, for $\tau \in S_r$, we have

$$\sum_{I\in[n]^r} L(e_{i_r})\cdots L(e_{i_1})L(\tau)L(e_{i_1}^*)\cdots L(e_{i_r}^*) = \frac{1}{(p-r)!} R\left(\sum_{\sigma\in S_p} \sigma^{-1}\tau\sigma\right).$$

3. DESCRIPTIONS OF COMMUTANTS

Using the multiplications and derivations defined in the previous section, we can describe some commutants of fundamental operators on tensor spaces.

3.1.

First we can prove the Schur–Weyl duality and its generalization by a simple calculation in $\mathcal{L}(V)$.

Theorem 3.1 For l = 0, 1, 2,..., we denote by $\mathscr{L}_l^{(q)} = \mathscr{L}_l^{(q)}(V)$ the set of all linear combinations of the following operators on $T_p^{(q)}(V)$:

$$L(v_l) \cdots L(v_1)L(\sigma)L(v_1^*) \cdots L(v_l^*).$$

Here v_i, v^*_j, and σ are elements of V, V^*, and S_{l+q}, respectively. Then we have the following assertions:

i. The inclusions $\mathscr{L}_0^{(q)} \subset \mathscr{L}_1^{(q)} \subset \cdots \subset \mathscr{L}_p^{(q)}$ hold as operators on $T_p^{(q)}(V)$. Moreover, we have $\mathscr{L}_l^{(q)} = \{0\}$ for $l > p$;

ii. Let R be the right multiplication by $\mathbb{C}S_{p+q}$ on $T_p^{(q)}(V)$. Then $R(\mathbb{C}S_{p+q})$ and $\mathscr{L}_p^{(q)}$ are mutual commutants of each other in $\mathrm{End}_{\mathbb{C}}(T_p^{(q)}(V))$.

Proof. To prove (i), we use the Euler type operator $A = \sum_{i=1}^n L(e_i)L(e_i^*)$ defined in (2.6). As a

consequence of Theorem 2.3, we have the relation for $v \in V$. Using this relation, we have

$$\sum_{k=1}^n L(v_l) \cdots L(v_1)L(e_k)L(\alpha(\sigma))L(e_k^*)L(v_1^*) \cdots L(v_l^*)$$
$$= L(v_l) \cdots L(v_1)AL(\sigma)L(v_1^*) \cdots L(v_l^*)$$
$$= (A - l)L(v_l) \cdots L(v_1)L(\sigma)L(v_1^*) \cdots L(v_l^*).$$

As operators on $T_p^{(q)}(V)$, this is equal to

$$(p-l)L(v_l)\cdots L(v_1)L(\sigma)L(v_1^*)\cdots L(v_l^*),$$

because $A|_{T_p^{(q)}(V)} = p\mathrm{id}_{T_p^{(q)}(V)}$. This calculation means the inclusion $\mathscr{L}_l^{(q)} \subset \mathscr{L}_{l+1}^{(q)}$ for $l < p$. The relation $\mathscr{L}_l^{(q)} = \{0\}$ for $l > p$ is obvious from Proposition 2.2. Thus (i) is proved.

To prove (ii), it suffices to show $R(\mathbb{C}S_{p+q})' = \mathscr{L}_p^{(q)}$ by the double commutant theorem (Theorem 3.3.7 in [3] Here we denote by $\mathcal{D}'$ the commutant of $\mathscr{D} \subset \mathrm{End}_{\mathbb{C}}(T_p^{(q)}(V))$. The inclusion $R(\mathbb{C}S_{p+q})' \supset \mathscr{L}_p^{(q)}$ is immediate from Proposition 2.1, so that we only have to show the reverse inclusion $R(\mathbb{C}S_{p+q})' \subset \mathscr{L}_p^{(q)}$. To show this, we consider a higher analogue of the Euler type operator A:

$$A_p = \frac{1}{p!}\sum_{I\in[n]^p} L(e_{i_p})\cdots L(e_{i_1})L(e_{i_1}^*)\cdots L(e_{i_p}^*). \quad (3.2)$$

Using (3.1) repeatedly, we can express this A_p as

$$A_p = \frac{1}{p!}(A-p+1)(A-p+2)\cdots(A-1)A,$$

so that $A_p\varphi = \varphi$ for $\varphi \in T_p^{(q)}(V)$. Thus, for any $f \in R(\mathbb{C}S_{p+q})'$, we have

$$\begin{aligned}
f(\varphi) &= f(A_p\varphi)\\
&= f\left(\frac{1}{p!}\sum_{I\in[n]^p} L(e_{i_p})\cdots L(e_{i_1})L(e_{i_1}^*)\cdots L(e_{i_p}^*)\varphi\right)\\
&= f\left(\frac{1}{p!}\sum_{I\in[n]^p} e_{i_p}\cdots e_{i_1}t_I\right)\\
&= \frac{1}{p!}\sum_{I\in[n]^p} f(e_{i_p}\cdots e_{i_1}t_I),
\end{aligned}$$

where we denote $L(e_{i_1}^*)\cdots L(e_{i_p}^*)\varphi \in \mathbb{C}S_{p+q}$ by t_I. Since $f \in R(\mathbb{C}S_{p+q})'$, we have

$$f(\varphi) = \frac{1}{p!}\sum_{I\in[n]^p} f(e_{i_p}\cdots e_{i_1})t_I = \frac{1}{p!}\sum_{I\in[n]^p} L(f(e_{i_p}\cdots e_{i_1}))L(e_{i_1}^*)\cdots L(e_{i_p}^*)\varphi,$$

namely,

$$f = \frac{1}{p!}\sum_{I\in[n]^p} L(f(e_{i_p}\cdots e_{i_1}))L(e_{i_1}^*)\cdots L(e_{i_p}^*).$$

This means that f is an element of $\mathscr{L}_p^{(q)}$.

When $q = 0$, Theorem 3.1 is equal to the Schur–Weyl duality, because we have the following relation.

Proposition 3.2 We have $\mathscr{L}_p^{(0)} = \pi(U(\mathfrak{gl}(V)))$.

Proof. As a consequence of Theorem 2.3, we have

$$[L(e_i)L(e_j^*), L(e_a)] = \delta_{ai}L(e_i).$$

Using this commutation relation, we have

$$[L(e_i)L(e_j^*), L(e_a)] = \delta_{ai}L(e_i)$$

$$= L(e_{i_1})L(e_{j_1}^*)\cdot L(e_{i_l})\cdots L(e_{i_2})L(e_{j_2}^*)\cdots L(e_{j_l}^*) + \text{an element of } \mathscr{L}_{l-1}^{(0)}$$

Repeating this, we see that any element of $\mathscr{L}_l^{(0)}$ can be expressed as a polynomial in $\pi(E_{ab}) = L(e_a)L(e_b^*),\ 1 \le a, b \le n$, $1 \le a, b \le n$. Thus we see the inclusion $\mathscr{L}_p^{(0)} \subset \pi(U(\mathfrak{gl}(V)))$.

Conversely, using the same commutation relation, we can express any element in $\pi(U(\mathfrak{gl}(V)))$ as a sum of elements in $\mathscr{L}_0^{(0)}, \mathscr{L}_1^{(0)}, \ldots$. Furthermore, this can be regarded as an element of $\mathscr{L}_p^{(0)}$ by Theorem 3.1 (i).

Remarks

1) When $p = 0$, Theorem 3.1 is equal to the fact that $L(\mathbb{C}S_q)$ and $R(\mathbb{C}S_q)$ are mutual commutants of each other in $\mathrm{End}_{\mathbb{C}}(\mathbb{C}S_q)$.
2) The results in this subsection are generalized to their q-analogues. Namely, we can construct q-analogues of $\overline{\overline{T}}(V)$ and $\mathcal{L}(V)$, which produce a simple proof of the q-Schur–Weyl duality between the quantum enveloping algebra $U_q(\mathfrak{gl}_n)$ and the Iwahori–Hecke algebra of type A. This result will be written somewhere else.
3) For any group G, every map $f: G \to G$ commuting with all right translations is equal to a left translation. This fact is proved quickly as follows. Let x be an element of G. Then we have $f(x) = f(ex)$ with the identity element e. Since f commutes with the right multiplication by x, this $f(ex)$ is equal to $f(e)x$. This means that f is equal to the left multiplication by $f(e)$. It should be noted that the proof of Theorem 3.1 is based on the same principle.

3.2.

We have a similar relation about operators on $\overline{\overline{T}}(V)$.

Theorem 3.3

Let R be the right multiplication by $\mathbb{C}S_\infty$ on $\overline{\overline{T}}(V)$. In addition, we denote by $\mathcal{M}_p = \mathcal{M}_p(V)$ the set of all operators on $\overline{\overline{T}}(V)$ in the form

$$\sum_{I\in[n]^p} L(\psi_I)L(e^*_{i_1})\cdots L(e^*_{i_p}),$$

where ψ_I' s are elements of $\overline{\overline{T}}(V)$. Moreover, we denote by $\mathcal{M} = \mathcal{M}(V)$ the set of all operators in the form $\sum_{p\geq 0} D_p$ with $D_p \in \mathcal{M}_p$. This is an infinite sum in general, but this actually acts as a finite sum, when it is applied to any element of $\overline{\overline{T}}(V)$. Then these $R(\mathbb{C}S_\infty)$ and $\mathcal{M}$ are mutual commutants of each other as operators on $\overline{\overline{T}}(V)$.

The inclusions $R(\mathbb{C}S_\infty)' \supset \mathcal{M}$ and $R(\mathbb{C}S_\infty) \subset \mathcal{M}'$ are obvious, so that it suffices to show the following two propositions.

Proposition 3.4 We have $\mathcal{M}' \subset R(\mathbb{C}S_\infty)$.

Proposition 3.5 We have $R(\mathbb{C}S_\infty)' \subset \mathcal{M}$.

To prove Proposition 3.5, we use the following lemma.

Lemma 3.6 For any $f \in R(\mathbb{C}S_\infty)'$, there exists $D_p \in \mathcal{M}_p$ such that $f(\varphi) = D_p \varphi$ for $\varphi \in \overline{T}_p(V)$.

Proof of Proposition 3.4 Assume that $f \in \mathcal{M}'$. Then we have

$$L(v^*)f(1) = f(L(v^*)1) = f(0) = 0,$$

so that $f(1)$ belongs to $\overline{T}_0(V) = \mathbb{C}S_\infty$. Moreover, we have $f(\varphi) = f(\varphi \cdot 1) = \varphi f(1)$ for $\varphi \in \overline{T}(V)$. From these we see $f \in R(\mathbb{C}S_\infty)$.

Proof of Lemma 3.6

In a way similar to the proof of Theorem 3.1 (ii), we have

$$f(\varphi) = f(A_p \varphi) = \frac{1}{p!} \sum_{I \in [n]^p} L(f(e_{i_p} \cdots e_{i_1}))L(e^*_{i_1}) \cdots L(e^*_{i_p})\varphi$$

for $\varphi \in \overline{T}_p(V)$. Here A_p is the analogue of the Euler operator defined in (3.2). This means the assertion.

Proof of Proposition 3.5

Fix an arbitrary $f \in R(\mathbb{C}S_\infty)'$. We take $D_0 \in \mathcal{M}_0$ such that $f(\varphi_0) = D_0\varphi_0$ for all $\varphi_0 \in \overline{T}_0(V)$. Next, we take $D_1 \in \mathcal{M}_1$ such that $(f - D_0)(\varphi_1) = D_1\varphi_1$ for all $\varphi_1 \in \overline{T}_1(V)$. In this way, we take $D_k \in \mathcal{M}_k$ for $k = 0, 1, 2, \ldots$ such that

$$(f - D_0 - \cdots - D_{r-1})(\varphi_r) = D_r \varphi_r$$

for all $\varphi_r \in \overline{T}_r(V)$. Here we used Lemma 3.6 and the fact $f - D_0 - \ldots - D_{r-1} \in R(\mathbb{C}S_\infty)'$. From this, we can deduce the following relation for $\varphi \in \bigoplus_{k=0}^{r} \overline{T}_k(V)$:

This is proved by induction on r by noting the relation $D_k\varphi_l = 0$ for $\varphi_l \in \overline{T}_l(V)$ when $k > l$. This means that $f = \sum_{k \geq 0} D_k \in \mathcal{M}$.

3.3.

We can also consider an analogue of the Howe duality. The general linear group $GL_n(\mathbb{C})$ naturally acts on the tensor product $W = \mathbb{C}^n \otimes \mathbb{C}^{n'}$ (canonically on the first component $\mathbb{C}^n$ and trivially on the second component $\mathbb{C}^{n'}$). This can be extended to the action ν on $T_p^{(q)}(W)$ naturally. Then we have the following theorem as an analogue of the $(GL_n(\mathbb{C}), GL_{n'}(\mathbb{C}))$ duality on the polynomial space $\mathcal{P}(W)$ due to Howe [5]

Theorem 3.7 Let Q_1 be the operator algebra in $\mathrm{End}_{\mathbb{C}}(T_p^{(q)}(\mathbb{C}^n \otimes \mathbb{C}^{n'}))$ generated by $\nu(GL_n(\mathbb{C}))$ and the right multiplication by S_{p+q}. Moreover, we denote by Q_2 the set of all linear combinations of the operators in the form

$$\Phi_{a_1,\ldots,a_p;b_1,\ldots,b_p}(\sigma) = \sum_{I\in[n]^p} L(w_{i_p a_p}) \cdots L(w_{i_1 a_1}) L(\sigma) L(w^*_{i_1 b_1}) \cdots L(w^*_{i_p b_p}).$$

Here w_{ij} and w^*_{ij} mean the canonical basis of $W = \mathbb{C}^n \otimes \mathbb{C}^{n'}$ and the dual basis, respectively. Moreover, σ is an element of S_{p+q}. Then Q_1 and Q_2 are mutual commutants of each other.

Proof. It is sufficient to show $\mathcal{Q}_1' = \mathcal{Q}_2$, because Q_1 is semisimple.

The proof of the inclusion $\mathcal{Q}_1' \supset \mathcal{Q}_2$ is plane. First, Q_2 commutes with the right multiplication by $\mathbb{C}S_{p+q}$, because $\mathcal{Q}_2 \subset \mathcal{L}_p^{(q)}(W)$. Secondly, Q_2 also commutes with $\nu(GL_n(\mathbb{C}))$ as seen the following calculation for $g \in GL_n(\mathbb{C})$:

$$\begin{aligned} &\nu(g)\Phi_{a_1,\ldots,a_p;b_1,\ldots,b_p}(\sigma)\nu(g^{-1}) \\ &= \sum_{I\in[n]^p} L(\tilde{g}w_{i_p a_p}) \cdots L(\tilde{g}w_{i_1 a_1}) L(\sigma) L({}^t\tilde{g}^{-1} w^*_{i_1 b_1}) \cdots L({}^t\tilde{g}^{-1} w^*_{i_p b_p}) \\ &= \Phi_{a_1,\ldots,a_p;b_1,\ldots,b_p}(\sigma). \end{aligned}$$

The first equality is a consequence of (2.1). Here we regard $g \in GL_n(\mathbb{C})$ as an element of $GL(W)$ through the action of $GL_n(\mathbb{C})$ on W, and denote this image by $\tilde{g}$.

The converse inclusion $\mathcal{Q}_1' \subset \mathcal{Q}_2$ is shown as follows. The first fundamental theorem of the invariant theory for tensor spaces (Theorem

4.3.1 in [3] states that all $GL_n(\mathbb{C})$-invariants in $W^{\otimes p} \otimes W^{*\otimes p}$ are expressed as a linear combination of elements in the form

$$\sum_{I\in[n]^p} w_{i_p a_p} \otimes \cdots \otimes w_{i_1 a_1} \otimes w^*_{i_{\sigma(1)} b_{\sigma(1)}} \otimes \cdots \otimes w^*_{i_{\sigma(p)} b_{\sigma(p)}}$$

.

Thus the following elements with $\tau \in S_{p+q}$ and $\sigma \in S_p$ expand all $GL_n(\mathbb{C})$-invariants in $W^{\otimes p} \otimes \mathbb{C}S_{p+q} \otimes W^{*\otimes p}$:

$$\Psi_{a_1,\dots,a_p;b_1,\dots,b_p}(\tau,\sigma) = \sum_{I\in[n]^p} w_{i_p a_p} \otimes \cdots \otimes w_{i_1 a_1} \otimes \tau \otimes w^*_{i_{\sigma(1)} b_{\sigma(1)}} \otimes \cdots \otimes w^*_{i_{\sigma(p)} b_{\sigma(p)}}$$

.

Here we consider the trivial action of $GL_n(\mathbb{C})$ on $\mathbb{C}S_{p+q}$. Next, we consider the linear map $F : W^{\otimes p} \otimes \mathbb{C}S_{p+q} \otimes W^{*\otimes p} \to \mathscr{L}_p^{(q)}(W)$ defined by the following correspondence:

$$w_p \otimes \cdots \otimes w_1 \otimes \tau \otimes w_1^* \otimes \cdots \otimes w_p^* \mapsto L(w_p)\cdots L(w_1)L(\tau)L(w_1^*)\cdots L(w_p^*)$$

.

This map is surjective and $GL(W)$-equivariant (hence, also $GL_n(\mathbb{C})$-equivariant) as seen from (2.1). Here $GL(W)$ acts on $\mathscr{L}_p^{(q)}(W)$ by conjugation. Thus all $GL_n(\mathbb{C})$-invariants in $\mathscr{L}_p^{(q)}(W)$ comes from $GL_n(\mathbb{C})$-invariants in $W^{\otimes p} \otimes \mathbb{C}S_{p+q} \otimes W^{*\otimes p}$. This inplies $\mathscr{Q}_1' \subset \mathscr{Q}_2$, because the image of $\Psi_{a_1,\dots,a_p;b_1,\dots,b_p}(\tau, \sigma)$ under F is equal to $\Phi_{a_1,\dots,a_p;b_1,\dots,b_p}(\tau\sigma)$.

The following relation is immediate from Theorem 3.7.

Corollary 3.8

For any central element C in the universal enveloping algebra $U(\mathfrak{gl}_n)$, the action of C on $T_p^{(q)}(\mathbb{C}^n \otimes \mathbb{C}^{n'})$ is equal to an operator in Q_2, and this operator is also central in Q_2.

In the next section, we will give an explicit description of this relation in terms of generators as an analogue of the Capelli identity.

4. CAPELLI IDENTITY ON $T(\mathbb{C}N \otimes \mathbb{C}^{N'})$ AND ITS APPLICATION TO INVARIANT THEORY

In this section, we give a Capelli type identity on $T(\mathbb{C}^n \otimes \mathbb{C}^{n'})$ using the multiplication and derivation operators defined in Section 2 (we will give a more general relation, namely a "higher version" in Section 9.2). This can be regarded as a description of the correspondence of invariant operators in Corollary 3.8.

The original Capelli identity played an important role in invariant theory. Our Capelli type identity also has an application to invariant theory. That is, we can determine the $SL_n(\mathbb{C})$-invariants in $T(\mathbb{C}^n \otimes \mathbb{C}^{n'})$ using this Capelli type identity. This application can be regarded as a noncommutative version of a classical result, namely the description of $SL_n(\mathbb{C})$-invariants in the polynomial algebra $(\mathbb{C}^n \otimes \mathbb{C}^{n'})$ using the original Capelli identity (Theorem 2.6.A in [31]).

Let us recall some central elements of $U(\mathfrak{gl}_n)$ called the Capelli elements [1,2,6,25]. Let E_{ij} be the standard basis of $\mathfrak{gl}_n$, and consider the matrix $E = (E_{ij})_{1\le i,j\le n}$ in $\mathrm{Mat}_n(U(\mathfrak{gl}_n))$. We consider the following element in $U(\mathfrak{gl}_n)$:

$$C_n = \text{column-det}(E + \mathrm{diag}(n-1, n-2, \ldots, 0)).$$

Here "column-det" means the "column-determinant." Namely, for a square matrix $X = (X_{ij})_{1\le i,j\le n}$ whose entries are not necessarily commutative, we put

$$\text{column-det}\, X = \sum_{\sigma\in S_n} \mathrm{sgn}(\sigma) X_{\sigma(1)1} X_{\sigma(2)2} \cdots X_{\sigma(n)n}.$$

This C_n is known to be central in $U(\mathfrak{gl}_n)$. We can generalize this to sums of minors:

$$C_r = \sum_{I\in\binom{[n]}{r}} \text{column-det}(E_{II} + \mathrm{diag}(r-1, r-2, \ldots, 0)) \tag{4.1}$$

Here we put $X_{IJ} = (X_{i_a j_b})_{1\le a,b\le r}$ for a matrix $X = (X_{ij})_{1\le i\le n, 1\le j\le n'}$ and sequences $I = (i_1,\ldots,i_r) \in [n]^r$, $J = (j_1,\ldots,j_r) \in [n']^r$. It is known that C_r is also central in $U(\mathfrak{gl}_n)$, and $C_1,\ldots, C_n$ generate the center of $U(\mathfrak{gl}_n)$. We call this C_r the "Capelli element" of degree r.

We work in the same situation as in Theorem 3.7. Namely, we consider the natural action of $GL_n(\mathbb{C})$ on $\mathbb{C}^n \otimes \mathbb{C}^{n'}$. This action can be extended to the actions on $T^{(q)}(\mathbb{C}^n \otimes \mathbb{C}^{n'})$ and moreover on $\bar{T}^{(q)}(\mathbb{C}^n \otimes \mathbb{C}^{n'})$. The infinitesimal action of $\mathfrak{gl}_n$ on $\bar{T}^{(q)}(\mathbb{C}^n \otimes \mathbb{C}^{n'})$ is expressed as follows (cf. Section 2.4):

$$\nu(E_{ij}) = \sum_{a=1}^{n'} L(w_{ia})L(w^*_{ja}) \qquad (4.2)$$

Here w_{ij} and w^*_{ij} mean the canonical basis of $\mathbb{C}^n \otimes \mathbb{C}^{n'}$ and its dual basis, respectively. It is convenient to consider the matrices Z and Z^* in $\mathrm{Mat}_{n,n'}(\mathcal{L}(\mathbb{C}^n \otimes \mathbb{C}^{n'}))$ defined by

$$Z = (L(w_{ij}))_{1\le i\le n, 1\le j\le n'}, \qquad Z^* = (L(w^*_{ij}))_{1\le i\le n, 1\le j\le n'}.$$

Then, we can rewrite (4.2) simply as $\nu(E) = Z\,{}^tZ^*$.

In this situation, we have the following analogue of the Capelli identity (cf. [6, 25]):

Theorem 4.1 For $1 \le r \le p$, we have

$$\nu(C_r) = \sum_{I\in\binom{[n]}{r}} \sum_{J\in\left(\binom{[n']}{r}\right)} \frac{1}{J!}\,\text{column-det}Z_{I^\circ J^\circ}\,\text{column-det}Z^*_{IJ}$$
$$= \frac{1}{r!^2} \sum_{I\in[n]^r} \sum_{J\in[n']^r} \text{column-det}Z_{I^\circ J^\circ}\,\text{column-det}Z^*_{IJ}.$$

Here we put $I^\circ = (i_r,\dots, i_1)$ and $J^\circ = (j_r,\dots, j_1)$ for $I = (i_1,\dots, i_r)$ and $J = (j_1,\dots, j_r)$.

The proof will be given soon in Section 4.3.

Recall the relation between the centers of $U(\mathfrak{gl}_n)$ and Q_2 in Corollary 3.8.

We can regard Theorem 4.1 as an explicit description of this relation in terms of generators. Indeed, the right-hand side of Theorem 4.1 belongs to Q_2(recall the relation $\mathcal{L}^{(q)}_r(\mathbb{C}^n \otimes \mathbb{C}^{n'}) \subset \mathcal{L}^{(q)}_p(\mathbb{C}^n \otimes \mathbb{C}^{n'})$ in Theorem 3.1). This is parallel to the fact that the original Capelli identity describes the correspondence of central elements of the universal enveloping algebras and invariant differential operators in the dual pair $(GL_n(\mathbb{C}), GL_{n'}(\mathbb{C}))$ (cf. Section 9.1; see also [5, 6, 25]).

In Section 9.2, we will give a more detailed description of this correspondence. Namely, we will describe the image of a basis of the center of $U(\mathfrak{gl}_n)$ as a vector space.

4.2.

We recall the exterior calculus to deal with noncommutative determinants. This is a key of the proof of Theorem 4.1 and serves as a prototype of the approach to (noncommutative) immanants in Sections 7–9. See [4, 7-13, 26-30] for the details and further developments of this approach to noncommutative determinants.

It is convenient to introduce the "symmetrized determinant" (or the "double determinant") as another noncommutative determinant. For an $n \times n$matrix X, we put

$$\text{symm-det}\, X = \frac{1}{n!} \sum_{\sigma,\sigma' \in S_n} \operatorname{sgn}(\sigma\sigma'^{-1}) X_{\sigma(1)\sigma'(1)} \cdots X_{\sigma(n)\sigma'(n)} .$$

We also consider a variant with n complex parameters $a_1,\ldots, a_n$:

$$\text{symm-det}(X; a_1, \ldots, a_n) = \frac{1}{n!} \sum_{\sigma,\sigma' \in S_n} \operatorname{sgn}(\sigma\sigma'^{-1}) X_{\sigma(1)\sigma'(1)}(a_1) \cdots X_{\sigma(n)\sigma'(n)}(a_n) .$$

Here $X_{ij}(u)$ means $X_{ij}(u) = X_{ij} + \delta_{ij} u$. Moreover, we put

$$\begin{aligned} \det_r(X; a_1, \ldots, a_r) &= \frac{1}{r!} \sum_{I \in [n]^r} \text{symm-det}(X_{II}; a_1, \ldots, a_r) \\ &= \sum_{I \in \binom{[n]}{r}} \text{symm-det}(X_{II}; a_1, \ldots, a_r). \end{aligned}$$

The second equality holds because symm-det(X_{II}; $a_1,\ldots, a_n$) = 0 for $I! \neq 1$ and

$$\text{symm-det}(X_{II}; a_1, \ldots, a_r) = \text{symm-det}(X_{\sigma(I)\sigma(I)}; a_1, \ldots, a_r)$$

for any $\sigma \in S_r$. We can also express this in terms of the column-determinant:

$$\det_r(X; a_1, \ldots, a_r) = \frac{1}{r!} \sum_{I \in [n]^r} \text{column-det}(X_{II} + 1_{II}\operatorname{diag}(a_1, \ldots, a_r)) .$$

This follows by a simple calculation.

These determinants have their own advantages. For example, the column-determinant is easily calculated, and the symmetrized determinant has some good invariances.

Remark. In [8-12], the matrix functions "column-det," "symm-det," and "det$_r$" are denoted by "det," "Det," and "Det$_r$," respectively.

These noncommutative determinants can be expressed in terms of the exterior calculus as follows. Let V be an n-dimensional ℂ-vector space and consider the exterior algebra $\Lambda(V)$ on V. Let $X = (X_{ij})$ be an $n \times n$ matrix whose entries are elements of an associative ℂ-algebra $\mathcal{A}$. From now on, we calculate in the extended algebra $\Lambda(V) \otimes \mathcal{A}$ in which the two algebras $\Lambda(V)$ and $\mathcal{A}$ commute with each other.

We fix a basis $e_1,\ldots, e_n$ of V and put $\xi_j = \sum_{i=1}^{n} e_i X_{ij} \in \Lambda(V) \otimes \mathcal{A}$. Then we have

$$\xi_{j_1} \cdots \xi_{j_r} = \sum_{I\in[n]^r} e_{i_1} \cdots e_{i_r} \text{column-det}\, X_{IJ} = r! \sum_{I\in\binom{[n]}{r}} e_{i_1} \cdots e_{i_r} \text{column-det}\, X_{IJ}.$$

Indeed, the first equality is seen by a straightforward calculation. The second equality is also immediate, because the column-determinant is alternating under the permutations of rows. This is generalized as follows:

$$\xi_{j_1}(a_1) \cdots \xi_{j_r}(a_r) = \sum_{I\in[n]^r} e_{i_1} \cdots e_{i_r} \text{column-det}(X_{IJ} + 1_{IJ}\text{diag}(a_1, \ldots, a_r))$$

$$= r! \sum_{I\in\binom{[n]}{r}} e_{i_1} \cdots e_{i_r} \text{column-det}(X_{IJ} + 1_{IJ}\text{diag}(a_1, \ldots, a_r)).$$

Here 1 means the unit matrix of size n, and we put $\xi_j(u) = \sum_{i=1}^{n} e_i X_{ij}(u)$.

To deal with the symmetrized determinant, we double the anticommuting generators. Namely, in addition to the basis $e_1,\ldots, e_n$ of V, we consider the dual basis $e^*_1,\ldots, e^*_n$ of V^* and work in $\Lambda(V \oplus V^*) \otimes \mathcal{A}$. We consider the following elements in $\Lambda(V \oplus V^*) \otimes \mathcal{A}$:

$$\tau = \sum_{i=1}^{n} e_i e_i^*, \quad \Xi = \sum_{i,j=1}^{n} e_i X_{ij} e_j^*, \quad \Xi(u) = \sum_{i,j=1}^{n} e_i X_{ij}(u) e_j^* = \sum_{j=1}^{n} \xi_j(u) e_j^* = \Xi + u\tau.$$

Then, a straightforward calculation shows us that

$$\begin{aligned}
&\Xi(a_1)\cdots\Xi(a_r)\\
&=\frac{1}{r!^2}\sum_{I,J\in[n]^r} e_{i_1}e^*_{j_1}\cdots e_{i_r}e^*_{j_r}\sum_{\sigma,\sigma'\in S_r}\mathrm{sgn}(\sigma)\mathrm{sgn}(\sigma')X_{i_{\sigma(1)}j_{\sigma'(1)}}(a_1)\cdots X_{i_{\sigma(r)}j_{\sigma'(r)}}(a_r)\\
&=\sum_{I,J\in\binom{[n]}{r}} e_{i_1}e^*_{j_1}\cdots e_{i_r}e^*_{j_r}\sum_{\sigma,\sigma'\in S_r}\mathrm{sgn}(\sigma)\mathrm{sgn}(\sigma')X_{i_{\sigma(1)}j_{\sigma'(1)}}(a_1)\cdots X_{i_{\sigma(r)}j_{\sigma'(r)}}(a_r).
\end{aligned}$$

It is useful to consider the symmetric bilinear form $\langle\cdot,\ \cdot\rangle$: $\Lambda(V\oplus V^*)\times\Lambda(V\oplus V^*)\to\mathbb{C}$ such that the following forms an orthonormal basis of $\Lambda(V\oplus V^*)$:

$$\left\{e_{i_1}\cdots e_{i_p}e^*_{j_1}\cdots e^*_{j_q}\ \middle|\ I\in\binom{[n]}{p},\ J\in\binom{[n]}{q},\ 0\le p,q\le n\right\}$$

Moreover, we extend this to the bilinear map

$$\langle\cdot,\cdot\rangle:\Lambda(V\oplus V^*)\times\Lambda(V\oplus V^*)\otimes\mathcal{A}\to\mathcal{A},\qquad \langle\varphi,\psi\otimes a\rangle=\langle\varphi,\psi\rangle a\ .$$

Here, φ and ψ are elements of $\Lambda(V\oplus V^*)$, and a is an element of $\mathcal{A}$. Using this bilinear map, we can express some noncommutative determinants as follows. These can be shown by simple calculations.

$$\langle e_{i_1}\cdots e_{i_r},\xi_{j_1}\cdots\xi_{j_r}\rangle=\text{column-det}X_{IJ}\ ,\ (4.4)$$

$$\langle e_{i_1}\cdots e_{i_r},\xi_{j_1}(a_1)\cdots\xi_{j_r}(a_r)\rangle=\text{column-det}(X_{IJ}+1_{IJ}\mathrm{diag}(a_1,\ldots,a_r))\ ,\ (4.5)$$

$$\frac{1}{r!}\langle\tau^{(r)},\Xi(a_1)\cdots\Xi(a_r)\rangle=\det{}_r(X;a_1,\ldots,a_r)\ .\ (4.6)$$

Here $x^{(r)}$ means the divided power $x^{(r)}=\frac{1}{r!}x^r$. Note that

$$\tau^{(r)}=\frac{1}{r!}\sum_{I\in[n]^r}e_{i_1}\cdots e_{i_r}e^*_{i_r}\cdots e^*_{i_1}=\sum_{I\in\binom{[n]}{r}}e_{i_1}\cdots e_{i_r}e^*_{i_r}\cdots e^*_{i_1}\ .$$

4.3.

As a preparation for the proof of Theorem 4.1, we note the following relation.

Proposition 4.2 For *I, J* ∈ [*n*] *r* and σ ∈*S* *r,* we have

$$\text{column-det}(E_{\sigma(I)\sigma'(J)} + 1_{\sigma(I)\sigma'(J)}\text{diag}(r-1, r-2, \ldots, 0))$$
$$= \text{sgn}(\sigma)\text{sgn}(\sigma')\text{column-det}(E_{IJ} + 1_{IJ}\text{diag}(r-1, r-2, \ldots, 0)).$$

In particular, this quantity vanishes when *I*! ≠ 1 or *J*! ≠ 1.

Proof For $\xi_j(u) = \sum_{i=1}^n e_i E_{ij}(u) \in \Lambda(V) \otimes U(\mathfrak{gl}_n)$**, we have the commutation relation**

$$\xi_{j_1}(u+1)\xi_{j_2}(u) = -\xi_{j_2}(u+1)\xi_{j_1}(u) \qquad \textbf{(4.7)}$$

by a direct calculation. Thus we have

$$\xi_{j_{\sigma(1)}}(r-1)\xi_{j_{\sigma(2)}}(r-2)\cdots\xi_{j_{\sigma(r)}}(0) = \text{sgn}(\sigma)\xi_{j_1}(r-1)\xi_{j_2}(r-2)\cdots\xi_{j_r}(0)$$

.

Combining this with (4.5), we see the alternating property under permutations of columns. In addition, the alternating property under permutations of rows holds in general.

The following is immediate from this and (4.3).

Corollary 4.3 We have

$$C_r = \det_r(E; r-1, r-2, \ldots, 0) = \frac{1}{r!}\langle \tau^{(r)}, \Xi(r-1)\Xi(r-2)\cdots\Xi(0)\rangle.$$

Here we define $\Xi(u) \in \Lambda(V \oplus V^*) \otimes U(\mathfrak{gl}_n)$ by $\Xi(u) = \sum_{i,j=1}^n e_i e_j^* E_{ij}(u) = \sum_{j=1}^n \xi_j(u) e_j^*$.

Noting this, let us prove Theorem 4.1. This proof is parallel to that of the original Capelli identity given in [9] (see also [29]).

Proof of Theorem 4.1

We work in $\Lambda(V \oplus V^*) \otimes \mathcal{L}(\mathbb{C}^n \otimes \mathbb{C}^{n'})$. We put

$$\Xi = \Xi_{\nu(E)} = \sum_{i,j=1}^{n} e_i e_j^* \nu(E_{ij}), \qquad \Xi(u) = \Xi_{\nu(E)}(u) = \sum_{i,j=1}^{n} e_i e_j^* \nu(E_{ij}(u))$$

,

so that

$$\Xi(0) = \Xi = \sum_{j=1}^{n'} \eta_j \eta_j^* \tag{4.8}$$

Here we define η_j and η^*_j by $\eta_j = \sum_{a=1}^{n} e_a L(w_{aj})$ and $\eta_j^* = \sum_{a=1}^{n} e_a^* L(w_{aj}^*)$. The following commutation relations are straightforward from Theorem 2.3:

$$\eta_i \eta_j = -\eta_j \eta_i s_1, \qquad \eta_i^* \eta_j^* = -s_1 \eta_j^* \eta_i^*, \qquad \eta_i^* \eta_j = -\eta_j s_1 \eta_i^* - \delta_{ij}\tau$$

. (4.9)

In particular, we have

$$\eta_{j_r} \cdots \eta_{j_1} \sigma = \eta_{j_{\sigma(r)}} \cdots \eta_{j_{\sigma(1)}}, \qquad \sigma^{-1} \eta_{j_1}^* \cdots \eta_{j_r}^* = \eta_{j_{\sigma(1)}}^* \cdots \eta_{j_{\sigma(r)}}^*.$$

(4.10)

Moreover, combining (4.8) and (4.9), we have

$$\Xi(u+1)\eta_j = \eta_j \Xi(u). \tag{4.11}$$

Using the relations (4.8) and (4.11) repeatedly, we have

$$
\begin{aligned}
&\Xi(r-1)\Xi(r-2)\cdots\Xi(2)\Xi(1)\Xi(0)\\
&= \sum_{1\le j_r\le n'} \Xi(r-1)\Xi(r-2)\cdots\Xi(2)\Xi(1)\eta_{j_r}\eta^*_{j_r}\\
&= \sum_{1\le j_r\le n'} \eta_{j_r}\Xi(r-2)\Xi(r-3)\cdots\Xi(1)\Xi(0)\eta^*_{j_r}\\
&= \sum_{1\le j_{r-1},j_r\le n'} \eta_{j_r}\Xi(r-2)\Xi(r-3)\cdots\Xi(1)\eta_{j_{r-1}}\eta^*_{j_{r-1}}\eta^*_{j_r}\\
&= \sum_{1\le j_{r-1},j_r\le n'} \eta_{j_r}\eta_{j_{r-1}}\Xi(r-3)\Xi(r-4)\cdots\Xi(0)\eta^*_{j_{r-1}}\eta^*_{j_r}\\
&\quad\vdots\\
&= \sum_{1\le j_1,\ldots,j_r\le n'} \eta_{j_r}\eta_{j_{r-1}}\cdots\eta_{j_1}\eta^*_{j_1}\cdots\eta^*_{j_{r-1}}\eta^*_{j_r}\\
&= \sum_{J\in\left(\binom{[n']}{r}\right)} \frac{r!}{J!}\eta_{j_r}\eta_{j_{r-1}}\cdots\eta_{j_1}\eta^*_{j_1}\cdots\eta^*_{j_{r-1}}\eta^*_{j_r}.
\end{aligned}
$$

The last equality is seen from (4.10). Taking the coupling with $\tau^{(r)}$ and and using (4.4) and Corollary 4.3, we have the assertion.

4.4.

Using Theorem 4.1 for the case $r = n$, we can determine the $SL_n(\mathbb{C})$-

invariants in $T(\mathbb{C}^n\otimes\mathbb{C}^{n'})$ and $\overline{T}(\mathbb{C}^n\otimes\mathbb{C}^{n'})$ (namely, we have the first fundamental theorem of invariant theory for $SL_n(\mathbb{C})$ in the tensor algebra).

Theorem 4.4 We put

$$\mathcal{J} = \left\{ \text{column-det}\, Y_{I_n^\circ, J^\circ} \,\middle|\, J \in \left(\binom{[n']}{n}\right) \right\}.$$

Here Y denotes the matrix $Y = (w_{ij})_{1\le i\le n,\ 1\le j\le n'}$ in $\mathrm{Mat}_{n,n'}(\overline{T}(\mathbb{C}^n\otimes\mathbb{C}^{n'}))$, and I_n means the sequence $I_n = (1, 2,\ldots, n)$. Then we have the following assertions:

i. The algebra $\overline{T}(\mathbb{C}^n\otimes\mathbb{C}^{n'})^{SL_n(\mathbb{C})}$ is generated by $\mathcal{J}$ and S_∞;

ii. The algebra $T(\mathbb{C}^n \otimes \mathbb{C}^{n'})^{SL_n(\mathbb{C})}$ is spanned by the elements in the form $A_1 \ldots A_k\sigma$. Here $A_1, \ldots, A_k$ are elements of $\mathfrak{J}$, and σ is an element of S_{rn}.

Proof For $\varphi \in \overline{T}_k(\mathbb{C}^n \otimes \mathbb{C}^{n'})^{SL_n(\mathbb{C})}$**, we have** $v(E_{ij})\varphi = \delta_{ij}\frac{k}{n}\varphi$**. This is immediate from the invariance of φ under the action of $SL_n(\mathbb{C})$. Noting this and the definition of the column-determinant, we have**

$$v(C_n)\varphi = \left(\frac{k}{n} + n - 1\right)\left(\frac{k}{n} + n - 2\right)\cdots\left(\frac{k}{n} + 0\right)\varphi.$$

On the other hand, from Theorem 4.1, we see

$$v(C_n)\varphi = \sum_{J \in \left(\left(\binom{|n'|}{n}\right)\right)} \frac{1}{J!}\text{column-det } X_{I_n^\circ J^\circ}\text{column-det } X^*_{I_n J}\varphi.$$

The quantity $X^*_{I_n J}\varphi$ is $SL_n(\mathbb{C})$-invariant, because the operator column-det $X^*_{I_n J}$ commutes with the action of $SL_n(\mathbb{C})$. Thus, we can express φ as $\varphi = \sum_{i=1}^{l} A_i\varphi_i$. Here A_i is an element of J, and φ_i is an element of $\overline{T}_{k-n}(\mathbb{C}^n \otimes \mathbb{C}^{n'})^{SL_n(\mathbb{C})}$. Repeating this argument, we see the assertion by induction on k.

5. OTHER NATURAL EXTENSIONS OF THE TENSOR ALGEBRA

Suggested by the operator algebra L(V) introduced in Section 2, we can consider similar algebras associated to two vector spaces V and W. We also consider the opposite algebra of $\overline{T}(V)$. These algebras are useful to treat the immanants in the next section.

5.1.

We consider the associative algebra defined by the following generators and relations:

$$\begin{aligned} &\textit{generators:}\ e_1, \ldots, e_m, f_1, \ldots, f_n, s_1, s_2, \ldots, \\ &\textit{relations:}\ e_b e_a = e_a e_b s_1, \quad f_b f_a = s_1 f_a f_b, \quad f_b e_a = e_a s_1 f_b + c_{ab}, \\ &s_i e_a = e_a s_{i+1}, \quad f_a s_i = s_{i+1} f_a, \\ &s_i^2 = 1, \quad s_i s_{i+1} s_i = s_{i+1} s_i s_{i+1}, \quad s_i s_j = s_j s_i \text{ when } |i-j| > 1 \end{aligned}$$

.

Here we fix arbitrary complex numbers c_{ab} for $1 \le a \le m$ and $1 \le b \le n$. Note that this $\mathscr{C}'_{m,n}$ is equal to $\mathscr{B}_n \simeq \mathscr{L}(V)$ defined in (2.3), when $m = n$ and $c_{ab} = \delta_{ab}$. It is also fundamental to consider the case $c_{ab} = 0$ for all a and b, and we denote $\mathscr{C}'_{m,n}$ by $\mathscr{C}_{m,n}$ in this case. It appears natural to regard $\mathscr{C}'_{m,n}$ as a "quantization" of $\mathscr{C}_{m,n}$.

Let us regard $e_1,\ldots, e_m$ and $f_1,\ldots, f_n$ as bases of two vector spaces V and W. Moreover, we consider a bilinear map $\langle .,\ .\rangle: W \times V \to \mathbb{C}$ determined by $\langle f_b, e_a\rangle = c_{ab}$. Then, as seen soon, the algebra $\mathscr{C}'_{m,n}$ can be identified with the vector space

$$\overline{T}'(V, W) = \bigoplus_{p,q \ge 0} \overline{T}'_{p,q}(V, W)$$

Here we put

$$\overline{T}'_{p,q}(V, W) = V^{\otimes p} \otimes_{\mathbb{C}S_p} \mathbb{C}S_\infty \otimes_{\mathbb{C}S_q} W^{\otimes q}$$

.

Proposition 5.1 The linear extension of the mapping

$$\overline{T}'(V, W) \to \mathscr{C}'_{m,n}, \qquad v_p \cdots v_1 t w_1 \cdots w_q \mapsto v_p \cdots v_1 t w_1 \cdots w_q$$

defines a linear isomorphism from $\overline{T}'(V, W)$ onto $\mathscr{C}'_{m,n}$. Here v_i, w_j, and t are elements of V, W, and $\mathbb{C}S_\infty$, respectively.

Proof By the definition of $\mathscr{C}'_{m,n}$, this correspondence induces a linear map from $\overline{T}'(V, W)\mathscr{C}'_{m,n}$ onto . Thus, it suffices to show that this is injective.

We consider a sufficiently large vector space U and its linear dual U^*. Then we can identify W and V with subspaces $W \subset U^*$ and $V \subset U$ such that the restriction of the natural coupling of U^* and U to $W \times V$ is equal to the bilinear map $\langle \cdot, \cdot \rangle$. As seen in Corollary 2.6, $\overline{T}'(U, U^*)$ is isomorphic to the operator algebra $\mathscr{L}(U)$. Moreover, we have a natural algebra homomorphism from $\mathscr{C}'_{m,n}$ to $\mathscr{B}_{\dim U}$ defined in (2.3). Similarly, we have an injective linear map from $\overline{T}'(V, W)$ to $\mathscr{L}(U)$. Thus we see that this correspondence between $\overline{T}'(V, W)$ and $\mathscr{C}'_{m,n}$ is injective.

We introduce an algebra structure on the vector space $\overline{T}'(V, W)$ through this identification with $\mathscr{C}'_{m,n}$.

Let us denote $\overline{T}'(V, W)$ and $\overline{T}'_{p,q}(V, W)$ by $\overline{T}(V, W)$ and $\overline{T}_{p,q}(V, W)$, respectively, when $c_{ab} = 0$ (namely, when the bilinear map $\langle \cdot, \cdot \rangle: W \times V \to \mathbb{C}$ is trivial). Then $\overline{T}(V, W) \simeq \mathscr{C}_{m,n}$ becomes a graded algebra, where $\overline{T}_{p,q}(V, W)$ is the homogeneous part of degree (p, q). Note that $\overline{T}'(V, W)$ has only a structure of filtered algebra in general.

The following proposition is easily seen.

Proposition 5.2 Let v and w be elements of V and W, respectively. Then, in the algebra $\overline{T}(V, W)$, the element vw is central, though wv is not central.

5.2.

Let us denote the opposite algebra of $\overline{T}(V)$ by $\overline{T}^0(V)$. Namely, we put

$$\overline{T}^\circ(V) = \bigoplus_{p \geq 0} \overline{T}^\circ_p(V) \qquad \text{with } \overline{T}^\circ_p(V) = \mathbb{C}S_\infty \otimes_{\mathbb{C}S_p} V^{\otimes p}$$

and define the multiplication on $\overline{T}^0(V)$ by the formula

$$\varphi\varphi' = (\sigma v_1 \cdots v_p)(\sigma' v'_1 \cdots v'_{p'}) = \sigma\alpha^p(\sigma') v_1 \cdots v_p v'_1 \cdots v'_{p'}$$

for $\varphi = \sigma v_1 \cdots v_p \in \overline{T}_p^\circ(V)$ and $\varphi' = \sigma' v_1' \cdots v_{p'}' \in \overline{T}_{p'}^\circ(V)$. Then, the following is a subalgebra of $\overline{T}^\circ(V)$ and can be regarded as the opposite algebra of $T^{(q)}(V)$:

$$T^{(q)\circ}(V) = \bigoplus_{p\geq 0} T_p^{(q)\circ}(V) \qquad \text{with } T_p^{(q)\circ}(V) = \mathbb{C}S_{p+q} \otimes_{\mathbb{C}S_p} V^{\otimes p}$$

The algebras $\overline{T}(V)$ and $\overline{T}^\circ(V)$ are isomorphic to special cases of the algebra $\overline{T}(V, W)$:

$$\overline{T}(V) \simeq \overline{T}(V, \{0\}), \qquad \overline{T}^\circ(W) \simeq \overline{T}(\{0\}, W).$$

5.3.

Let us consider some natural bilinear maps. First, we define the bilinear map

$$\langle \cdot, \cdot \rangle : \overline{T}_p^\circ(V^*) \times \overline{T}_p(V) \to \mathbb{C}S_\infty$$

by

$$\langle \sigma' v_1^* \cdots v_p^*, v_p \cdots v_1 \sigma \rangle = \sigma' L(v_1^*) \cdots L(v_p^*) v_p \cdots v_1 \sigma.$$

It is easily seen that this is well defined. Namely, for $\varphi^* \in \overline{T}_p^\circ(V^*)$, $\varphi \in \overline{T}_p(V)$, and σ, $\sigma' \in S_\infty$, we have $\langle \sigma' \varphi^*, \varphi\sigma \rangle = \sigma' \langle \varphi^*, \varphi \rangle \sigma$.

Restricting this to $T_p^{(0)\circ}(V^*) \times T_p^{(0)}(V) = T_p^\circ(V^*) \times T_p(V)$, we obtain a map

$$\langle \cdot, \cdot \rangle : T_p^\circ(V^*) \times T_p(V) \to \mathbb{C}S_p$$

Here $T_p^{(0)\circ}(V^*)$ is isomorphic to $T_p(V^*)$, but we denote this by $T_p^\circ(V^*)$ to emphasize that we regard this as a subspace of $\overline{T}^\circ(V^*)$. Composing this with the irreducible character χ_λ of S_p determined by a partition λ of p, we have a bilinear form. Namely, we define

$$\langle \cdot, \cdot \rangle_\lambda : T_p^\circ(V^*) \times T_p(V) \to \mathbb{C}$$

by $\langle x^*, x\rangle_\lambda = \chi_\lambda(\langle x^*, x\rangle)$

The map $\langle \cdot, \cdot \rangle$ can be generalized to a bilinear map

$$\diamond : \overline{T}_{p,q}(U, V^*) \times \overline{T}_{q,r}(V, W) \to \overline{T}_{p,r}(U, W)$$

defined by

$$(u_p \cdots u_1 \sigma v_1^* \cdots v_q^*) \diamond (v_q \cdots v_1 \sigma' w_1 \cdots w_r) = u_p \cdots u_1 \sigma \langle v_1^* \cdots v_q^*, v_q \cdots v_1 \rangle \sigma' w_1 \cdots w_r.$$

It is easily seen that this multiplication ◊ is well defined and associative.

Suggested by the relation (4.6), we also define $\langle\cdot\rangle : \overline{T}_{p,p}(V, V^*) \to \mathbb{C}S_\infty$ by

$$\langle \Phi \rangle = \frac{1}{p!} \sum_{I \in [n]^p} e_{i_1}^* \cdots e_{i_p}^* \diamond \Phi \diamond e_{i_p} \cdots e_{i_1}.$$

When $\langle \Phi \rangle \in \mathbb{C}S_p$, we denote $\chi_\lambda(\langle \Phi \rangle)$ simply by $\langle \Phi \rangle_\lambda$.

5.4.

We can use these bilinear maps to provide some central elements in $\mathbb{C}S_p$. Let us give some preliminaries to the discussion in Sections 7 and 8.

First, the following is easily seen from the definition of $\langle \cdot \rangle$.

Proposition 5.3 When $\Phi \in \overline{T}_{p,p}(V, V^*)$, the element $\langle\Phi\rangle \in \mathbb{C}S_\infty$ commutes with any element in S_p.

Let us consider $\varphi_J \in T_p(V)$ and $\varphi_J^* \in T_p^\circ(V^*)$ indexed by $J \in [n]^p$ satisfying

$$\varphi_J \sigma = \varphi_{\sigma(J)}, \qquad \sigma^{-1} \varphi_J^* = \varphi_{\sigma(J)}^* \quad . (5.1)$$

For example, $\varphi_J = v_{j_p} \cdots v_{j_1}$ and $\varphi_J^* = v_{j_1}^* \cdots v_{j_p}^*$ satisfy these relations for any $v_1, \ldots, v_n \in V$ and $v_1^*, \ldots, v_n^* \in V^*$. For these, we have the following proposition.

Proposition 5.4 Let and satisfy (5.1).

i. We have $\sigma^{-1}\langle \varphi_I^*, \varphi_J\rangle\sigma' = \langle \varphi_{\sigma(I)}^*, \varphi_{\sigma'(J)}\rangle$.

ii. We have $\sum_{J\in[n]^p}\langle \varphi_J^*, \varphi_J\rangle = \sum_{J\in[n]^p}\langle \varphi_J \varphi_J^*\rangle$. In particular, this quantity is central in $\mathbb{C}S_p$.

Proof The proof of (i) is easy. We can deduce (ii) from

$$\langle \varphi_J \varphi_J^*\rangle = \frac{1}{p!}\sum_{\sigma\in S_p}\sigma^{-1}\langle \varphi_J^*, \varphi_J\rangle\sigma = \frac{1}{p!}\sum_{\sigma\in S_p}\langle \varphi_{\sigma(J)}^*, \varphi_{\sigma(J)}\rangle$$

Here we used (i) and the following relation for $\varphi \in T_p(V)$ and $\varphi^* \in T^0{}_p(V^*)$:

$$\langle \varphi\varphi^*\rangle = \frac{1}{p!}\sum_{\sigma\in S_p}\sigma^{-1}\langle \varphi^*, \varphi\rangle\sigma$$

6. IMMANANTS

The algebras $\bar{T}(V)$, $\bar{T}^0(V)$, and $\bar{T}(V, V^*)$ are useful to treat a matrix function called the “immanant.” In this section, we recall this notion and see its basic properties. We also introduce the notion of “preimmanants.”

6.1.

The immanant is a natural generalization of the determinant and the permanent. Let $\mathcal{A}$ be an associative $\mathbb{C}$-algebra, and consider an $n \times n$ matrix $X = (X_{ij})_{1\le i,j\le n}$ in $\mathrm{Mat}_n(\mathcal{A})$. When $\mathcal{A}$ is commutative, we define the immanant by

$$\mathrm{imm}_\lambda X = \sum_{\sigma\in S_n}\chi_\lambda(\sigma)X_{\sigma(1)1}\cdots X_{\sigma(n)n} = \sum_{\sigma\in S_n}\chi_\lambda(\sigma^{-1})X_{1\sigma(1)}\cdots X_{n\sigma(n)} \tag{6.1}$$

Here λ is a partition of n, and χ_λ is the irreducible character of S_n determined by λ. This is equal to the determinant and the permanent when $\lambda = (1^n)$ and $\lambda = (n)$, respectively:

$$\mathrm{imm}_{(1^n)} = \det, \qquad \mathrm{imm}_{(n)} = \mathrm{per}$$

When $\mathcal{A}$ is not commutative, the two expressions in (6.1) do not coincide in general, so that we should consider the following two functions independently:

$$\text{column-imm}_\lambda X = \sum_{\sigma \in S_n} \chi_\lambda(\sigma) X_{\sigma(1)1} \cdots X_{\sigma(n)n},$$

$$\text{row-imm}_\lambda X = \sum_{\sigma \in S_n} \chi_\lambda(\sigma^{-1}) X_{1\sigma(1)} \cdots X_{n\sigma(n)}$$

Additionally, we consider

$$\text{double-imm}_\lambda X = \frac{\chi_\lambda(1)}{n!} \sum_{\sigma, \sigma' \in S_n} \chi_\lambda(\sigma)\chi_\lambda(\sigma'^{-1}) X_{\sigma(1)\sigma'(1)} \cdots X_{\sigma(n)\sigma'(n)},$$

$$\text{symm-imm}_\lambda X = \frac{1}{n!} \sum_{\sigma, \sigma' \in S_n} \chi_\lambda(\sigma\sigma'^{-1}) X_{\sigma(1)\sigma'(1)} \cdots X_{\sigma(n)\sigma'(n)}.$$

These names are parallel to those of the noncommutative determinants in Section 4 and [7-13]. These four functions are all equal to $\mathrm{imm}_\lambda X$, when $\mathcal{A}$ is commutative.

Proposition 6.1 If $\mathcal{A}$ is commutative, we have

$$\text{column-imm}_\lambda X = \text{row-imm}_\lambda X = \text{double-imm}_\lambda X = \text{symm-imm}_\lambda X$$

This coincidence is easy from the following two relations of the irreducible characters:

$$|S_n| \frac{\chi_\lambda(\sigma)}{\chi_\lambda(1)} = \sum_{\sigma' \in S_n} \chi_\lambda(\sigma\sigma'^{-1})\chi_\lambda(\sigma'), \qquad \chi_\lambda(\sigma) = \chi_\lambda(\sigma^{-1})$$

The first relation holds for general finite groups, and the second one is deduced from the fact that the irreducible representations of S_n are all over $\mathbb{R}$.

Proof of Proposition 6.1 We can show the equality between "double-imm" and "column-imm" as follows:

$$\sum_{\sigma,\sigma'\in S_n} \chi_\lambda(\sigma)\chi_\lambda(\sigma'^{-1})X_{\sigma(1)\sigma'(1)}\cdots X_{\sigma(n)\sigma'(n)}$$

$$= \sum_{\sigma,\sigma'\in S_n} \chi_\lambda(\sigma\sigma')\chi_\lambda(\sigma'^{-1})X_{\sigma\sigma'(1)\sigma'(1)}\cdots X_{\sigma\sigma'(n)\sigma'(n)}$$

$$= \sum_{\sigma,\sigma'\in S_n} \chi_\lambda(\sigma\sigma')\chi_\lambda(\sigma'^{-1})X_{\sigma(1)1}\cdots X_{\sigma(n)n}$$

$$= \sum_{\sigma\in S_n} \frac{n!}{\chi_\lambda(1)}\chi_\lambda(\sigma)X_{\sigma(1)1}\cdots X_{\sigma(n)n}.$$

Here we replaced σ by $\sigma\sigma'$ in the first equality. Moreover, we changed the order of X_{ij}'s in the second equality. The third equality is from the first relation of (6.2).

The proofs of the other equalities are similar, so that we omit them.

The following relation follows from the fact that χ_λ is a class function.

Proposition 6.2 Even if $\mathcal{A}$ is noncommutative, we have

$$\text{symm-imm}_\lambda\sigma^{-1}X\sigma = \text{symm-imm}_\lambda X$$ **for $X \in \mathrm{Mat}_n(\mathcal{A})$ and $\sigma \in S_n$. Here we define the left and right actions of $\sigma \in S_n$ on $\mathrm{Mat}_n(\mathcal{A})$ by**

$$\sigma(X_{ij})_{1\le i,j\le n} = (X_{\sigma^{-1}(i)j})_{1\le i,j\le n}, \qquad (X_{ij})_{1\le i,j\le n}\sigma = (X_{i\sigma(j)})_{1\le i,j\le n}.$$

In particular, we have $\mathrm{imm}_\lambda\sigma^{-1}X\sigma = \mathrm{imm}_\lambda X$, when $\mathcal{A}$ is commutative.

6.2.

The Cauchy–Binet identity is known for determinants. Namely, when the entries of $n \times n'$ matrix X and $n' \times n''$ matrix Y are commutative, we have

$$\det(XY)_{IK} = \sum_{J\in\binom{[n']}{p}} \det X_{IJ} \det Y_{JK}.$$

We have a similar relation for immanants.

Proposition 6.3

Consider two matrices $X \in \mathrm{Mat}_{n,n'}(\mathcal{A})$ and $Y \in \mathrm{Mat}_{n',n''}(\mathcal{A})$. When $\mathcal{A}$ is commutative, we have the following relation for $I \in [n]^p$ and $K \in [n'']^p$:

$$\mathrm{imm}_\lambda (XY)_{IK} = \frac{\chi_\lambda(1)}{p!} \sum_{J\in[n']^p} \mathrm{imm}_\lambda X_{IJ} \mathrm{imm}_\lambda Y_{JK}.$$

The higher Capelli identity given in [21] can be regarded as a noncommutative analogue of this relation (we will discuss this in Section 9).

Proof of Proposition 6.3 This is seen from the following calculation:

$$\begin{aligned}
&\text{double-imm}_\lambda (XY)_{IK} \\
&= \frac{\chi_\lambda(1)}{p!} \sum_{\sigma,\sigma'\in S_p} \chi_\lambda(\sigma)\chi_\lambda(\sigma'^{-1})(XY)_{i_{\sigma(1)}k_{\sigma'(1)}} \cdots (XY)_{i_{\sigma(p)}k_{\sigma'(p)}} \\
&= \frac{\chi_\lambda(1)}{p!} \sum_{\sigma,\sigma'\in S_p} \sum_{J\in[n']^p} \chi_\lambda(\sigma)\chi_\lambda(\sigma'^{-1}) X_{i_{\sigma(1)}j_1} Y_{j_1 k_{\sigma'(1)}} \cdots X_{i_{\sigma(p)}j_p} Y_{j_p k_{\sigma'(p)}} \\
&= \frac{\chi_\lambda(1)}{p!} \sum_{J\in[n']^p} \sum_{\sigma\in S_p} \chi_\lambda(\sigma) X_{i_{\sigma(1)}j_1} \cdots X_{i_{\sigma(p)}j_p} \sum_{\sigma'\in S_p} \chi_\lambda(\sigma'^{-1}) Y_{j_1 k_{\sigma'(1)}} \cdots Y_{j_p k_{\sigma'(p)}} \\
&= \frac{\chi_\lambda(1)}{p!} \sum_{J\in[n']^p} \text{column-imm}_\lambda X_{IJ} \text{row-imm}_\lambda Y_{JK}.
\end{aligned}$$

6.3.

Next we introduce a fundamental matrix function expressed as a sum of immanants. This is invariant under the conjugation by $GL_n(\mathbb{C})$, and this invariance follows from Proposition 6.3 when the matrix entries are commutative.

We assume that $\mathcal{A}$ is not necessarily commutative. A simple calculation shows

$$\sum_{I\in[n]^p} \text{column-imm}_\lambda X_{II} = \sum_{I\in[n]^p} \text{row-imm}_\lambda X_{II} = \sum_{I\in[n]^p} \text{symm-imm}_\lambda X_{II}$$

Let us denote by "$\mathrm{imm}_{\lambda,p} X$" this quantity divided by $p!$. We can also express this as

$$\mathrm{imm}_{\lambda,p} X = \sum_{I\in\left(\binom{[n]}{p}\right)} \frac{1}{I!} \text{symm-imm}_\lambda X_{II}$$

as seen from Proposition 6.2. We can regard this function "$\mathrm{imm}_{\lambda,p}$" as the counterpart of "$\det_p$" given in Section 4. This function is invariant under the conjugation by $GL_n(\mathbb{C})$:

Proposition 6.4 For any $g \in GL_n(\mathbb{C})$, we have $\mathrm{imm}_{\lambda,p} X = \mathrm{imm}_{\lambda,p}\, gXg^{-1}$.

This proposition is immediate from Proposition 6.3, when $\mathcal{A}$ is commutative. We will prove the noncommutative case later as a corollary of Proposition 6.7 (iii).

For $X \in \mathrm{Mat}_n(\mathbb{C})$, this quantity $\mathrm{imm}_{\lambda,p} X$ is equal to the Schur polynomial in eigenvalues of X. We will discuss this issue in Section 6.6.

6.4.

Let us consider the following quantities determined by $X \in \mathrm{Mat}_n(\mathcal{A})$:

$$\text{column-preimm}\, X = \sum_{\sigma \in S_n} \sigma X_{\sigma(1)1} \cdots X_{\sigma(n)n}$$

$$\text{row-preimm}\, X = \sum_{\sigma \in S_n} X_{1\sigma(1)} \cdots X_{n\sigma(n)} \sigma^{-1},$$

$$\text{symm-preimm}\, X = \frac{1}{n!} \sum_{\sigma, \sigma' \in S_n} \sigma X_{\sigma(1)\sigma'(1)} \cdots X_{\sigma(n)\sigma'(n)} \sigma'^{-1}$$

.

These are elements of the algebra $\mathbb{C}S_n \otimes \mathcal{A}$, in which the two algebras $\mathcal{A}$ and $\otimes S_n$ commute with each other. In this article, we call these $\mathbb{C}S_n \otimes \mathcal{A}$-valued matrix functions the "preimmanants" or the "predeterminants," because the immanants and the determinants can be obtained from these by applying the character χ_λ and the signature function:

$$\text{column-imm}_\lambda X = \chi_\lambda(\text{column-preimm}\, X),$$

$$\text{row-imm}_\lambda X = \chi_\lambda(\text{row-preimm}\, X),$$

$$\text{symm-imm}_\lambda X = \chi_\lambda(\text{symm-preimm}\, X).$$

Here we define the linear map $\chi_\lambda\colon \mathbb{C}S_n \otimes \mathcal{A} \to \mathcal{A}$ by $\chi_\lambda(t \otimes a) = \chi_\lambda(t)a$. We also put

$$\text{column-imm}^\circ X = (\text{column-imm} X)^\circ,$$

$$\text{row-imm}^\circ X = (\text{row-imm} X)^\circ,$$

$$\text{symm-imm}^\circ X = (\text{symm-imm} X)^\circ.$$

Here we define the antiautomorphism $t \to t^\circ$ on $\mathbb{C}S_n$ by $\sigma^\circ = \sigma i^{-1}$ for $\sigma \in S_n$. Moreover, we extend this to a linear transformation on $\mathbb{C}S_n \otimes \mathcal{A}$ by $(t \otimes a)^\circ = t^\circ \otimes a$.

When $\mathcal{A}$ is commutative, we have

$$\text{column-preimm}\, X = \text{row-preimm}\, X = \text{symm-preimm}\, X,$$

$$\text{column-preimm}^\circ X = \text{row-preimm}^\circ X = \text{symm-preimm}^\circ X$$

by a simple calculation. Let us denote these two quantities by

respectively. These two are connected by the transposition:

preimm X, preimm^{0}X,

respectively. These two are connected by the transposition: $\text{preimm}^t X = \text{preimm}^0 X$

For the actions of S_n on $\text{Mat}_n(\mathcal{A})$ given in Proposition 6.2, the following relations hold.

Proposition 6.5 We have

$$\text{column-preimm}(\sigma X) = \sigma(\text{column-preimm}\, X),$$

$$\text{row-preimm}(X\sigma) = (\text{row-preimm}\, X)\sigma,$$

$$\text{symm-preimm}(\sigma X \sigma') = \sigma(\text{symm-preimm}\, X)\sigma',$$

$$\text{column-preimm}^\circ(\sigma X) = (\text{column-preimm}^\circ X)\sigma,$$

$$\text{row-preimm}^\circ(X\sigma) = \sigma(\text{row-preimm}^\circ X),$$

$$\text{symm-preimm}^\circ(\sigma X \sigma') = \sigma'(\text{symm-preimm}^\circ X)\sigma.$$

We can regard Proposition 6.2 as a corollary of this proposition.

6.5.

Under the assumption of commutativity, Cauchy–Binet type identities still hold for the preimmanants:

Proposition 6.6 When $\mathcal{A}$ is commutative, we have the following relations for $X \in \text{Mat}_{n,n'}(\mathcal{A})$, $Y \in \text{Mat}_{n',n''}(\mathcal{A})$ and $I \in [n]^p$, $K \in [n'']^p$:

$$\begin{aligned}\operatorname{preimm}(XY)_{IK} &= \frac{1}{p!}\sum_{J\in[n']^p} \operatorname{preimm} X_{IJ} \operatorname{preimm} Y_{JK} \\ &= \sum_{J\in\left(\left(\binom{[n']}{p}\right)\right)} \frac{1}{J!}\operatorname{preimm} X_{IJ} \operatorname{preimm} Y_{JK}.\end{aligned}$$

$$\begin{aligned}\operatorname{preimm}^{\circ}(XY)_{IK} &= \frac{1}{p!}\sum_{J\in[n']^p} \operatorname{preimm}^{\circ} Y_{JK} \operatorname{preimm}^{\circ} X_{IJ} \\ &= \sum_{J\in\left(\left(\binom{[n']}{p}\right)\right)} \frac{1}{J!}\operatorname{preimm}^{\circ} Y_{JK} \operatorname{preimm}^{\circ} X_{IJ}.\end{aligned}$$

The proof is almost the same as that of Proposition 6.3, so that we omit it.

In Section 9, we will obtain noncommutative analogues of this proposition, which are regarded as translations of the higher Capelli identities.

Proposition 6.6 does not hold in general if $\mathcal{A}$ is noncommutative, but we can consider an analogue of $\operatorname{imm}_{\lambda,p}$ without the assumption of commutativity.

Proposition 6.7 Even if $\mathcal{A}$ is noncommutative, the following assertions hold:

We have

$$\begin{aligned}\sum_{I\in[n]^p} \text{column-preimm}\, X_{II} &= \sum_{I\in[n]^p} \text{row-preimm}\, X_{II} \\ &= \sum_{I\in[n]^p} \text{symm-preimm}\, X_{II} \\ &= \sum_{I\in[n]^p} \text{column-preimm}^{\circ} X_{II} \\ &= \sum_{I\in[n]^p} \text{row-preimm}^{\circ} X_{II} \\ &= \sum_{I\in[n]^p} \text{symm-preimm}^{\circ} X_{II}.\end{aligned}$$

We denote by "$\operatorname{preimm}_p X$" this quantity divided by $p!$.

i. The quantity $\operatorname{preimm}_p X$ is commutative with any element in $\mathbb{C}S_p$.

ii. The quantity $\mathrm{preimm}_p X$ is invariant under the conjugation by $GL_n(\mathbb{C})$.

Proof of Proposition 6.7 (i) and (ii) A simple calculation shows us

$$\sum_{I\in[n]^p} \text{column-preimm}\, X_{II} = \sum_{I\in[n]^p} \text{row-preimm}\, X_{II} = \sum_{I\in[n]^p} \text{symm-preimm}\, X_{II}$$

.

This quantity commutes with any element in $\mathbb{C}S_p$. Indeed, we have

$$\sigma^{-1}(\text{symm-preimm}\, X_{II})\sigma = \text{symm-preimm}\,(\sigma^{-1}X_{II}\sigma) = \text{symm-preimm}\, X_{\sigma(I)\sigma(I)}$$

using Proposition 6.5. Since we have $t = t^{\circ}$ for any central element t in $\mathbb{C}S_p$, we conclude (i) and (ii).

We will show Proposition 6.7 (iii) using the algebra $\bar{T}$(V, V*) in Section 7.

6.6.

The immanant and the preimmanant have interesting relations with symmetric functions. Let X be a matrix in $\mathrm{Mat}_n(\mathbb{C})$. Then, as mentioned in Section 1.4 of [21] or Section 2 of [23], the following relation holds.

Proposition 6.8 We have

$$\mathrm{imm}_{\lambda,p} X = s_\lambda(a_1, \ldots, a_n).$$

Here s_λ is the Schur polynomial, and $a_1,\ldots, a_n$ are the eigenvalues of X.

Proof It suffices to consider the case when X is triangular, because both sides are invariant under conjugations. In this case, the assertion follows from the relations

$$s_\lambda = \sum_{\mu\vdash p} K_{\lambda\mu} m_\lambda, \qquad \sum_{\sigma\in S_\mu} \chi_\lambda(\sigma) = |S_\mu| K_{\lambda\mu}$$

,

where the first relation is well known formula (for Example (7.35) in [24]) and the second relation is equivalent

to $\langle \mathrm{Ind}_{S_\mu}^{S_p} 1, \chi_\lambda \rangle = K_{\lambda\mu}$ **(Proposition 7.18.7 in [24]). Here m_λ is the monomial symmetric polynomial, $K_{\lambda\mu}$ is the Kostka number, and S_μ is the Young subgroup $S_{\mu 1} \times \ldots \times S_{\mu l}$ of S_p determined by $\mu = (\mu_1,\ldots, \mu_l) \vdash p$.**

Next, let us interpret the preimmanant in the context of symmetric functions. We denote by Class(S_p) the set of all class functions on S_p. This can be identified with the center $Z\mathbb{C}S_p$ of $\mathbb{C}S_p$ through the canonical inner product on Class(S_p). Moreover, through the Frobenius characteristic map, we can identify Class(S_p) with Symm$_p$, the set of all homogeneous symmetric functions in $x_1, x_2, \ldots$ of degree p (Section 7.18 of [24]). For example, the following three elements correspond to each other through this identification:

$$\tilde{s}_\lambda \in Z\mathbb{C}S_p, \qquad \chi_\lambda \in \mathrm{Class}(S_p), \qquad s_\lambda \in \mathrm{Symm}_p$$

Here we put $\tilde{s}_\lambda = \frac{1}{p!}\sum_{\sigma\in S_p}\chi_\lambda(\sigma)\sigma$ for $\lambda \vdash p$. Then the following lemma is easy.

Lemma 6.9

We have $t = \sum_{\lambda\vdash p}\chi_\lambda(t)\tilde{s}_\lambda$ for any central element t in $\mathbb{C}S_p$.

Applying this, we have

$$\mathrm{preimm}_p X = \sum_{\lambda\vdash p}\mathrm{imm}_{\lambda,p}X\cdot\tilde{s}_\lambda = \sum_{\lambda\vdash p}s_\lambda(a_1,\ldots,a_n)\tilde{s}_\lambda.$$

We can regard this relation as the counterpart of the Cauchy identity (Theorem 7.12.1 in [24]) or the irreducible decomposition of the regular representation of $S_p \times S_p$ on $\mathbb{C}S_p$ through the identification $Z\mathbb{C}S_p \approx$ Class(S_p) $\approx$ Symm$_p$.

We can generalize (6.3) as follows.

Theorem 6.10

We have

$$\mathrm{preimm}_p X = \sum_{\lambda\vdash p}s_\lambda(a_1,\ldots,a_n)\tilde{s}_\lambda = \sum_{\lambda\vdash p}m_\lambda(a_1,\ldots,a_n)\tilde{h}_\lambda$$
$$= \sum_{\lambda\vdash p}\frac{p!}{z_\lambda}p_\lambda(a_1,\ldots,a_n)\tilde{p}_\lambda.$$

Here, m_λ and p_λ are the monomial symmetric polynomial and the power sum symmetric polynomial, respectively. Moreover, $\tilde{h}_\lambda$ and $\tilde{p}_\lambda$ in $Z\mathbb{C}S_p$ denote the counterparts of the complete homogeneous symmetric function h_λ and the power sum symmetric function p_λ in Symm$_p$. Namely, we put

$$\tilde{h}_\lambda = \frac{1}{|S_p||S_\lambda|} \sum_{\sigma \in S_p} \sum_{\tau \in S_\lambda} \sigma^{-1} \tau \sigma, \qquad \tilde{p}_\lambda = \frac{z_\lambda}{p!} \sum_{\sigma \in \mathrm{Conj}_\lambda} \sigma.$$

Here $\mathrm{Conj}_\lambda \subset S_p$ is the conjugacy class corresponding to λ, and z_λ means $z_\lambda = p!/|\mathrm{Conj}_\lambda|$. These $\tilde{h}_\lambda$ and $\tilde{p}_\lambda$ correspond to $\mathrm{Ind}_{S_\lambda}^{S_p} 1$ and $1_{\mathrm{Conj}\lambda}$ in Class(S_p), respectively. Here $1_{\mathrm{Conj}\lambda}$ is the characteristic function of Conj_λ:

$$1_{\mathrm{Conj}_\lambda}(\sigma) = \begin{cases} 1, & \sigma \in \mathrm{Conj}_\lambda, \\ 0, & \sigma \notin \mathrm{Conj}_\lambda. \end{cases}$$

7. IMMANANTS AND EXTENSIONS OF THE TENSOR ALGEBRA

In Section 4, we used the exterior algebra in order to treat noncommutative determinants conveniently. Replacing the exterior algebra by the algebras $\overline{T}(V)$, $\overline{T}^0(V)$, and $\overline{T}(V, V^*)$, we can treat the noncommutative immanants similarly. We can regard this approach as a kind of generating function method. This method plays an important role to treat the quantum immanant in Section 8.

7.1.

Before going to the main subject, we note a natural construction of extensions of bilinear maps. Let U_1, U_2, and U_3 be three $\mathbb{C}$-vector spaces, and consider a bilinear map $\mathbb{C}\cdot$, $\langle \cdot, \cdot \rangle$: $U_1 \times U_2 \to U_3$. We consider another $\mathbb{C}$-vector space $\mathcal{A}$, and define bilinear maps

$$\langle \cdot, \cdot \rangle_R : U_1 \times U_2 \otimes \mathcal{A} \to U_3 \otimes \mathcal{A}, \qquad \langle \cdot, \cdot \rangle_L : U_1 \otimes \mathcal{A} \times U_2 \to U_3 \otimes \mathcal{A}$$

by $\langle u_1, u_2 \otimes a \rangle_R = \langle u_1, u_2 \rangle \otimes a$ and $\langle u_1 \otimes a, u_2 \rangle_L = \langle u_1, u_2 \rangle \otimes a$.

Let us denote these $\langle \cdot, \cdot \rangle_R$ and $\langle \cdot, \cdot \rangle_L$ by the same symbol $\langle \cdot, \cdot \rangle$, when no confusion arises. All bilinear maps which we use in this section are defined from the bilinear maps given in Section 5.3 in this way.

7.2.

First, let us see how to treat the column-immanant and the row-immanant.

Let $\mathcal{A}$ be an associative $\mathbb{C}$ -algebra, and fix $X = (X_{ij})_{1 \leq i, j \leq n} \in \mathrm{Mat}_n(\mathcal{A})$. We consider an n-dimensional $\mathbb{C}$-vector space V and work in the extended algebra $\overline{T}(V) \otimes \mathcal{A}$ in which the two algebras $\overline{T}(V)$ and $\mathcal{A}$ commute with each other.

Fix a basis $e_1, \ldots, e_n$ of V and its dual basis $e^*_1, \ldots, e^*_n$. We consider $\xi_j = \sum_{i=1}^{n} e_i X_{ij} \in \overline{T}(V) \otimes \mathcal{A}$. Then we have

$$\begin{aligned}\langle e^*_{i_1} \cdots e^*_{i_p}, \xi_{j_p} \cdots \xi_{j_1} \rangle &= \langle e^*_{i_1} \cdots e^*_{i_p}, \sum_{K \in [n]^p} e_{k_p} \cdots e_{k_1} X_{k_p j_p} \cdots X_{k_1 j_1} \rangle \\ &= \sum_{\sigma \in S_p} \frac{1}{I!} \langle e^*_{i_1} \cdots e^*_{i_p}, e_{i_{\sigma(p)}} \cdots e_{i_{\sigma(1)}} \rangle X_{i_{\sigma(p)} j_p} \cdots X_{i_{\sigma(1)} j_1} \\ &= \sum_{\sigma \in S_p} s_I \sigma X_{i_{\sigma(p)} j_p} \cdots X_{i_{\sigma(1)} j_1} \\ &= \sum_{\sigma \in S_p} \sigma X_{i_{\sigma(p)} j_p} \cdots X_{i_{\sigma(1)} j_1} \\ &= \text{column-preimm}\, X_{I^\circ J^\circ}.\end{aligned}$$

Here, we see the third equality, because (2.2) implies

$$\langle e^*_{i_1} \cdots e^*_{i_p}, e_{i_{\sigma(p)}} \cdots e_{i_{\sigma(1)}} \rangle = L(e^*_{i_1}) \cdots L(e^*_{i_p}) e_{i_{\sigma(p)}} \cdots e_{i_{\sigma(1)}} = I! s_I \sigma$$

.

Moreover, we see the fourth equality, because $s_I = \frac{1}{I!} \sum_{\sigma' \in (S_p)_I} \sigma'$ and

$$\sum_{\sigma \in S_p} \sigma' \sigma X_{i_{\sigma(p)} j_p} \cdots X_{i_{\sigma(1)} j_1} = \sum_{\sigma \in S_p} \sigma X_{i_{\sigma(p)} j_p} \cdots X_{i_{\sigma(1)} j_1}$$

for $\sigma' \in (S_p)_I$.

We have a similar calculation for $\xi^*_i = \sum_{j=1}^{n} X_{ij} e^*_j \in \overline{T}^\circ(V^*) \otimes \mathcal{A}$. Thus, the following lemma holds.

Lemma 7.1

For $I, J \in [n]^p$, we have

$$\text{column-preimm}\, X_{I^\circ J^\circ} = \langle e^*_{i_1} \cdots e^*_{i_p}, \xi_{j_p} \cdots \xi_{j_1} \rangle,$$
$$\text{column-imm}_\lambda X_{I^\circ J^\circ} = \langle e^*_{i_1} \cdots e^*_{i_p}, \xi_{j_p} \cdots \xi_{j_1} \rangle_\lambda,$$
$$\text{row-preimm}\, X_{IJ} = \langle \xi^*_{i_1} \cdots \xi^*_{i_p}, e_{j_p} \cdots e_{j_1} \rangle,$$
$$\text{row-imm}_\lambda X_{IJ} = \langle \xi^*_{i_1} \cdots \xi^*_{i_p}, e_{j_p} \cdots e_{j_1} \rangle_\lambda.$$

Similarly,
for $\gamma^*_j = \sum_{i=1}^n X_{ij} e^*_i \in \overline{T}^\circ(V^*) \otimes \mathscr{A}$ and
$\gamma_i = \sum_{j=1}^n e_j X_{ij} \in \overline{T}(V) \otimes \mathscr{A}$, we have the following lemma.

Lemma 7.2 For *I, J* ∈ [*n*]^p, we have

$$\text{column-preimm}^\circ X_{IJ} = \langle \gamma^*_{j_p} \cdots \gamma^*_{j_1}, e_{i_1} \cdots e_{i_p} \rangle,$$
$$\text{column-imm}_\lambda X_{IJ} = \langle \gamma^*_{j_p} \cdots \gamma^*_{j_1}, e_{i_1} \cdots e_{i_p} \rangle_\lambda,$$
$$\text{row-preimm}^\circ X_{I^\circ J^\circ} = \langle e^*_{j_1} \cdots e^*_{j_p}, \gamma_{i_p} \cdots \gamma_{i_1} \rangle,$$
$$\text{row-imm}_\lambda X_{I^\circ J^\circ} = \langle e^*_{j_1} \cdots e^*_{j_p}, \gamma_{i_p} \cdots \gamma_{i_1} \rangle_\lambda.$$

7.3. To treat the symmetrized immanant, the algebra $\overline{T}$(V, V*) is useful. We consider the following element of $\overline{T}(V, V^*) \otimes \mathscr{A}$:

$$\Xi = \sum_{i,j=1}^n e_i X_{ij} e^*_j = \sum_{j=1}^n \xi_j e^*_j = \sum_{i=1}^n e_i \xi^*_i.$$

Then, we have

$$\begin{aligned}
&e^*_{i_1}\cdots e^*_{i_p}\diamond \Xi^p \diamond e_{j_p}\cdots e_{j_1}\\
&=\sum_{K,L\in[n]^p} e^*_{i_1}\cdots e^*_{i_p}\diamond e_{k_p}\cdots e_{k_1}X_{k_1l_1}\cdots X_{k_pl_p}e^*_{l_1}\cdots e^*_{l_p}\diamond e_{j_p}\cdots e_{j_1}\\
&=\sum_{\sigma,\sigma'\in S_p}\frac{1}{I!J!}e^*_{i_1}\cdots e^*_{i_p}\diamond e_{i_{\sigma(p)}}\cdots e_{i_{\sigma(1)}}X_{i_{\sigma(1)}j_{\sigma'(1)}}\cdots X_{i_{\sigma(p)}j_{\sigma'(p)}}e^*_{j_{\sigma'(1)}}\cdots e^*_{j_{\sigma'(p)}}\diamond e_{j_p}\cdots e_{j_1}\\
&=\sum_{\sigma,\sigma'\in S_p} s_I\sigma X_{i_{\sigma(1)}j_{\sigma'(1)}}\cdots X_{i_{\sigma(p)}j_{\sigma'(p)}}\sigma'^{-1}s_J\\
&=\sum_{\sigma,\sigma'\in S_p}\sigma X_{i_{\sigma(1)}j_{\sigma'(1)}}\cdots X_{i_{\sigma(p)}j_{\sigma'(p)}}\sigma'^{-1}\\
&=p!\text{symm-preimm}\, X_{IJ}.
\end{aligned}$$

Here, we used Proposition 5.2 to show the first equality. Thus the following lemma holds.

Lemma 7.3

For $I, J \in [n]^p$, we have

$$\text{symm-preimm}\, X_{IJ} = e^*_{i_1}\cdots e^*_{i_p}\diamond \Xi^{(p)}\diamond e_{j_p}\cdots e_{j_1},$$

$$\text{symm-imm}_\lambda X_{IJ} = \chi_\lambda(e^*_{i_1}\cdots e^*_{i_p}\diamond \Xi^{(p)}\diamond e_{j_p}\cdots e_{j_1}).$$

Here, $x^{(k)}$ denotes the divided power: $x^{(k)} = \frac{1}{k!}x^k$.

Moreover, we can express preimm_p and $\text{imm}_{\lambda,p}$ as follows.

Lemma 7.4

We have

$$\text{preimm}_p X = \langle \Xi^{(p)}\rangle, \qquad \text{imm}_{\lambda,p}X = \frac{1}{p!}\sum_{I\in[n]^p}\text{symm-imm}_\lambda X_{II} = \langle \Xi^{(p)}\rangle_\lambda.$$

We can also express "symm-preimm°" using the following element of $\overline{T}(V, V^*)\otimes \mathscr{A}$:

$$\Gamma = \sum_{i,j=1}^{n} e_jX_{ij}e^*_i = \sum_{i=1}^{n}\gamma_i e^*_i = \sum_{j=1}^{n} e_j\gamma^*_j.$$

Lemma 7.5 For $I, J \in [n]^p$, we have

$$\text{symm-preimm}^{\circ} X_{IJ} = e^*_{j_1} \cdots e^*_{j_p} \diamond \Gamma^{(p)} \diamond e_{i_p} \cdots e_{i_1},$$

$$\text{symm-imm}^{\circ}_{\lambda} X_{IJ} = \chi_{\lambda}(e^*_{j_1} \cdots e^*_{j_p} \diamond \Gamma^{(p)} \diamond e_{i_p} \cdots e_{i_1}).$$

Moreover, we have

$$\text{preimm}_p X = \langle \Gamma^{(p)} \rangle, \qquad \text{imm}_{\lambda,p} X = \frac{1}{p!} \sum_{I \in [n]^p} \text{symm-imm}^{\circ}_{\lambda} X_{II} = \langle \Gamma^{(p)} \rangle_{\lambda}.$$

Using Lemma 7.4, we can show Proposition 6.7 (iii), namely the invariance of preimm_p under the conjugation by $GL_n(\mathbb{C})$. This also concludes Proposition 6.4.

**Proof of Proposition 6.7 (iii) Fix $g = (g_{ij}) \in GL_n(\mathbb{C})$, and put $\tilde{e}_j = \sum_{i=1}^n e_i g_{ij}$, so that $\tilde{e}_1, \ldots, \tilde{e}_n$ form a basis of V. We can express the dual basis $\tilde{e}^*_1, \ldots, \tilde{e}^*_n$ as ,
$\tilde{e}^*_i = \sum_{j=1}^n e^*_j g^{ij}$ where g^{ij} is the (i, j)th entries of g^{-1}. We put**

$$\Xi_{gXg^{-1}} = \sum_{i,j=1}^{n} e_i (gXg^{-1})_{ij} e^*_j, \qquad \tilde{\Xi}_X = \sum_{i,j=1}^{n} \tilde{e}_i X_{ij} \tilde{e}^*_j.$$

Then, we can easily see that these are equal: $\Xi_{gXg^{-1}} = \tilde{\Xi}_X$. Thus we have

$$\text{preimm}_p gXg^{-1} = \langle \Xi^{(p)}_{gXg^{-1}} \rangle = \langle \tilde{\Xi}^{(p)} \rangle = \langle \Xi^{(p)} \rangle = \text{preimm}_p X.$$

Indeed, $\langle \cdot \rangle$ does not depend on basis chosen. This means the assertion.

Besides Proposition 6.7 (iii), we can easily see various relations in Section 6 from the lemmas in this section. For example, Proposition 6.5 is immediate from these lemmas and the properties of $\langle \cdot, \cdot \rangle$ and $\langle \cdot, \cdot \rangle$. Proposition 6.7 (ii) is also easy from Proposition 5.3.

This method is more effective against the quantum immanants as seen in the next section.

8. QUANTUM IMMANANTS

The quantum immanants are a generalization of the Capelli elements and form a basis of the center of the universal enveloping algebra $U(\mathfrak{gl}_n)$ as a vector space. These were first introduced in [21] and have been studied by using the R-matrix method, fusion procedure, and representation theory of the Yangian $Y(\mathfrak{gl}_n)$ [16, 17, 20-23]. In this section, we employ the expressions of immanants given in the previous section instead of these standard methods and show various relations for the quantum immanants. Most of these relations were already given in [21] (and the others can also be shown by the same method). Thus it is not that our approach is stronger than established ones, but the author thinks that our method has its own advantage as a natural analogue of the usual generating function method. In other words, our approach is regarded as an advanced version of the exterior calculus in the study of the Capelli type identities in [4, 7-13, 26-30] (and also in Section 4 of this article).

It is also interesting that the notion of "quantum preimmanant" naturally appears.

8.1.

First, we recall some notions and facts of from the representation theory of the symmetric group. See [14, 21] for the details.

We identify a partition with the corresponding Young diagram. For a cell (i, j) of a Young diagram, we call the number $j - i$ the content of the cell. For example, for the Young diagram corresponding to (4, 3, 1), each cell has the following value as its content:

$$\begin{matrix} 0 & 1 & 2 & 3 \\ -1 & 0 & 1 & \\ -2 & & & \end{matrix}$$

Let $\mathrm{STab}(\lambda)$ be the set of all standard tableaux of shape λ. For $T \in \mathrm{STab}(\lambda)$, we put $c_T(k) = j - i$, when the kth cell is (i, j). For example, we have $c_T(1) = 0$, $c_T(2) = -1$, and $c_T(3) = 1$ for the tableau

$$T = \begin{matrix} 1 & 3 \\ 2 & \end{matrix} \in \mathrm{STab}(2, 1).$$

We consider the Young orthogonal form ρ_λ of the irreducible representation of S_p determined by a partition $\lambda \vdash p$ [14, 21]. We denote by v_T the vector in the Young orthogonal basis corresponding to $T \in \mathrm{STab}(\lambda)$. Moreover, let $\rho_\lambda(\sigma)_{TT'}$ be the (T, T')th matrix entry of $\rho_\lambda(\sigma)$. Namely, we define this by

$$\rho_\lambda(\sigma)v_T = \sum_{T' \in \mathrm{STab}(\lambda)} \rho_\lambda(\sigma)_{TT'} v_{T'}.$$

We can describe the matrix entries of $\rho_\lambda(s_i)$ as follows. We put $r_T(i) = c_T(i+1) - c_T(i)$. Then, for $T \in \mathrm{STab}(\lambda)$, we have

$$\rho_\lambda(\sigma)v_T = \sum_{T' \in \mathrm{STab}(\lambda)} \rho_\lambda(\sigma)_{TT'} v_{T'}.$$

when $s_i T$ is non-standard (in this case, $r_T(i)$ is equal to ± 1). Here we consider the natural action of permutation in S_p on a tableau. When $T' = s_i T$ is standard, we have

$$\begin{pmatrix} \rho_\lambda(s_i)v_T & \rho_\lambda(s_i)v_{T'} \end{pmatrix} = \begin{pmatrix} v_T & v_{T'} \end{pmatrix} \begin{pmatrix} r^{-1} & \sqrt{1-r^{-2}} \\ \sqrt{1-r^{-2}} & -r^{-1} \end{pmatrix},$$

where $r = r_T(i)$. Using these and the recurrence formula (8.1), we can prove Proposition 8.1 by induction on i.

Finally, we consider the Jucys–Murphy elements $x_k \in \mathbb{C}S_p$ [15, 21]:

$$x_k = \sum_{i=1}^{k-1} (i\ k) = (1\ k) + (2\ k) + \cdots + (k-1\ k).$$

For $k = 1$, we put $x_1 = 0$. Then $\rho_\lambda(x_i)$ is diagonal, and its entries are expressed in terms of contents.

Proposition 8.1 We have $\rho_\lambda(x_i)v_T = c_T(i)v_T$, namely, $\rho_\lambda(x_i)_{TT'} = \delta_{TT'} c_T(i)$.

This can be proved by using the recurrence formula

$$x_{i+1} = s_i x_i s_i + s_i \tag{8.1}$$

and the description of the matrix entries of $\rho_\lambda(s_i)$ above.

8.2.

The main objects of this section are the following elements in the universal enveloping algebra $U(\mathfrak{gl}_n)$:

$$G_\lambda = \frac{\chi_\lambda(1)}{p!} \sum_{J\in[n]^p} \sum_{\sigma\in S_p} \rho_\lambda(\sigma)_{TT} E_{j_{\sigma(1)}j_1}(c_T(1)) E_{j_{\sigma(2)}j_2}(c_T(2)) \cdots E_{j_{\sigma(p)}j_p}(c_T(p))$$

$$= \frac{\chi_\lambda(1)}{p!} \sum_{J\in[n]^p} \sum_{\sigma\in S_p} \rho_\lambda(\sigma^{-1})_{TT} E_{j_p j_{\sigma(p)}}(c_T(p)) \cdots E_{j_2 j_{\sigma(2)}}(c_T(2)) E_{j_1 j_{\sigma(1)}}(c_T(1)),$$

$$G_\lambda^\circ = \frac{\chi_\lambda(1)}{p!} \sum_{J\in[n]^p} \sum_{\sigma\in S_p} \rho_\lambda(\sigma^{-1})_{TT} E_{j_{\sigma(p)}j_p}(-c_T(p)) \cdots E_{j_{\sigma(2)}j_2}(-c_T(2)) E_{j_{\sigma(1)}j_1}(-c_T(1)$$

$$= \frac{\chi_\lambda(1)}{p!} \sum_{J\in[n]^p} \sum_{\sigma\in S_p} \rho_\lambda(\sigma)_{TT} E_{j_1 j_{\sigma(1)}}(-c_T(1)) E_{j_2 j_{\sigma(2)}}(-c_T(2)) \cdots E_{j_p j_{\sigma(p)}}(-c_T(p)).$$

Here λ is a partition of p, and T is a standard tableau of shape λ. We will see that these G_λ and G^0_λ are central in $U(\mathfrak{gl}_n)$, and these expressions do not depend on $T \in \mathrm{STab}(\lambda)$. These can also be expressed as

$$G_\lambda = \chi_\lambda(G_p), \qquad G_\lambda^\circ = \chi_\lambda(G_p^\circ),$$

where we define G_p and G_p and $G_p^\circ \in \mathbb{C}S_p \otimes U(\mathfrak{gl}_n)$ as follows (these are also central):

$$G_p = \frac{1}{p!} \sum_{J\in[n]^p} \sum_{\sigma\in S_p} \sigma E_{j_{\sigma(1)}j_1}(x_1) E_{j_{\sigma(2)}j_2}(x_2) \cdots E_{j_{\sigma(p)}j_p}(x_p)$$

$$= \frac{1}{p!} \sum_{J\in[n]^p} \sum_{\sigma\in S_p} E_{j_p j_{\sigma(p)}}(x_p) \cdots E_{j_2 j_{\sigma(2)}}(x_2) E_{j_1 j_{\sigma(1)}}(x_1)\sigma^{-1},$$

$$G_p^\circ = \frac{1}{p!} \sum_{J\in[n]^p} \sum_{\sigma\in S_p} E_{j_{\sigma(p)}j_p}(-x_p) \cdots E_{j_{\sigma(2)}j_2}(-x_2) E_{j_{\sigma(1)}j_1}(-x_1)\sigma^{-1}$$

$$= \frac{1}{p!} \sum_{J\in[n]^p} \sum_{\sigma\in S_p} \sigma E_{j_1 j_{\sigma(1)}}(-x_1) E_{j_2 j_{\sigma(2)}}(-x_2) \cdots E_{j_p j_{\sigma(p)}}(-x_p).$$

We will show the equivalence of these various expressions later.

We can regard G_λ and G^o_λ (resp., G_p and G^0_p) as modifications of $\mathrm{imm}_{\lambda,p} X$(resp., $\mathrm{preimm}_p X$). Indeed, the highest terms of these elements with respect to the canonical filtration of $U(\mathfrak{gl}_n)$ are equal to $\mathrm{imm}_{\lambda,p} E$ and $\mathrm{preimm}_p E$, respectively. To be more specific, G_p and G^0_p are obtained by modifying $\frac{1}{p!}\sum_I \mathrm{preimm}\, X_{II}$ and $\frac{1}{p!}\sum_I \mathrm{preimm}\, X_{II}$, respectively. It is interesting that two different modifications naturally appear.

The second central element G^0_λ of $U(\mathfrak{gl}_n)$ is known as the "quantum immanant" [21] (this is denoted by $\mathbb{S}_\lambda$ in [21]). Thus, it is natural to call G^0_p the "quantum preimmanant." As seen in Section 9, we have beautiful Capelli type identities for the quantum immanant G^0_λ (the higher Capelli identities). This fact seems to indicate that G^0_p is more fundamental than G_p, though "preimm " is simpler than "preimm°" in Proposition 6.6.

It is easily seen that these two are related by the automorphism of $U(\mathfrak{gl}_n)$ defined by $E_{ij} \to -E_{ji}$ as follows: $G_p \mapsto (-)^p G^\circ_p$.

Remark. When $\lambda = (1^p)$, we can express G_λ and G°_λ as follows:

$$G_{(1^p)} = \frac{1}{p!} \sum_{J \in [n]^p} \text{column-det}(E_{JJ} - \text{diag}(0, 1, \ldots, n-1)),$$

$$G^\circ_{(1^p)} = \frac{1}{p!} \sum_{J \in [n]^p} \text{column-det}(E_{JJ} + \text{diag}(n-1, n-2, \ldots, 0)).$$

In particular, $G^\circ_{(p)}$ is equal to the Capelli element C_p seen in Section 4. Similarly, we can express $G_{(p)}$ and $G^\circ_{(p)}$ in terms of column-permanent ($G^\circ_{(p)}$ was first given and studied in [19]).

Let us express these elements using the algebras $\overline{\overline{T}}(V)$, $\overline{\overline{T}}^\circ(V)$, and $\overline{\overline{T}}(V, V)$. Under these expressions, we can easily show various relations for these elements.

8.3.

Let V be an n-dimensional $\mathbb{C}$-vector space with basis $e_1, \ldots, e_n$, and consider $\xi_j = \sum_{i=1}^n e_i E_{ij} \in \overline{T}(V) \otimes U(\mathfrak{gl}_n)$. Then we have the commutation relation

$$\begin{aligned}\xi_j \xi_l &= \sum_{i,k=1}^{n} e_i e_k E_{ij} E_{kl} \\ &= \sum_{i,k=1}^{n} e_i e_k (E_{kl} E_{ij} + E_{il}\delta_{kj} - E_{kj}\delta_{il}) \\ &= \xi_l \xi_j \cdot (1\ 2) + \xi_l e_j - e_l \xi_j,\end{aligned} \tag{8.2}$$

namely,

$$\xi_j(\xi_l + e_l \cdot (1\ 2)) = \xi_l(\xi_j + e_j \cdot (1\ 2)) \cdot (1\ 2).$$

This can be generalized as follows.

Lemma 8.2 We put $\xi_j(u) = \sum_{i=1}^{n} e_i E_{ij}(u) = \xi_j + e_j u$ **. Then we have**

$$\xi_i(y_l)\xi_j(y_{l+1}) = \xi_j(y_l)\xi_i(y_{l+1})s_1.$$

Here we define $y_k \in \mathbb{C}S_p$ as follows (for $k = 1$, we put $y_1 = 0$):

$$y_k = \sum_{i=1}^{k-1}(1\ i+1) = (1\ 2) + (1\ 3) + \cdots + (1\ k).$$

This relation can be regarded as an analogue of (4.7) in the exterior algebra. Before proving Lemma 8.2, we consider $x_k^\circ \in \mathbb{C}S_p$ defined by

$$x_k^\circ = \sum_{i=1}^{k-1}(p-i+1\ p-k+1).$$

For $k = 1$, we put $x^\circ_1 = 0$. The elements x_k, x°_k, and y_k are connected by the relations

$$\varepsilon^{-1} x_i \varepsilon = x_i^\circ, \qquad \alpha^{p-i}(y_i) = x_i^\circ,$$

where ϵ is the following element in S_p:

$$\varepsilon = \begin{pmatrix} 1 & 2 & \ldots & p-1 & p \\ p & p-1 & \ldots & 2 & 1 \end{pmatrix}.$$

Proof of Lemma 8.2

By a simple calculation, we have

$$(\xi_i + e_i y_l)(\xi_j + e_j y_{l+1})$$
$$= \xi_i\xi_j + e_i e_j \alpha(y_l) y_{l+1} + e_i \xi_j \alpha(y_l) + \xi_i e_j y_{l+1}$$

$$= \{\xi_j\xi_i s_1 + \xi_j e_i - e_j\xi_i\} + e_j e_i s_1\alpha(y_l)y_{l+1} + \xi_j e_i s_1\alpha(y_l) + e_j\xi_i s_1 y_{l+1}$$
$$= \xi_j\xi_i s_1 + e_j e_i s_1\alpha(y_l)y_{l+1} + \xi_j e_i s_1(\alpha(y_l) + s_1) + e_j\xi_i s_1(y_{l+1} - s_1). \tag{8.4}$$

Here, in the second equality, we used (8.2). Since we have

$$s_1\alpha(y_l)s_1 = y_{l+1} - s_1, \quad s_1 y_{l+1} s_1 = \alpha(y_l) + s_1, \quad \alpha(y_l)y_{l+1} = y_{l+1}\alpha(y_l),$$
$$s_1\alpha(y_l)y_{l+1} = (y_{l+1} - s_1)s_1 y_{l+1} = y_{l+1}s_1(y_{l+1} - s_1) = y_{l+1}\alpha(y_l)s_1,$$

the last line of (8.4) is equal to

$$\xi_j\xi_i s_1 + e_j e_i\alpha(y_l)y_{l+1}s_1 + \xi_j e_i y_{l+1}s_1 + e_j\xi_i\alpha(y_l)s_1 = (\xi_j + e_j y_l)(\xi_i + e_i y_{l+1})s_1 .$$

This means the assertion.

We see the following from Lemma 8.2, because $\sigma\xi_i(y_{k+1}) = \xi_i(y_{k+1})\alpha(\sigma)$ for $\sigma \in S_k$.

Corollary 8.3 For $\sigma \in S_p$, we have

$$\xi_{i_p}(y_1)\xi_{i_{p-1}}(y_2)\cdots\xi_{i_1}(y_p)\sigma = \xi_{i_{\sigma(p)}}(y_1)\xi_{i_{\sigma(p-1)}}(y_2)\cdots\xi_{i_{\sigma(1)}}(y_p) .$$

Namely, the element $\varphi_J = \xi_{j_p}(y_1)\xi_{j_{p-1}}(y_2)\cdots\xi_{j_1}(y_p)$ satisfies the relation (5.1).

Recalling Proposition 5.4(ii), we put

$$
\begin{aligned}
G^{IJ} &= \langle e_{i_1}^* \cdots e_{i_p}^*, \xi_{j_p}(y_1)\xi_{j_{p-1}}(y_2) \cdots \xi_{j_1}(y_p) \rangle, \\
G_p &= \frac{1}{p!} \sum_{J \in [n]^p} G^{JJ} \\
&= \frac{1}{p!} \sum_{J \in [n]^p} \langle e_{j_1}^* \cdots e_{j_p}^*, \xi_{j_p}(y_1)\xi_{j_{p-1}}(y_2) \cdots \xi_{j_1}(y_p) \rangle \\
&= \frac{1}{p!} \sum_{J \in [n]^p} \langle \xi_{j_p}(y_1)\xi_{j_{p-1}}(y_2) \cdots \xi_{j_1}(y_p) e_{j_1}^* \cdots e_{j_p}^* \rangle \\
&= \frac{1}{p!} \left\langle \sum_{I, J \in [n]^p} e_{i_p} \cdots e_{i_1} E_{i_p j_p}(x_1^\circ) \cdots E_{i_1 j_1}(x_p^\circ) e_{j_1}^* \cdots e_{j_p}^* \right\rangle.
\end{aligned}
$$

The last equality is seen from the second relation of (8.3). Moreover, we put

$$
G_\lambda = \chi_\lambda(G_p) = \frac{1}{p!} \sum_{J \in [n]^p} \chi_\lambda(G^{JJ}).
$$

We aim to prove the following theorem.

Theorem 8.4 The element G_p is central in $\mathbb{C}S_p \otimes U(\mathfrak{gl}_n)$.

Corollary 8.5 The element G_λ is central in $U(\mathfrak{gl}_n)$.

First, the following is a consequence of Proposition 5.4 (ii).

Proposition 8.6 The element G_p commutes with any element in $\mathbb{C}S_p$.

Thus it suffices to show the commutativity with elements of $U(\mathfrak{gl}_n)$. For this, we note the following relation. This is seen by a straightforward calculation (see [13, 21] for the details of the proof).

Lemma 8.7 The matrix E satisfies the following relation for any $g \in GL_n(\mathbb{C})$:

$$
\mathrm{Ad}(g)E = {}^t g \cdot E \cdot {}^t g^{-1}.
$$

Here $\mathrm{Ad}(g)E$ means the matrix $(\mathrm{Ad}(g)E_{ij})_{1 \le i, j \le n}$.

Combining this lemma and Proposition, we can show the invariance of G_p under the conjugation by $GL_n(\mathbb{C})$ in a way similar to that of Proposition 6.7 (iii).

Theorem 8.8 We have $\mathrm{Ad}(g)G_p = G_p$ for any $g \in GL_n(\mathbb{C})$. In particular, G_p commutes with any element of $\mathfrak{gl}_n$.

Theorem 8.4 follows from this and Proposition 8.6.

Proof of Theorem 8.8 Using Lemma 8.7, we have

$$\begin{aligned}
&\mathrm{Ad}(g) \sum_{I,J\in[n]^p} e_{i_p} \cdots e_{i_1} E_{i_p j_p}(x_1^\circ) \cdots E_{i_1 j_1}(x_p^\circ) e_{j_1}^* \cdots e_{j_p}^* \\
&= \sum_{I,J\in[n]^p} e_{i_p} \cdots e_{i_1} ({}^t g \cdot E \cdot {}^t g^{-1})_{i_p j_p}(x_1^\circ) \cdots ({}^t g \cdot E \cdot {}^t g^{-1})_{i_1 j_1}(x_p^\circ) e_{j_1}^* \cdots e_{j_p}^* \\
&= \sum_{I,J\in[n]^p} \tilde{e}_{i_p} \cdots \tilde{e}_{i_1} E_{i_p j_p}(x_1^\circ) \cdots E_{i_1 j_1}(x_p^\circ) \tilde{e}_{j_1}^* \cdots \tilde{e}_{j_p}^*.
\end{aligned}$$

Here we put $\tilde{e}_i = \sum_{j=1}^n e_j g_{ij}$ and $\tilde{e}_i^* = \sum_{i=1}^n e_i^* g^{ij}$, so that these are dual bases of each other. Applying ⟨•⟩ to this and using Proposition, we obtain the assertion, because ⟨•⟩ does not depend on basis chosen.

Let us find other expressions of G_p. From Proposition 5.4 (i), we see

$$G^{\sigma(I)\sigma'(J)} = \sigma^{-1} G^{IJ} \sigma'.$$

Noting this, we can express G^{IJ} and G_p as follows.

Proposition 8.9 We have

$$G^{IJ} = \sum_{\sigma\in S_p} \sigma E_{i_{\sigma(p)} j_p}(x_1^\circ) E_{i_{\sigma(p-1)} j_{p-1}}(x_2^\circ) \cdots E_{i_{\sigma(1)} j_1}(x_p^\circ)$$

$$= \sum_{\sigma\in S_p} \sigma E_{i_{\sigma(1)}j_1}(x_1)E_{i_{\sigma(2)}j_2}(x_2)\cdots E_{i_{\sigma(p)}j_p}(x_p),$$

$$G_p = \frac{1}{p!^2}\sum_{I\in[n]^p}\sum_{\sigma,\sigma'\in S_p} \sigma E_{i_{\sigma(p)}i_{\sigma'(p)}}(x_1^\circ)\cdots E_{i_{\sigma(1)}i_{\sigma'(1)}}(x_p^\circ)\sigma'^{-1}$$

$$= \frac{1}{p!^2}\sum_{I\in[n]^p}\sum_{\sigma,\sigma'\in S_p} \sigma E_{i_{\sigma(1)}i_{\sigma'(1)}}(x_1)\cdots E_{i_{\sigma(p)}i_{\sigma'(p)}}(x_p)\sigma'^{-1}.$$

Proof Using the second relation of (8.3), we have

$$\varphi_J = \sum_{I\in[n]^p} e_{i_p}e_{i_{p-1}}\cdots e_{i_1}E_{i_pj_p}(x_1^\circ)E_{i_{p-1}j_{p-1}}(x_2^\circ)\cdots E_{i_1j_1}(x_p^\circ)$$

.

The first expression of G^{IJ} follows from this in a way similar to the proof of Lemma 7.1. Moreover, using (8.5) and the first relation of (8.3), we have

$$G^{IJ} = \varepsilon G^{\varepsilon(I)\varepsilon(J)}\varepsilon^{-1}$$

$$= \varepsilon\sum_{\sigma\in S_p} \sigma E_{i_{\varepsilon\sigma(p)}j_{\varepsilon(p)}}(x_1^\circ)E_{i_{\varepsilon\sigma(p-1)}j_{\varepsilon(p-1)}}(x_2^\circ)\cdots E_{i_{\varepsilon\sigma(1)}j_{\varepsilon(1)}}(x_p^\circ)\varepsilon^{-1}$$

$$= \sum_{\sigma\in S_p} \varepsilon\sigma\varepsilon^{-1}E_{i_{\varepsilon\sigma(p)}j_{\varepsilon(p)}}(x_1)E_{i_{\varepsilon\sigma(p-1)}j_{\varepsilon(p-1)}}(x_2)\cdots E_{i_{\varepsilon\sigma(1)}j_{\varepsilon(1)}}(x_p).$$

Replacing σ with $\varepsilon^{-1}\sigma\varepsilon$, we have

$$G^{IJ} = \sum_{\sigma\in S_p} \sigma E_{i_{\sigma\varepsilon(p)}j_{\varepsilon(p)}}(x_1)E_{i_{\sigma\varepsilon(p-1)}j_{\varepsilon(p-1)}}(x_2)\cdots E_{i_{\sigma\varepsilon(1)}j_{\varepsilon(1)}}(x_p)$$

$$= \sum_{\sigma\in S_p} \sigma E_{i_{\sigma(1)}j_1}(x_1)E_{i_{\sigma(2)}j_2}(x_2)\cdots E_{i_{\sigma(p)}j_p}(x_p).$$

This means the second expression of G^{IJ}. The expressions of G_p are shown similarly.

We can express G_λ simply in terms of contents of the Young diagram λ as follows.

Theorem 8.10 For any $T \in$ STab(λ), we have

$$G_\lambda = \frac{\chi_\lambda(1)}{p!} \sum_{I\in[n]^p} \sum_{\sigma\in S_p} \rho_\lambda(\sigma)_{TT} E_{i_{\sigma(1)}i_1}(c_T(1)) \cdots E_{i_{\sigma(p)}i_p}(c_T(p))$$

$$= \frac{\chi_\lambda(1)}{p!} \langle \Xi(c_T(1)) \cdots \Xi(c_T(p)) \rangle_T$$

$$= \frac{1}{p!} \langle \Xi(c_T(1)) \cdots \Xi(c_T(p)) \rangle_\lambda .$$

In particular, these expressions do not depend on T. Here we define $\Xi(u)$ by $\sum_{i,j=1}^n e_i E_{ij}(u) e_j^* = \Xi + u\tau$, and we put $\langle \Phi \rangle_T = \rho_\lambda(\langle\langle \Phi \rangle\rangle)_{TT}$ *for* $\Phi \in T_{p,p}(V, V^*)$.

Proof **Note that $\rho_\lambda(A)$ is a "scalar matrix" (namely, $\rho_\lambda(A)_{TT'} = \delta_{TT'} \frac{1}{\chi_\lambda(1)} \chi_\lambda(A)$) by Schur's lemma, when $A \in \mathbb{C}S_p \otimes \mathcal{A}$ commutes with any element in $\mathbb{C}S_p$. Thus we see the last equality of the assertion, because $\langle \Xi(c_T(1)) \ldots \Xi(c_T(p)) \rangle$ commutes with any element in $\mathbb{C}S_p$. Similarly, we have $G_\lambda = \chi_\lambda(1)\rho_\lambda(G_p)_{TT}$. Noting this and the relation $\rho_\lambda(x_i)_{TT} = c_T(i)$ in Proposition 8.1, we can express G_λ as**

$$G_\lambda = \frac{\chi_\lambda(1)}{p!} \sum_{I\in[n]^p} \sum_{\sigma\in S_p} \rho_\lambda(\sigma E_{i_{\sigma(1)}i_1}(x_1) \cdots E_{i_{\sigma(1)}i_p}(x_p))_{TT}$$

$$= \frac{\chi_\lambda(1)}{p!} \sum_{I\in[n]^p} \sum_{\sigma\in S_p} \rho_\lambda(\sigma)_{TT} E_{i_{\sigma(1)}i_1}(c_T(1)) \cdots E_{i_{\sigma(p)}i_p}(c_T(p))$$

This can be rewritten as

$$G_\lambda = \frac{\chi_\lambda(1)}{p!} \sum_{J\in[n]^p} \left\langle e_{j_p}^* \cdots e_{j_1}^*, \xi_{j_1}(c_T(1)) \cdots \xi_{j_p}(c_T(p)) \right\rangle_T$$

$$= \frac{\chi_\lambda(1)}{p!} \sum_{I\in[n]^p} \left\langle \xi_{j_1}(c_T(1)) \cdots \xi_{j_p}(c_T(p)) e_{j_p}^* \cdots e_{j_1}^* \right\rangle_T$$

$$= \frac{\chi_\lambda(1)}{p!} \langle \Xi(c_T(1)) \cdots \Xi(c_T(p)) \rangle_T .$$

Here we put $\langle \varphi^*, \varphi \rangle_T = \rho_\lambda(\langle \varphi^*, \varphi \rangle)_{TT}$ for $\varphi^* \in T_p^\circ(V^*)$ and $\varphi \in T_p(V)$.

8.4.

By a similar calculation, we can obtain expressions of G_λ corresponding to the row-immanant as follows. We consider $\xi_i^*(u) = \sum_{j=1}^n E_{ij}(u) e_j^*$ in $\overline{T}^\circ(V^*) \otimes U(\mathfrak{gl}_n)$. Then the following relation holds.

Lemma 8.11 We have

$$\xi_i^*(y_{l+1})\xi_j^*(y_l) = s_1 \xi_j^*(y_{l+1})\xi_i^*(y_l)$$

and in particular

$$\sigma^{-1} \xi_{i_1}^*(y_p)\xi_{i_2}^*(y_{p-1}) \cdots \xi_{i_p}^*(y_1) = \xi_{i_{\sigma(1)}}^*(y_p)\xi_{i_{\sigma(2)}}^*(y_{p-1}) \cdots \xi_{i_{\sigma(p)}}^*(y_1)$$

Noting this, we put

$$\begin{aligned} G'^{IJ} &= \langle \xi_{i_1}^*(y_p)\xi_{i_2}^*(y_{p-1}) \cdots \xi_{i_p}^*(y_1), e_{j_p} \cdots e_{j_1} \rangle \\ &= \sum_{\sigma \in S_p} E_{i_1 j_{\sigma(1)}}(x_p^\circ) E_{i_2 j_{\sigma(2)}}(x_{p-1}^\circ) \cdots E_{i_p j_{\sigma(p)}}(x_1^\circ)\sigma^{-1} \\ &= \sum_{\sigma \in S_p} E_{i_p j_{\sigma(p)}}(x_p) E_{i_{p-1} j_{\sigma(p-1)}}(x_{p-1}) \cdots E_{i_1 j_{\sigma(1)}}(x_1)\sigma^{-1} \end{aligned}$$

And

$$G'_p = \frac{1}{p!} \sum_{I \in [n]^p} G'^{II}, \qquad G'_\lambda = \chi_\lambda(G'_p)$$

These G'_p and G_λ' are central in $\mathbb{C}S_p \otimes U(\mathfrak{gl}_n)$ and $U(\mathfrak{gl}_n)$, respectively. We can also express these as

$$G'_p = \frac{1}{p!^2} \sum_{I \in [n]^p} \sum_{\sigma, \sigma' \in S_p} \sigma E_{i_{\sigma(p)} i_{\sigma'(p)}}(x_p) \cdots E_{i_{\sigma(1)} i_{\sigma'(1)}}(x_1)\sigma'^{-1} \tag{8.6}$$

$$\begin{aligned}G'_\lambda &= \frac{\chi_\lambda(1)}{p!}\sum_{I\in[n]^p}\sum_{\sigma\in S_p} E_{i_p i_{\sigma(p)}}(c_T(p))\cdots E_{i_1 i_{\sigma(1)}}(c_T(1))\rho_\lambda(\sigma^{-1})_{TT}\\ &= \frac{\chi_\lambda(1)}{p!}\langle\Xi(c_T(p))\cdots\Xi(c_T(1))\rangle_T\\ &= \frac{1}{p!}\langle\Xi(c_T(p))\cdots\Xi(c_T(1))\rangle_\lambda .\end{aligned} \tag{8.7}$$

Hence we have $G_\lambda = G'_\lambda$, because $\Xi(c_T(p))\ldots\Xi(c_T(1)) = \Xi(c_T(1))\ldots\Xi(c_T(p))$. Noting Lemma 6.9, we also have $G_p = G'_p$

Remark

It seems not easy to see this equality $G_p = G'_p$ without using the expressions of G_λ and G'_λ in terms of contents.

8.5.

We can deal with $G^o{}_\lambda$ and $G^{o\prime}{}_\lambda$ similarly. We define $\gamma_j^*(u) \in \overline{T}^\circ(V^*)\otimes U(\mathfrak{gl}_n)$ and $\gamma_j(u) \in \overline{T}(V)\otimes U(\mathfrak{gl}_n)$ by

$$\gamma_j^*(u) = \sum_{i=1}^n E_{ij}(u)e_i^*, \qquad \gamma_i(u) = \sum_{j=1}^n e_j E_{ij}(u).$$

Then we have the following commutation relations.

Lemma 8.12 We have

$$\gamma_i^*(-y_{l+1})\gamma_j^*(-y_l) = s_l\gamma_j^*(-y_{l+1})\gamma_i^*(-y_l), \qquad \gamma_i(-y_l)\gamma_j(-y_{l+1}) = \gamma_j(-y_l)\gamma_i(-y_{l+1})s_l$$

and in particular

$$\begin{aligned}\sigma^{-1}\gamma_{j_1}^*(-y_p)\gamma_{j_2}^*(-y_{p-1})\cdots\gamma_{j_p}^*(-y_1) &= \gamma_{j_{\sigma(1)}}^*(-y_p)\gamma_{j_{\sigma(2)}}^*(-y_{p-1})\cdots\gamma_{j_{\sigma(p)}}^*(-y_1),\\ \gamma_{i_p}(-y_1)\gamma_{i_{p-1}}(-y_2)\cdots\gamma_{i_1}(-y_p)\sigma &= \gamma_{i_{\sigma(p)}}(-y_1)\gamma_{i_{\sigma(p-1)}}(-y_2)\cdots\gamma_{i_{\sigma(1)}}(-y_p).\end{aligned}$$

Noting this, we put

$$G^{\circ IJ} = \langle\gamma_{j_1}^*(-y_p)\gamma_{j_2}^*(-y_{p-1})\cdots\gamma_{j_p}^*(-y_1), e_{i_p}\cdots e_{i_1}\rangle$$

$$= \sum_{\sigma\in S_p} E_{i_{\sigma(1)}j_1}(-x_p^\circ)E_{i_{\sigma(2)}j_2}(-x_{p-1}^\circ)\cdots E_{i_{\sigma(p)}j_p}(-x_1^\circ)\sigma^{-1}$$

$$= \sum_{\sigma\in S_p} E_{i_{\sigma(p)}j_p}(-x_p)E_{i_{\sigma(p-1)}j_{p-1}}(-x_{p-1})\cdots E_{i_{\sigma(1)}j_1}(-x_1)\sigma^{-1}.$$

$$G^{\circ\prime IJ} = \langle e_{j_1}^* \cdots e_{j_p}^*, \gamma_{i_p}(-y_1)\gamma_{i_{p-1}}(-y_2)\cdots\gamma_{i_1}(-y_p)\rangle$$

$$= \sum_{\sigma\in S_p} \sigma E_{i_p j_{\sigma(p-1)}}(-x_1^\circ)E_{i_{p-1}j_{\sigma(p-2)}}(-x_2^\circ)\cdots E_{i_1 j_{\sigma(1)}}(-x_p^\circ)$$

$$= \sum_{\sigma\in S_p} \sigma E_{i_1 j_{\sigma(1)}}(-x_1)E_{i_2 j_{\sigma(2)}}(-x_2)\cdots E_{i_p j_{\sigma(p)}}(-x_p)$$

and moreover

$$G_p^\circ = \frac{1}{p!}\sum_{I\in[n]^p} G^{\circ II}, \qquad G_p^{\circ\prime} = \frac{1}{p!}\sum_{I\in[n]^p} G^{\circ\prime II}, \qquad G_\lambda^\circ = \chi_\lambda(G_p^\circ), \qquad G_\lambda^{\circ\prime} = \chi_\lambda(G_p^{\circ\prime}).$$

Then we have

$$G_p^\circ = \frac{1}{p!^2}\sum_{J\in[n]^p}\sum_{\sigma,\sigma'\in S_p} \sigma' E_{j_{\sigma(p)}j_{\sigma'(p)}}(-x_p)\cdots E_{j_{\sigma(1)}j_{\sigma'(1)}}(-x_1)\sigma^{-1}, \tag{8.8}$$

$$G_p^{\circ\prime} = \frac{1}{p!^2}\sum_{J\in[n]^p}\sum_{\sigma,\sigma'\in S_p} \sigma' E_{j_{\sigma(1)}j_{\sigma'(1)}}(-x_1)\cdots E_{j_{\sigma(p)}j_{\sigma'(p)}}(-x_p)\sigma^{-1}, \tag{8.9}$$

$$\begin{aligned} G_\lambda^\circ &= \frac{\chi_\lambda(1)}{p!}\sum_{J\in[n]^p}\sum_{\sigma\in S_p} E_{j_{\sigma(p)}j_p}(-c_T(p))\cdots E_{j_{\sigma(1)}j_1}(-c_T(1))\rho_\lambda(\sigma^{-1})_{TT} \\ &= \frac{\chi_\lambda(1)}{p!}\langle\Xi(-c_T(p))\Xi(-c_T(p-1))\cdots\Xi(-c_T(1))\rangle_T \\ &= \frac{1}{p!}\langle\Xi(-c_T(p))\Xi(-c_T(p-1))\cdots\Xi(-c_T(1))\rangle_\lambda. \end{aligned} \tag{8.10}$$

$$\begin{aligned} G_\lambda^{\circ\prime} &= \frac{\chi_\lambda(1)}{p!} \sum_{J\in[n]^p} \sum_{\sigma\in S_p} \rho_\lambda(\sigma)_{TT} E_{j_1 j_{\sigma(1)}}(-c_T(1)) \cdots E_{j_p j_{\sigma(p)}}(-c_T(p)) \\ &= \frac{\chi_\lambda(1)}{p!} \langle \Xi(-c_T(1))\Xi(-c_T(2)) \cdots \Xi(-c_T(p)) \rangle_T \\ &= \frac{1}{p!} \langle \Xi(-c_T(1))\Xi(-c_T(2)) \cdots \Xi(-c_T(p)) \rangle_\lambda . \end{aligned} \tag{8.11}$$

Combining these, we see that $G_\lambda^\circ = G_\lambda^{\circ\prime}$ and in particular $G_p^\circ = G_p^{\circ\prime}$. The quantity $G_\lambda^\circ = G_\lambda^{\circ\prime}$ is known as the "quantum immanants" [21, 23]. Thus it is natural to call $G_p^\circ = G_p^{\circ\prime}$ the "quantum preimmanant."

We can also deduce these relations from the results in Sections 8.3 and 8.4 by applying the automorphism of $U(gl_n)$defined by $E_{ij} \to -E_{ji}$.

Most of various relations in this Section were already given in [21Okounkov, A. (1996).Quantum immanants and higher Capelli identities.*Transform. Groups* 1 (1): 99– 126. [Google Scholar]]. However, it is interesting that we can deduce these relations by simple calculations using our algebras.

8.6.

Combining these relations, we can also express G_λ° as

$$G_\lambda^\circ = \sum_{I\in\left(\binom{[n]}{p}\right)^\circ} \frac{1}{I!} \sum_{\sigma\in S_p} \rho_\lambda(\sigma E_{i_1 i_{\sigma(1)}}(-x_1) \cdots E_{i_p i_{\sigma(p)}}(-x_p)). \tag{8.12}$$

Here we put

$$\left(\binom{[n]}{p}\right)^\circ = \{(i_1, \ldots, i_p) \in [n]^p \mid i_1 \geq \cdots \geq i_n\}.$$

This expression was given in [21]. Using this expression, we can calculate the eigenvalue of G_λ° on the irreducible representation π_μ of $\mathfrak{gl}_n$ determined by a partition μ. The result is as follows (see Section 3.7 of [21] for the details of this calculation):

$$\pi_\mu(G_\lambda) = \sum_T \prod_\alpha (\lambda_{T(\alpha)} - c_\alpha)$$

Here T runs over all reverse semistandard tableaux of shape λ with entries in {1,..., n}, and α runs over all cells in the Young tableau μ. Moreover, T(α) means the number in α, and c_α means the content of the cell α.

Similarly, G_λ is expressed as

$$G_\lambda = \sum_{I \in \left(\binom{[n]}{p}\right)} \frac{1}{I!} \sum_{\sigma \in S_p} \rho_\lambda(\sigma E_{i_{\sigma(1)} i_1}(x_1) \cdots E_{i_{\sigma(p)} i_p}(x_p)), \tag{8.13}$$

and we can calculate its eigenvalue as follows:

$$\pi_\mu(G_\lambda) = \sum_T \prod_\alpha (\lambda_{T(\alpha)} + c_\alpha)$$

The proof is almost the same. Here T runs over all semistandard tableaux of shape λ with entries in {1,..., n}, and α runs over all cells in the Young tableau μ.

On one hand, these expressions (8.12) and (8.13) can be regarded as the counterparts of the definition of the Capelli elements (4.1), and we can calculate the eigenvalues easily under these expressions. On the other hand, the expressions of G_p, G_λ, and G°_p, G°_λ given in Proposition 8.9, Theorem 8.10, and (8.6), (8.7), (8.8), (8.10), (8.9), (8.11) are corresponding to the expression of the Capelli elements given in Corollary 4.3. Under these expressions, we can show the centrality of these elements easily.

Similar phenomena on contrastive expressions are also known in the universal enveloping algebras $U(\mathfrak{o}_n)$ and $U(\mathfrak{sp}_n)$. First, central elements of $U(\mathfrak{o}_n)$ expressed in terms of the column-determinant were given in [30], nd we can easily calculate their eigenvalues. These elements can also be expressed in terms of the symmetrized determinant, and we can easily show their centrality under this expression [11]. Similar central elements are given in $U(\mathfrak{sp}_n)$ in terms of permanents [12]. It is natural to expect good bases of the centers of $U(\mathfrak{o}_n)$ and $U(\mathfrak{sp}_n)$ which contains these elements and can be regarded as analogues of the quantum immanants. The author hopes that our extensions of the tensor algebra are useful to construct such bases.

9. HIGHER CAPELLI IDENTITY AND ITS ANALOGUE ON TENSOR ALGEBRAS

The method in the previous section is also useful to show a Capelli type identity for the quantum immanants, namely, the "higher Capelli identity."

Several proofs are already known for this identity [17, 20-22], and the proof given in [17] is particularly simple. Our proof is similarly simple and parallel to the easy proof of the original Capelli identities using the exterior calculus given in [11] or [9] (cf. Section 4 in this article).

Furthermore, we give the "higher version" of Theorem 4.1 in $\mathcal{L}(\mathbb{C}^n \otimes \mathbb{C}^{n'})$. We can also assemble them in terms of the quantum preimmanants.

9.1.

We work in the following situation. The general linear group $GL_n(\mathbb{C})$ naturally acts on $\mathbb{C}^n \otimes \mathbb{C}^{n'}$ as in Section 4 and, moreover, the space $(\mathbb{C}^n \otimes \mathbb{C}^{n'})$ of all polynomial functions on $\mathbb{C}^n \otimes \mathbb{C}^{n'}$. The infinitesimal action ν is expressed as follows:

$$\nu(E_{ij}) = \sum_{k=1}^{n'} x_{ik} \partial_{jk}.$$

Here x_{ij} is the standard coordinate of $\mathbb{C}^n \otimes \mathbb{C}^{n'}$, and ∂_{ij} means the partial differentiation $\partial_{ij} = \frac{\partial}{\partial x_{ij}}$. Let us express this relation as $\nu(E) = X\,{}^t\partial$ using the following matrices $E \in \mathrm{Mat}_n(U(\mathfrak{gl}_n))$ and $X, \partial \in \mathrm{Mat}_{n,n'}(\mathcal{PD}(\mathbb{C}^n \otimes \mathbb{C}^{n'}))$:

$$\nu(U(\mathfrak{gl}_n)^{GL_n(\mathbb{C})}) = \mathcal{PD}(\mathbb{C}^n \otimes \mathbb{C}^{n'})^{GL_n(\mathbb{C}) \times GL_{n'}(\mathbb{C})} = \nu'(U(\mathfrak{gl}_{n'})^{GL_{n'}(\mathbb{C})})$$

Then we have the following relation. This is called the "higher Capelli identity" [21].

Theorem 9.1

We have

$$\nu(G_p^{\circ}) = \frac{1}{p!^2} \sum_{I \in [n]^p, K \in [n']^p} \operatorname{preimm} X_{IK} \operatorname{preimm}^{\circ} \partial_{IK}$$

This relation can be regarded as a generalization of the Capelli identity and has an interpretation in the dual pair theory as follows (see also [5, 6, 9, 10, 18, 21, 25]). The general linear group $GL_{n'}(\mathbb{C})$ acts on $\mathcal{P}(\mathbb{C}^n \otimes \mathbb{C}^{n'})$ naturally, and these two actions of $GL_n(\mathbb{C})$ and $GL_{n'}(\mathbb{C})$ form the dual pair. This tells us the relation

$$\nu(U(\mathfrak{gl}_n)^{GL_n(\mathbb{C})}) = \mathcal{PD}(\mathbb{C}^n \otimes \mathbb{C}^{n'})^{GL_n(\mathbb{C}) \times GL_{n'}(\mathbb{C})} = \nu'(U(\mathfrak{gl}_{n'})^{GL_{n'}(\mathbb{C})})$$

Here ν' means the infinitesimal action of $GL_{n'}(\mathbb{C})$. Theorem 9.1 can be regarded as a beautiful description of this relation in terms of bases as vector spaces (from the symmetry of ν and ν', we see that the quantity in this theorem is also equal to $\nu'(G_\lambda^\circ)$). Note that we can check that the right hand side of Theorem 9.1 is $GL_n(\mathbb{C}) \times GL_{n'}(\mathbb{C})$-invariant using the Cauchy–Binet type formula (Proposition 6.3).

Furthermore, we can rewrite this in terms of the quantum preimmanants.

Theorem 9.2

We have

$$\nu(G_p^\circ) = \frac{1}{p!^2} \sum_{I\in[n]^p,\, K\in[n']^p} \operatorname{preimm} X_{IK} \operatorname{preimm}^\circ \partial_{IK}$$

We can also regard these relations as noncommutative analogues of the Cauchy–Binet type identities in Section 6 (Propositions 6.3 and 6.6).

These theorems are shown as follows. This proof is parallel to that of the original Capelli identity given in [9, 29] in the exterior calculus.

Proof of Theorems 9.1 and 9.2 Put $V = \mathbb{C}^n$ and consider a basis $e_1, \ldots, e_n$ of V and its dual basis $e^*_1, \ldots, e^*_n$. We consider the following elements in $\overline{T}^\circ(V^*) \otimes \mathscr{PD}(\mathbb{C}^n \otimes \mathbb{C}^{n'})$:

$$\eta_k^* = \sum_{i=1}^{n} x_{ik} e_i^*, \qquad \gamma_j^* = \sum_{i=1}^{n} \nu(E_{ij}) e_i^* = \sum_{i=1}^{n}\sum_{k=1}^{n'} x_{ik}\partial_{jk} e_i^* = \sum_{k=1}^{n'} \eta_k^* \partial_{jk}$$

Moreover, we put

$$\gamma_j^*(u) = \sum_{i=1}^{n} \nu(E_{ij}(u)) e_i^* = \gamma_j^* + u e_j^*.$$

A simple calculation tells us $\nu(E_{ij}) x_{ka} = x_{ka} \nu(E_{ij}) + x_{ia} \delta_{kj}$. From this, we see the commutation relation

$$\gamma_j^*(-y_{k+1})\eta_k^* = s_1 \eta_k^* \gamma_j^*(-y_k)$$

Using this and the relation $\gamma_j^*(-y_1) = \gamma_j^* = \sum_{k=1}^{n'} \eta_k^* \partial_{jk}$ repeatedly, we have

$$\gamma_{j_1}^*(-y_p)\cdots\gamma_{j_{p-1}}^*(-y_2)\gamma_{j_p}^*(-y_1) = \sum_{k_p=1}^{n'} \gamma_{j_1}^*(-y_p)\cdots\gamma_{j_{p-1}}^*(-y_2)\eta_{k_p}^*\partial_{j_p k_p}$$
$$= \sum_{k_p=1}^{n'} \eta_{k_p}^*\gamma_{j_1}^*(-y_{p-1})\cdots\gamma_{j_{p-1}}^*(-y_1)\partial_{j_p k_p}$$
$$\vdots$$
$$= \sum_{k_1,\ldots,k_p=1}^{n'} \eta_{k_p}^*\cdots\eta_{k_1}^*\partial_{j_1 k_1}\cdots\partial_{j_p k_p}.$$

Thus, we have

$$\sum_{J\in[n]^p} e_{j_p}\cdots e_{j_1}\gamma_{j_1}^*(-y_p)\cdots\gamma_{j_p}^*(-y_1)$$
$$= \sum_{I,J\in[n]^p}\sum_{K\in[n']^p} e_{j_p}\cdots e_{j_1}x_{i_p k_p}\cdots x_{i_1 k_1}\partial_{j_1 k_1}\cdots\partial_{j_p k_p}e_{i_1}^*\cdots e_{i_p}^*$$

in $\overline{T}(V, V^*)\otimes\mathscr{PD}(\mathbb{C}^n\otimes\mathbb{C}^{n'})$. Applying $\langle\bullet\rangle$ and dividing by $p!$, we obtain

$$v(G_p^\circ) = \frac{1}{p!^2}\sum_{I\in[n]^p}\sum_{K\in[n']^p}\sum_{\sigma,\sigma'\in S_p}\sigma' x_{i_{\sigma(p)}k_p}\cdots x_{i_{\sigma(1)}k_1}\partial_{i_{\sigma'(1)}k_1}\cdots\partial_{i_{\sigma'(p)}k_p}\sigma^{-1}.$$

The left-hand side is invariant under the antiautomorphism $t\to t^\circ$given in Section 6.4, because . $G_p^\circ\in Z\mathbb{C}S_p\otimes U(\mathfrak{gl}_n)$ Hence, we have

$$v(G_p^\circ) = \frac{1}{p!^2}\sum_{I\in[n]^p}\sum_{K\in[n']^p}\sum_{\sigma,\sigma'\in S_p}\sigma x_{i_{\sigma(p)}k_p}\cdots x_{i_{\sigma(1)}k_1}\partial_{i_{\sigma'(1)}k_1}\cdots\partial_{i_{\sigma'(p)}k_p}\sigma'^{-1}.$$

This means Theorem 9.2. Moreover, applying χ_λ to (9.1), we have

$$v(G_\lambda^\circ) = \frac{1}{p!^2}\sum_{I\in[n]^p}\sum_{K\in[n']^p}\sum_{\sigma,\sigma'\in S_p}\chi_\lambda(\sigma'\sigma^{-1})x_{i_{\sigma(p)}k_p}\cdots x_{i_{\sigma(1)}k_1}\partial_{i_{\sigma'(1)}k_1}\cdots\partial_{i_{\sigma'(p)}k_p}$$

As seen from the first relation of (6.2), the right-hand side is equal to

$$\frac{\chi_\lambda(1)}{p!^2}\sum_{I\in[n]^p}\sum_{K\in[n']^p}\sum_{\sigma\in S_p}\chi_\lambda(\sigma^{-1})x_{i_{\sigma(p)}k_p}\cdots x_{i_{\sigma(1)}k_1}\sum_{\sigma'\in S_p}\chi_\lambda(\sigma')\partial_{i_{\sigma'(1)}k_1}\cdots\partial_{i_{\sigma'(p)}k_p}.$$

This means Theorem 9.1.

Remark

The right-hand side of Theorem 9.1 is also equal to

$$\frac{1}{p!}\sum_{I\in[n]^p}\sum_{K\in[n']^p}\sum_{\sigma\in S_p}\chi_\lambda(\sigma)x_{i_{\sigma(p)}k_p}\cdots x_{i_{\sigma(1)}k_1}\partial_{i_1k_1}\cdots\partial_{i_pk_p}.$$

9.2.

In the same situation as Section 4, we have the following relation as an analogue of the higher Capelli identity.

Theorem 9.3

We have

$$\nu(G_\lambda^\circ)=\frac{\chi(1)}{p!^2}\sum_{I\in[n]^p,K\in[n']^p}\text{column-imm}_\lambda X_{I^\circ K^\circ}\text{column-imm}_\lambda X_{IK}^*.$$

We can regard this as the "higher version" of Theorem 4.1. Namely this describes the relation between the centers of $U(\mathfrak{gl}_n)$ and Q_2 in Corollary 3.8 in terms of a basis of the center of $U(\mathfrak{gl}_n)$.

Moreover, we have the following relation for the quantum preimmanants.

Theorem 9.4

We have

$$\nu(G_p^\circ)=\frac{1}{p!^2}\sum_{I\in[n]^p,K\in[n']^p}\text{column-preimm}\,X_{I^\circ K^\circ}\text{column-preimm}^\circ X_{IK}^*.$$

Proof of Theorems 9.3 and 9.4

The proof is almost the same as that of Theorem 9.1. We consider the following elements in $\overline{T}^\circ(V^*)\otimes\mathcal{L}(\mathbb{C}^n\otimes\mathbb{C}^{n'})$:

$$\eta_k^*=\sum_{i=1}^n L(w_{ik})e_i^*,\qquad \gamma_j^*=\sum_{i=1}^n \nu(E_{ij})e_i^*=\sum_{i=1}^n\sum_{k=1}^{n'}L(w_{ik})L(w_{jk}^*)e_i^*=\sum_{k=1}^{n'}\eta_k^*L(w_{jk}^*),$$

$$\gamma_j^*(u)=\sum_{i=1}^n \nu(E_{ij}(u))e_i^*=\gamma_j^*+ue_j^*.$$

By a direct calculation, we have

$$v(E_{ij})L(w_{ka}) = L(w_{ka})v(E_{ij}) + L(w_{ia})\delta_{kj}.$$

From this, we see the commutation relation

$$\gamma_j^*(-y_{k+1})\eta_k^* = s_1\eta_k^*\gamma_j^*(-y_k).$$

Here s_i and y_k belong to $\overline{T}^\circ(V^*)$ (these should be distinguished from $L(s_i)$ and $L(y_k)$ in $\mathcal{L}(\mathbb{C}$ n $\otimes$ $\mathbb{C}$ n')). Using this and the relation $\gamma_j^*(-y_1) = \gamma_j^* = \sum_{k=1}^{n'} \eta_k^* L(w_{jk}^*)$ repeatedly, we have

$$\gamma_{j_1}^*(-y_p)\cdots\gamma_{j_p}^*(-y_1) = \sum_{K\in[n']^p} \eta_{k_p}^*\cdots\eta_{k_1}^* L(w_{j_1k_1}^*)\cdots L(w_{j_pk_p}^*).$$

Thus, we have

$$\sum_{I,J\in[n]^p} e_{j_p}\cdots e_{j_1}\gamma_{j_1}^*(-y_p)\cdots\gamma_{j_p}^*(-y_1)$$

$$= \sum_{I,J\in[n]^p} e_{j_p}\cdots e_{j_1}L(w_{i_pk_p})\cdots L(w_{i_1k_1})L(w_{j_1k_1}^*)\cdots L(w_{j_pk_p}^*)e_{i_1}^*\cdots e_{i_p}^*$$

in $\overline{T}(V, V^*) \otimes \mathscr{L}(\mathbb{C}^n \otimes \mathbb{C}^{n'})$. The remainder is the same as the proof of Theorems 9.1 and 9.2.

ACKNOWLEDGMENTS

This research was partially supported by JSPS Grant-in-Aid for Young Scientists (B) 20740020.

The author is grateful to Professor T\^oru Umeda for fruitful discussions. He is also grateful to Professor Alexander Molev for pointing out some developments in the study of the quantum immanants and the higher Capelli identities. Finally he thanks to the referee for valuable comments which improved this article.

NOTES

Communicated by V. A. Artamonov.

REFERENCES

1. Capelli, A. (1887). Über die Zurückführung der Cayley'schen operation Ω auf gewöhnliche Polar-Operationen. *Math. Ann.* 29: 331 – 338.
2. Capelli, A. (1890). Sur les opérations dans la théorie des formes algébriques.*Math. Ann.* 37: 1 – 37.
3. Goodman, R., Wallach, N. R. (2003). *Representations and Invariants of the Classical Groups*. Cambridge Univ. Press.
4. Hashimoto, T. (2008). A central element in the universal enveloping algebra of type D_n via minor summation formula of Pfaffians. *J. Lie Theory* 18: 581 – 594.
5. Howe, R. (1989). Remarks on classical invariant theory. *Trans. Amer. Math. Soc.*313: 539 – 570.
6. Howe, R., Umeda, T. (1991). The Capelli identity, the double commutant theorem, and multiplicity-free actions. *Math. Ann.* 290: 565 – 619.
7. Itoh, M. (2000). Capelli elements for the orthogonal Lie algebras. *J. Lie Theory*10: 463 – 489.
8. Itoh, M. (2001). A Cayley–Hamilton theorem for the skew Capelli elements. *J. Algebra* 242: 740 – 761.
9. Itoh, M. (2004). Capelli identities for the dual pair (O_M, Sp_N). *Math. Z.* 246:125 – 154.
10. Itoh, M. (2005). Capelli identities for reductive dual pairs. *Adv. Math.* 194: 345 –397.
11. Itoh, M. (2007). Two determinants in the universal enveloping algebras of the orthogonal Lie algebras. *J. Algebra* 314: 479 – 506.
12. Itoh, M. (2009). Two permanents in the universal enveloping algebras of the symplectic Lie algebras. *Internat. J. Math.* 20: 339 – 368.
13. Itoh, M., Umeda, T. (2001). On central elements in the universal enveloping algebras of the orthogonal Lie algebras. *Compositio Math.* 127: 333 – 359.
14. James, G., Kerber, A. (1981). *The Representation Theory of the Symmetric Group*. Encyclopedia of Mathematics and Its Applications, Vol. 16. Addison-Wesley.
15. Jucys, A.-A. A. (1974). Symmetric polynomials and the center of the symmetric group ring. *Rep. Mathematical Phys.* 5 (1): 107 – 112.
16. Molev, A. (1998). *Factorial Supersymmetric Schur Functions and Super Capelli Identities*. Kirillov's Seminar on Representation Theory, Amer. Math. Soc. Transl. Ser. 2, 181. Providence, RI: Amer. Math. Soc., pp. 109–137.
17. Molev, A. (2007). *Yangians and Classical Lie Algebras*. Mathematical Surveys and Monographs, 143. Providence, RI: American Mathematical Society.
18. Molev, A., Nazarov, M. (1999). Capelli identities for classical Lie algebras.*Math. Ann.* 313: 315 – 357.
19. Nazarov, M. (1991). Quantum Berezinian and the classical Capelli identity. *Lett. Math. Phys.* 21: 123 – 131.
20. Nazarov, M. (1998). *Yangians and Capelli Identities*. Kirillov's Seminar on Representation Theory, Amer. Math. Soc. Transl. Ser. 2, 181. Providence, RI:Amer. Math. Soc., pp. 139–163.
21. Okounkov, A. (1996). Quantum immanants and higher Capelli identities.*Transform. Groups* 1 (1): 99 – 126.
22. Okounkov, A. (1996). Young basis, Wick formula, and higher Capelli identities.*Internat. Math. Res. Notices* 17: 817 – 839.
23. Okounkov, A., Olshanski, G. (1998). *Shifted Schur Functions*. Algebra i Analiz 9(2) (Russian); English version in St. Petersburg Math. J. 9:239–300.
24. Stanley, R. P. (1999). *Enumerative Combinatorics*. Vol. 2. Cambridge Studies in Advanced Mathematics, 62. Cambridge: Cambridge University Press.
25. Umeda, T. (1994). The Capelli identities, a century after. *Sūgaku* 46:206–227 (in Japanese); English transl. In: Nomizu, K., ed. *Selected Papers on Harmonic Analysis, Groups, and Invariants*. Vol. 183. AMS Translations, Series 2, pp. 51–78.
26. Umeda, T. (1998). Newton's formula for $\mathfrak{gl}_n$. *Proc. Amer. Math. Soc.* 126:3169–3175.

27. Umeda, T. (2000). On Turnbull identity for skew symmetric matrices. *Proc. Edinburgh Math. Soc.* (2) 43:379–393.
28. Umeda, T. (2003). Application of Koszul complex to Wronski relations for $U(\mathfrak{gl}_n)$. *Comment. Math. Helv.* 78: 663 – 680.
29. Umeda, T. (2008). On the proof of the Capelli identities. *Funkcialaj Ekvacioj* 51:1 – 15.
30. Wachi, A. (2006). Central elements in the universal enveloping algebras for the split realization of the orthogonal Lie algebras. *Lett. Math. Phys.* 77: 155 – 168.
31. Weyl, H. (1946). *The Classical Groups. Their Invariants and Representations.* Princeton: Princeton Univ. Press.

CHAPTER 11

Group Inverse of 2 × 2 Block Matrices over Minkowski Space M

Dandapany Krishnaswamy, Tasaduq Hussain Khan

Department of Mathematics, Annamalai University, Annamalai Nagar, India

ABSTRACT

Necessary and sufficient conditions for the existence of the group inverse of the block matrix $\begin{pmatrix} P & P \\ Q & 0 \end{pmatrix}$ in Minkowski Space are studied, where $P^{\sim}$,$Q^{\sim}$ are both square and $(Q)\geq$ .rank (P) The representation of this group inverse and some related additive results are also given.

Keywords: Block Matrix, Group Inverse, Minkowski Adjoint, Minkowski Space

1. INTRODUCTION

Let F be a skew field and $F^{n\times n}(C)$ be the set of all matrices over F. For $A \in F^{n\times n}$ the matrix $A \in F^{n\times n}$ is said to be the group inverse of A, if $AXA = A$, $ZAX = X$, $AX = XA$

and is denoted by $X = A^{\#}$, and is unique by [1].

The generalized inverse of block matrix has important applications in statistical probability, mathematical programming, game theory, control theory etc. and for references see [2] [3] [4]. The research on the existence and the representation of the group inverse for block matrices in Euclidean space has been done in wide range. For the literature of the group inverse of block matrix in Euclidean space, see [5] - [11].

In [12] the existence of anti-reflexive with respect to the generalized reflection anti- symmetric matrix $P^{\sim}$ and solution of the matrix equation $AXB = C$ in Minkowski space M is given. In [13] necessary and sufficient condition for the existence of Re-nnd solution has been established of the matrix equation $AXA^{\sim} = C$ where $A \in C^{n \times n}$ and $C \in C^{n \times n}$. In [14] partitioned matrix $M^{\sim}$ in Minkowski space M was taken of the form $M^{\sim} = \begin{bmatrix} A^{\sim} & -C^{\sim}G_1 \\ -G_1B^{\sim} & D^{*} \end{bmatrix}$ to yield a formula for the inverse of $M^{\sim}$ in terms of the Schur complement of.

In this paper P^{*} and $P^{\sim}$ denote the conjugate transpose and Minkowski adjoint of a matrix P respectively. $I_n$denotes the identity matrix of order $n \times n$. Minkowski Space M is an indefinite inner product space in which the metric matrix associated with the indefinite inner product is denoted by G and is defined as

$G = \begin{bmatrix} 1 & 0 \\ 0 & -I_{n-1} \end{bmatrix}$ *satisfying* $G^2 = I_n$ *and* $G^{*} = G$.

G is called the Minkowski metric matrix. In case $u \in \mathbb{C}^n$, indexed as $u = (u_0, u_1, \cdots, u_{n-1})$, G is called the Minkowski metric tensor and is defined as $Gu = (u_0, -u_1, \cdots, -u_{n-1})$ [12]. For any $P \in C^{n \times n}$, the Minkowski adjoint of P denoted by $P^{\sim}$ is defined as $P^{\sim} = GP^{*}G$ where P^{*} is the usual Hermitian adjoint and G the Minkowski metric matrix of order n. We establish the necessary and sufficient condition for the existence and the representation of the group inverse of a block matrix $\begin{pmatrix} P^{\sim} & P^{\sim} \\ Q^{\sim} & 0 \end{pmatrix}$ or $\begin{pmatrix} P^{\sim} & Q^{\sim} \\ P^{\sim} & 0 \end{pmatrix}$ in Minkowski space, where $P^{\sim}, Q^{\sim} \in K^{n \times n}$ rank $(Q^{\sim}) \geq$.rank $(P^{\sim})$ We also give a sufficient condition for $(P^{\sim} Q^{\sim})$ to be similar to $(Q^{\sim} P^{\sim})$.

2. LEMMAS

Lemma 1. Let $P, Q \in F^{n \times n}(C)$. If

Rank (*P*) = r, rank (*Q*) = *rank* ((*PQ*) = *rank* (*QP*), then there are unitary matrices *A*, *B* ∈ $F^{n \times n}$ such that

$$P^{\sim} = B^{\sim} \begin{pmatrix} I_r & 0 \\ 0 & 0 \end{pmatrix} A^{\sim}, \quad Q^{\sim} = \left(A^*\right)^{\sim} \begin{pmatrix} Q_1^* & -Q_1^* X \\ -YQ_1^* & YQ_1^* X \end{pmatrix} \left(B^*\right)^{\sim}$$

where $Q_1^* \in F^{r \times r}, X \in F^{r \times (n-r)}$ and $Y \in F^{(n-r) \times r}$.

Proof. Since *rank (P) = r*, there are two unitary matrices $A, B \in F^{n \times n}(C)$ such that

$$P = A \begin{pmatrix} I_r & 0 \\ 0 & 0 \end{pmatrix} B, \qquad Q = B^* \begin{pmatrix} Q_1 & Q_2 \\ Q_3 & Q_4 \end{pmatrix} A^*$$

where

$$Q_1 \in F^{r \times r}, Q_2 \in F^{r \times (n-r)}, Q_3 \in F^{(n-r) \times r}, Q_4 \in F^{(n-r) \times (n-r)}.$$

Now

$$P^* = B^* \begin{pmatrix} I_r & 0 \\ 0 & 0 \end{pmatrix} A^* \Rightarrow P^{\sim} = GB^* G^2 \begin{pmatrix} I_r & 0 \\ 0 & 0 \end{pmatrix} G^2 A^* G = B^{\sim} \begin{pmatrix} I_r & 0 \\ 0 & 0 \end{pmatrix} A^{\sim}$$

and

$$Q = B^* \begin{pmatrix} Q_1 & Q_2 \\ Q_3 & Q_4 \end{pmatrix} A^*$$

$$Q^* = \left(A^*\right)^* \begin{pmatrix} Q_1^* & Q_3^* \\ Q_2^* & Q_4^* \end{pmatrix} \left(B^*\right)^*$$

$$Q^{\sim} = G\left(A^*\right)^* G^2 \begin{pmatrix} Q_1^* & Q_3^* \\ Q_2^* & Q_4^* \end{pmatrix} G^2 (B^*)^* G = \left(A^*\right)^{\sim} \begin{pmatrix} Q_1^* & -Q_3^* \\ -Q_2^* & Q_4^* \end{pmatrix} \left(B^*\right)^{\sim}$$

From *rank* = (*Q*) = *rank* (*PQ*) we have

$$Q_2^* = YQ_1^*, \qquad Q_4^* = YQ_3^*, \qquad Y \in F^{(n-r) \times r}$$

and from *rank* (*Q*) = *rank* (*QP*) we get

$$Q_3^* = Q_1^* X, \qquad Q_4^* = Q_2^* X = Y Q_1^* X, \qquad X \in F^{r\times(n-r)}$$

So,

$$Q^{\sim} = (A^*)^{\sim} \begin{pmatrix} Q_1^* & -Q_1^* X \\ -YQ_1^* & YQ_1^* X \end{pmatrix} (B^*)^{\sim}$$

Lemma 2. Let

$$P \in F^{r\times r}, Q \in F^{(n-r)\times r}, M = \begin{pmatrix} P^{\sim} & 0 \\ Q^{\sim} & 0 \end{pmatrix} \in F^{n\times n}(C)$$

Then the group inverse of M exists in *M* if and only if the group inverse of *P*~ exists in *M* and rank (*P* ~). If the group inverse of *M** exists in *M*,then

$$M^{\#} = \begin{pmatrix} (P^{\sim})^{\#} & 0 \\ Q\left((P^{\sim})^{\#}\right)^2 & 0 \end{pmatrix}$$

Proof. Since $M = \begin{pmatrix} P^{\sim} & 0 \\ Q^{\sim} & 0 \end{pmatrix}$, suppose group inverse of *P*~ exists in *M* and $rank(P^{\sim}) = rank\begin{pmatrix} P^{\sim} \\ Q^{\sim} \end{pmatrix}$. Now

$$rank(M) = rank\begin{pmatrix} P^{\sim} & 0 \\ Q^{\sim} & 0 \end{pmatrix} = rank\begin{pmatrix} P^{\sim} \\ Q^{\sim} \end{pmatrix} = rank(P^{\sim}).$$

But $rank(P^{\sim}) = rank(P^{\sim})^2$ because $(P^{\sim})^{\#}$ exists $\Rightarrow rank(M) = rank(M^2)$.

There- fore $(M^{\sim})^{\#}$ exists in *M*.

Conversely, suppose the group inverse of M exists in *M*, then it satisfies the following conditions: 1) *MM* #*M* = *M* 2) *M* #*MM*# = *M*# and 3).*MM* #*M*#*M*.Also

$$rank(M) = rank\begin{pmatrix} P^{\sim} & 0 \\ Q^{\sim} & 0 \end{pmatrix} = rank\begin{pmatrix} P^{\sim} \\ Q^{\sim} \end{pmatrix} \Rightarrow rank(P^{\sim}) = rank\begin{pmatrix} P^{\sim} \\ Q^{\sim} \end{pmatrix}.$$

Let $M^{\#} = X = \begin{pmatrix} \left(P^{\sim}\right)^{\#} & 0 \\ Q^{\sim}\left(\left(P^{\sim}\right)^{\#}\right)^{2} & 0 \end{pmatrix}$ then,

1)

$$MXM = \begin{pmatrix} P^{\sim} & 0 \\ Q^{\sim} & 0 \end{pmatrix}\begin{pmatrix} \left(P^{\sim}\right)^{\#} & 0 \\ Q^{\sim}\left(\left(P^{\sim}\right)^{\#}\right)^{2} & 0 \end{pmatrix}\begin{pmatrix} P^{\sim} & 0 \\ Q^{\sim} & 0 \end{pmatrix}$$

$$= \begin{pmatrix} P^{\sim}\left(P^{\sim}\right)^{\#}P^{\sim} & 0 \\ Q^{\sim}\left(P^{\sim}\right)^{\#}P^{\sim} & 0 \end{pmatrix} = \begin{pmatrix} P^{\sim} & 0 \\ Q^{\sim} & 0 \end{pmatrix}$$

$$MXM = M$$

2)

$$XMX = \begin{pmatrix} \left(P^{\sim}\right)^{\#} & 0 \\ Q^{\sim}\left(\left(P^{\sim}\right)^{\#}\right)^{2} & 0 \end{pmatrix}\begin{pmatrix} P^{\sim} & 0 \\ Q^{\sim} & 0 \end{pmatrix}\begin{pmatrix} \left(P^{\sim}\right)^{\#} & 0 \\ Q^{\sim}\left(\left(P^{\sim}\right)^{\#}\right)^{2} & 0 \end{pmatrix}$$

$$= \begin{pmatrix} \left(P^{\sim}\right)^{\#}P^{\sim}\left(P^{\sim}\right)^{\#} & 0 \\ Q^{\sim}\left(\left(P^{\sim}\right)^{\#}\right)^{2}P^{\sim}\left(P^{\sim}\right)^{\#} & 0 \end{pmatrix} = \begin{pmatrix} \left(P^{\sim}\right)^{\#} & 0 \\ Q^{\sim}\left(\left(P^{\sim}\right)^{\#}\right)^{2} & 0 \end{pmatrix}$$

$$XMX = X$$

3)

$$MX = \begin{pmatrix} P^{\sim} & 0 \\ Q^{\sim} & 0 \end{pmatrix}\begin{pmatrix} \left(P^{\sim}\right)^{\#} & 0 \\ Q^{\sim}\left(\left(P^{\sim}\right)^{\#}\right)^{2} & 0 \end{pmatrix} = \begin{pmatrix} P^{\sim}\left(P^{\sim}\right)^{\#} & 0 \\ Q^{\sim}\left(P^{\sim}\right)^{\#} & 0 \end{pmatrix}$$

$$XM = \begin{pmatrix} \left(P^{\sim}\right)^{\#} & 0 \\ Q^{\sim}\left(\left(P^{\sim}\right)^{\#}\right)^{2} & 0 \end{pmatrix}\begin{pmatrix} P^{\sim} & 0 \\ Q^{\sim} & 0 \end{pmatrix} = \begin{pmatrix} P^{\sim}\left(P^{\sim}\right)^{\#} & 0 \\ Q^{\sim}\left(P^{\sim}\right)^{\#} & 0 \end{pmatrix}$$

$$XM = MX.$$

Lemma 3. Let $P \in F^{r\times r}$, $Q \in F^{r\times(n-r)}$ and $M = \begin{pmatrix} P^{\sim} & Q^{\sim} \\ 0 & 0 \end{pmatrix} \in F^{n\times n}(C)$. Then the group inverse of M exists in M if and only if the group inverse of $P^{\sim}$ exists in M and $\left(P^{\sim}\right) = rank\left(P^{\sim} \quad Q^{\sim}\right)$. If the group inverse of M exists in M, then,

$$M^{\#} = \begin{pmatrix} \left(P^{\sim}\right)^{\#} & \left(\left(P^{\sim}\right)^{\#}\right)^{2} Q^{\sim} \\ 0 & 0 \end{pmatrix}$$

Proof. The proof is same as Lemma 2.

Lemma 4. Let $P, Q \in F^{n\times n}(C)$. If

Rank (P~) rank (Q~) = rank (P~ Q~) = (Q~ P~)

then the following conclusions hold:

1) $P^{\sim} Q^{\sim} \left(P^{\sim} Q^{\sim}\right)^{\#} P^{\sim} = P^{\sim}$

2) $P^{\sim} \left(Q^{\sim} P^{\sim}\right)^{\#} Q^{\sim} P^{\sim} = P^{\sim}$

3) $Q^{\sim} P^{\sim} \left(Q^{\sim} P^{\sim}\right)^{\#} P^{\sim} = Q^{\sim}$

4) $Q^{\sim} \left(P^{\sim} Q^{\sim}\right)^{\#} P^{\sim} = Q^{\sim} P^{\sim} \left(Q^{\sim} P^{\sim}\right)^{\#}$

5) $P^{\sim} \left(Q^{\sim} P^{\sim}\right)^{\#} = \left(P^{\sim} Q^{\sim}\right)^{\#} P^{\sim}$

Proof. Suppose rank (P) = r, then by Lemma 1 we have

$$P^{\sim} = B^{\sim} \begin{pmatrix} I_r & 0 \\ 0 & 0 \end{pmatrix} A^{\sim}, \qquad B^{\sim} = \left(A^{*}\right)^{\sim} \begin{pmatrix} Q_1^{*} & -Q_1^{*} X \\ -Y Q_1^{*} & Y Q_1^{*} X \end{pmatrix} \left(B^{*}\right)^{\sim}$$

where $Q_1^{*} \in F^{r\times r}, X \in F^{r\times(n-r)}, Y \in F^{(n-r)\times r}$. Then

$$P^{\sim} Q^{\sim} = B^{\sim} \begin{pmatrix} Q_1^{*} & -Q_1^{*} X \\ 0 & 0 \end{pmatrix} \left(B^{*}\right)^{\sim}, \quad Q^{\sim} P^{\sim} = \left(A^{*}\right)^{\sim} \begin{pmatrix} Q_1^{*} & 0 \\ -Y Q_1^{*} & 0 \end{pmatrix} A^{\sim}$$

Since rank (P~) = (Q~) we have that Q_1^{*} is invertible. By using Lemma 2 and 3 we get

$$\left(P^{\sim}Q^{\sim}\right)^{\#} = B^{\sim}\begin{pmatrix} \left(Q_1^{*}\right)^{-1} & -\left(Q_1^{*}\right)^{-1}X \\ 0 & 0 \end{pmatrix}\left(B^{*}\right)^{\sim}$$

$$\left(Q^{\sim}P^{\sim}\right)^{\#} = \left(A^{*}\right)^{\sim}\begin{pmatrix} \left(Q_1^{*}\right)^{-1} & 0 \\ -Y\left(Q_1^{*}\right)^{-1} & 0 \end{pmatrix}A^{\sim}$$

Then, 1)

$$P^{\sim}Q^{\sim}\left(P^{\sim}Q^{\sim}\right)^{\#}P^{\sim}$$

$$= B^{\sim}\begin{pmatrix} Q_1^{*} & -Q_1^{*}X \\ 0 & 0 \end{pmatrix}\left(B^{*}\right)^{\sim}B^{\sim}\begin{pmatrix} \left(Q_1^{*}\right)^{-1} & -\left(Q_1^{*}\right)^{-1}X \\ 0 & 0 \end{pmatrix}\left(B^{*}\right)^{\sim}B^{\sim}\begin{pmatrix} I_r & 0 \\ 0 & 0 \end{pmatrix}A^{\sim}$$

$$= B^{\sim}\begin{pmatrix} Q_1^{*}\left(Q_1^{*}\right)^{-1} & -Q_1^{*}\left(Q_1^{*}\right)^{-1}X \\ 0 & 0 \end{pmatrix}\begin{pmatrix} I_r & 0 \\ 0 & 0 \end{pmatrix}A^{\sim}$$

$$= B^{\sim}\begin{pmatrix} I_r & 0 \\ 0 & 0 \end{pmatrix}A^{\sim}$$

$$= P^{\sim}$$

Similarly we can prove 2) - 5).

3. MAIN RESULTS

Theorem 1.

Let $M = \begin{pmatrix} P & P \\ Q & 0 \end{pmatrix}$ where $P^{\sim}, Q^{\sim} \in F^{n\times n}(C), rank\left(Q^{\sim}\right) \geq rank\left(P^{\sim}\right) = r$, then

1) The group inverse of M exists in M if and only if $rank\left(P^{\sim}\right) = rank\left(Q^{\sim}\right) = rank\left(P^{\sim}Q^{\sim}\right) = rank\left(Q^{\sim}P^{\sim}\right)$.

2) If the group inverse of M exists in M, then $M^{\#} = \begin{pmatrix} M_{11} & M_{12} \\ M_{21} & M_{22} \end{pmatrix}$, where

$$M_{11} = \left(P^{\sim}Q^{\sim}\right)^{\#} P^{\sim} - \left(P^{\sim}Q^{\sim}\right)^{\#}\left(P^{\sim}\right)^{2}\left(Q^{\sim}P^{\sim}\right)^{\#} Q^{\sim},$$
$$M_{12} = \left(P^{\sim}Q^{\sim}\right)^{\#} P^{\sim},$$
$$M_{21} = \left(Q^{\sim}P^{\sim}\right)^{\#} Q^{\sim} - Q^{\sim}\left(P^{\sim}Q^{\sim}\right)^{\#}\left(P^{\sim}\right)^{2}\left(Q^{\sim}P^{\sim}\right)^{\#}$$
$$+ Q^{\sim}\left(P^{\sim}Q^{\sim}\right)^{\#} P^{\sim}\left(P^{\sim}Q^{\sim}\right)^{\#}\left(P^{\sim}\right)^{2}\left(Q^{\sim}P^{\sim}\right)^{\#} Q^{\sim},$$
$$M_{22} = -Q^{\sim}\left(P^{\sim}Q^{\sim}\right)^{\#}\left(P^{\sim}\right)^{2}\left(Q^{\sim}P^{\sim}\right)^{\#}.$$

Proof. 1) Given $rank\left(Q^{\sim}\right) \geq rank\left(P^{\sim}\right) = r$. Suppose $rank\left(P^{\sim}\right) = rank\left(Q^{\sim}\right)$ then,

$rank\left(P^{\sim}\right)^{2} = rank\left(P^{\sim}Q^{\sim}\right)$. We know that

$rank\left(P^{\sim}Q^{\sim}\right) = rank\left(P^{\sim}\right)$ so, $rank\left(P^{\sim}\right)^{2} = rank\left(P^{\sim}\right)$.

Therefore the group inverse of *M* exists. Now we show that the condition is ne- cessary,

$$rank(M) = rank\begin{pmatrix} P^{\sim} & P^{\sim} \\ Q^{\sim} & 0 \end{pmatrix} = rank\begin{pmatrix} 0 & P^{\sim} \\ Q^{\sim} & 0 \end{pmatrix} = rank\left(P^{\sim}\right) + rank\left(Q^{\sim}\right).$$

$$rank(M)^{2} = rank\begin{pmatrix} \left(P^{\sim}\right)^{2} + P^{\sim}Q^{\sim} & \left(P^{\sim}\right)^{2} \\ Q^{\sim}P^{\sim} & Q^{\sim}P^{\sim} \end{pmatrix} = rank\begin{pmatrix} P^{\sim}Q^{\sim} & \left(P^{\sim}\right)^{2} \\ 0 & Q^{\sim}P^{\sim} \end{pmatrix}.$$

Since the group inverse of M exists in *M* if and only if rank $(M) = (M)^2$,rank , we have

$$rank\left(P^{\sim}\right) + rank\left(Q^{\sim}\right) = rank(M)^{2}$$
$$\leq rank\left(P^{\sim}Q^{\sim}\right) + rank\begin{pmatrix} \left(P^{\sim}\right)^{2} \\ Q^{\sim}P^{\sim} \end{pmatrix}$$
$$\leq rank\left(P^{\sim}Q^{\sim}\right) + rank\left(P^{\sim}\right).$$

Also

$$rank(P^{\sim}) + rank(Q^{\sim}) = rank(M)^2$$
$$\leq rank\left(P^{\sim}Q^{\sim} \quad (P^{\sim})^2\right) + rank(Q^{\sim}P^{\sim})$$
$$\leq rank(Q^{\sim}P^{\sim}) + rank(P^{\sim}).$$

Then $rank(P^{\sim}) \leq rank(P^{\sim}Q^{\sim})$ and $rank(Q^{\sim}) \leq rank(Q^{\sim}P^{\sim})$. Therefore,

$$rank(Q^{\sim}) = rank(P^{\sim}Q^{\sim}) = rank(Q^{\sim}P^{\sim}).$$

From

$$rank(Q^{\sim}) = rank(P^{\sim}Q^{\sim}) \leq rank(P^{\sim})$$

and

$$rank(P^{\sim}) = rank(P^{\sim}Q^{\sim}) \leq rank(Q^{\sim}),$$

we have

$$rank(P^{\sim}) = rank(Q^{\sim})$$

Since

$$rank(P^{\sim}) + rank(Q^{\sim}) \leq rank\left(P^{\sim}Q^{\sim} \quad (P^{\sim})^2\right) + rank(Q^{\sim}P^{\sim})$$

and

$$rank\left(P^{\sim}Q^{\sim} \quad (P^{\sim})^2\right) \leq rank(P^{\sim}) \leq rank\left(P^{\sim}Q^{\sim} \quad (P^{\sim})^2\right),$$

we get

$$rank\left(P^{\sim}Q^{\sim} \quad (P^{\sim})^2\right) = rank(P^{\sim}).$$

Thus

$$rank\left(P^{\sim}Q^{\sim} \quad (P^{\sim})^2\right) = rank(P^{\sim}Q^{\sim}).$$

Then there exists a matrix $U^{\sim} \in F^{n \times n}(\mathcal{C})$ such that $P^{\sim}Q^{\sim}U^{\sim} = \left(P^{\sim}\right)^2$. Then

$$rank(M)^2 = rank\begin{pmatrix} P^{\sim}Q^{\sim} & 0 \\ 0 & Q^{\sim}P^{\sim} \end{pmatrix} = rank\left(P^{\sim}Q^{\sim}\right) + rank\left(Q^{\sim}P^{\sim}\right).$$

So, we get

$$rank\left(P^{\sim}\right) = rank\left(Q^{\sim}\right) = rank\left(P^{\sim}Q^{\sim}\right) = rank\left(Q^{\sim}P^{\sim}\right).$$

2) Let $X = \begin{pmatrix} M_{11} & M_{12} \\ M_{21} & M_{22} \end{pmatrix}$, we will prove that the matrix X satisfies the conditions of

the group inverse in $\mathcal{M}$. Firstly we compute

$$MX = \begin{pmatrix} P^{\sim}M_{11} + P^{\sim}M_{21} & P^{\sim}M_{12} + P^{\sim}M_{22} \\ Q^{\sim}M_{11} & Q^{\sim}M_{12} \end{pmatrix}$$

$$XM = \begin{pmatrix} M_{11}P^{\sim} + M_{12}Q^{\sim} & M_{11}P^{\sim} \\ M_{21}Q^{\sim} + M_{22}Q^{\sim} & M_{21}P^{\sim} \end{pmatrix}$$

Applying Lemma 4 1), 2) and 5) we have

$$\begin{aligned} P^{\sim}M_{11} + P^{\sim}M_{21} &= P^{\sim}\left(P^{\sim}Q^{\sim}\right)^{\#}P^{\sim} - P^{\sim}\left(P^{\sim}Q^{\sim}\right)^{\#}\left(P^{\sim}\right)^2\left(Q^{\sim}P^{\sim}\right)^{\#}Q^{\sim} + P^{\sim}\left(Q^{\sim}P^{\sim}\right)^{\#}Q^{\sim} \\ &\quad - P^{\sim}Q^{\sim}\left(P^{\sim}Q^{\sim}\right)^{\#}\left(P^{\sim}\right)^2\left(Q^{\sim}P^{\sim}\right)^{\#} \\ &\quad + P^{\sim}Q^{\sim}\left(P^{\sim}Q^{\sim}\right)^{\#}P^{\sim}\left(P^{\sim}Q^{\sim}\right)^{\#}\left(P^{\sim}\right)^2\left(Q^{\sim}P^{\sim}\right)^{\#}Q^{\sim} \\ &= P^{\sim}\left(P^{\sim}Q^{\sim}\right)^{\#}P^{\sim} + P^{\sim}\left(Q^{\sim}P^{\sim}\right)^{\#}Q^{\sim} - P^{\sim}P^{\sim}\left(Q^{\sim}P^{\sim}\right)^{\#} \\ &= P^{\sim}\left(Q^{\sim}P^{\sim}\right)^{\#}Q^{\sim} \\ M_{11}P^{\sim} + M_{12}Q^{\sim} &= \left(P^{\sim}Q^{\sim}\right)^{\#}\left(P^{\sim}\right)^2 - \left(P^{\sim}Q^{\sim}\right)^{\#}\left(P^{\sim}\right)^{\#}\left(Q^{\sim}P^{\sim}\right)^{\#}Q^{\sim}P^{\sim} + \left(P^{\sim}Q^{\sim}\right)^{\#}P^{\sim}Q^{\sim} \\ &= \left(P^{\sim}Q^{\sim}\right)^{\#}\left(P^{\sim}\right)^2 - \left(P^{\sim}Q^{\sim}\right)^{\#}P^{\sim}P^{\sim} + P^{\sim}\left(Q^{\sim}P^{\sim}\right)^{\#}Q^{\sim} \\ &= P^{\sim}\left(Q^{\sim}P^{\sim}\right)^{\#}Q^{\sim} \end{aligned}$$

$$P^{\sim}M_{12}+P^{\sim}M_{22}=P^{\sim}\left(P^{\sim}Q^{\sim}\right)^{\#}P^{\sim}-P^{\sim}Q^{\sim}\left(P^{\sim}Q^{\sim}\right)^{\#}\left(P^{\sim}\right)^{2}\left(Q^{\sim}P^{\sim}\right)^{\#}$$
$$=P^{\sim}\left(P^{\sim}Q^{\sim}\right)^{\#}P^{\sim}-P^{\sim}P^{\sim}\left(Q^{\sim}P^{\sim}\right)^{\#}$$
$$=P^{\sim}\left(P^{\sim}Q^{\sim}\right)^{\#}P^{\sim}-P^{\sim}\left(P^{\sim}Q^{\sim}\right)^{\#}P^{\sim}$$
$$=0$$
$$M_{11}P^{\sim}=\left(P^{\sim}Q^{\sim}\right)^{\#}\left(P^{\sim}\right)^{2}-\left(P^{\sim}Q^{\sim}\right)^{\#}\left(P^{\sim}\right)^{2}\left(Q^{\sim}P^{\sim}\right)^{\#}Q^{\sim}P$$
$$=\left(P^{\sim}Q^{\sim}\right)^{\#}\left(P^{\sim}\right)^{2}-\left(P^{\sim}Q^{\sim}\right)^{\#}\left(P^{\sim}\right)^{2}$$
$$=0$$

$$Q^{\sim}M_{11}=Q^{\sim}\left(P^{\sim}Q^{\sim}\right)^{\#}P^{\sim}-Q^{\sim}\left(P^{\sim}Q^{\sim}\right)^{\#}\left(P^{\sim}\right)^{2}\left(Q^{\sim}P^{\sim}\right)^{\#}Q^{\sim}$$
$$M_{21}P^{\sim}+M_{22}Q^{\sim}=\left(Q^{\sim}P^{\sim}\right)^{\#}Q^{\sim}P^{\sim}-Q^{\sim}\left(P^{\sim}Q^{\sim}\right)^{\#}\left(P^{\sim}\right)^{2}\left(Q^{\sim}P^{\sim}\right)P^{\sim}$$
$$+Q^{\sim}\left(P^{\sim}Q^{\sim}\right)^{\#}P^{\sim}\left(P^{\sim}Q^{\sim}\right)^{\#}\left(P^{\sim}\right)^{2}\left(Q^{\sim}P^{\sim}\right)^{\#}Q^{\sim}P^{\sim}$$
$$-Q^{\sim}\left(P^{\sim}Q^{\sim}\right)^{\#}\left(P^{\sim}\right)^{2}\left(Q^{\sim}P^{\sim}\right)^{\#}Q^{\sim}$$
$$=Q^{\sim}\left(P^{\sim}Q^{\sim}\right)^{\#}P^{\sim}-Q^{\sim}\left(P^{\sim}Q^{\sim}\right)^{\#}\left(P^{\sim}\right)^{2}\left(Q^{\sim}P^{\sim}\right)^{\#}P^{\sim}$$
$$+Q^{\sim}\left(P^{\sim}Q^{\sim}\right)^{\#}P^{\sim}P^{\sim}\left(Q^{\sim}P^{\sim}\right)^{\#}P^{\sim}$$
$$-Q^{\sim}\left(P^{\sim}Q^{\sim}\right)^{\#}\left(P^{\sim}\right)^{2}\left(Q^{\sim}P^{\sim}\right)^{\#}Q^{\sim}$$
$$=Q^{\sim}\left(P^{\sim}Q^{\sim}\right)^{\#}P^{\sim}-Q^{\sim}\left(P^{\sim}Q^{\sim}\right)^{\#}\left(P^{\sim}\right)^{2}\left(Q^{\sim}P^{\sim}\right)^{\#}Q^{\sim}$$
$$Q^{\sim}M_{12}=Q^{\sim}\left(P^{\sim}Q^{\sim}\right)^{\#}P^{\sim}$$
$$M_{21}P^{\sim}=Q^{\sim}\left(P^{\sim}Q^{\sim}\right)P^{\sim}-Q^{\sim}\left(P^{\sim}Q^{\sim}\right)^{\#}\left(P^{\sim}\right)^{2}\left(Q^{\sim}P^{\sim}\right)^{\#}P^{\sim}$$
$$+Q^{\sim}\left(P^{\sim}Q^{\sim}\right)^{\#}P^{\sim}\left(P^{\sim}Q^{\sim}\right)^{\#}\left(P^{\sim}\right)^{2}\left(Q^{\sim}P^{\sim}\right)^{\#}Q^{\sim}P^{\sim}$$
$$=Q^{\sim}\left(P^{\sim}Q^{\sim}\right)P^{\sim}-Q^{\sim}\left(P^{\sim}Q^{\sim}\right)^{\#}\left(P^{\sim}\right)^{2}\left(Q^{\sim}P^{\sim}\right)^{\#}P^{\sim}$$
$$+Q^{\sim}\left(Q^{\sim}P^{\sim}\right)^{\#}P^{\sim}P^{\sim}\left(Q^{\sim}P^{\sim}\right)^{\#}P^{\sim}$$
$$=Q^{\sim}\left(P^{\sim}Q^{\sim}\right)P^{\sim}$$

$$\Rightarrow MX = XM = \begin{pmatrix} P^{\sim}\left(Q^{\sim}P^{\sim}\right)^{\#}Q^{\sim} & 0 \\ Q^{\sim}\left(P^{\sim}Q^{\sim}\right)^{\#}P^{\sim} - Q^{\sim}\left(P^{\sim}Q^{\sim}\right)^{\#}\left(P^{\sim}\right)^{2}\left(Q^{\sim}P^{\sim}\right)^{\#}Q^{\sim} & Q^{\sim}\left(P^{\sim}Q^{\sim}\right)P^{\sim} \end{pmatrix}$$

$$XMX = \begin{pmatrix} M_{11} & M_{12} \\ M_{21} & M_{22} \end{pmatrix}\begin{pmatrix} P^{\sim}\left(Q^{\sim}P^{\sim}\right)^{\#}Q^{\sim} & 0 \\ Q^{\sim}\left(P^{\sim}Q^{\sim}\right)^{\#}P^{\sim} - Q^{\sim}\left(P^{\sim}Q^{\sim}\right)^{\#}\left(P^{\sim}\right)^{2}\left(Q^{\sim}P^{\sim}\right)^{\#}Q^{\sim} & Q^{\sim}\left(P^{\sim}Q^{\sim}\right)P^{\sim} \end{pmatrix}$$

$$= \begin{pmatrix} Y_{11} & Y_{12} \\ Y_{21} & Y_{22} \end{pmatrix}$$

$$MXM = \begin{pmatrix} P^{\sim} & P^{\sim} \\ Q^{\sim} & 0 \end{pmatrix}\begin{pmatrix} P^{\sim}\left(Q^{\sim}P^{\sim}\right)^{\#}Q^{\sim} & 0 \\ Q^{\sim}\left(P^{\sim}Q^{\sim}\right)^{\#}P^{\sim} - Q^{\sim}\left(P^{\sim}Q^{\sim}\right)^{\#}\left(P^{\sim}\right)^{2}\left(Q^{\sim}P^{\sim}\right)^{\#}Q^{\sim} & Q^{\sim}\left(P^{\sim}Q^{\sim}\right)P^{\sim} \end{pmatrix}$$

$$= \begin{pmatrix} X_{11} & X_{12} \\ X_{21} & 0 \end{pmatrix}$$

$$\begin{aligned} Y_{11} &= \left(P^{\sim}Q^{\sim}\right)^{\#}\left(P^{\sim}\right)^{2}\left(Q^{\sim}P^{\sim}\right)^{\#}Q^{\sim} - \left(P^{\sim}Q^{\sim}\right)^{\#}\left(P^{\sim}\right)^{2}\left(Q^{\sim}P^{\sim}\right)^{\#}Q^{\sim}P^{\sim}\left(Q^{\sim}P^{\sim}\right)^{\#}Q^{\sim} \\ &\quad + \left(P^{\sim}Q^{\sim}\right)^{\#}P^{\sim}Q^{\sim}\left(P^{\sim}Q^{\sim}\right)^{\#}P^{\sim} - \left(P^{\sim}Q^{\sim}\right)^{\#}P^{\sim}Q^{\sim}\left(P^{\sim}Q^{\sim}\right)^{\#}\left(P^{\sim}\right)^{2}\left(Q^{\sim}P^{\sim}\right)^{\#}Q^{\sim} \\ &= \left(P^{\sim}Q^{\sim}\right)P^{\sim} - \left(P^{\sim}Q^{\sim}\right)^{\#}\left(P^{\sim}\right)^{2}\left(Q^{\sim}P^{\sim}\right)^{\#}Q^{\sim} \\ &= M_{11} \end{aligned}$$

$$Y_{12} = \left(P^{\sim}Q^{\sim}\right)^{\#}P^{\sim}Q^{\sim}\left(P^{\sim}Q^{\sim}\right)^{\#}P^{\sim} = \left(P^{\sim}Q^{\sim}\right)^{\#}P^{\sim} = M_{12}$$

$$\begin{aligned} Y_{21} &= \left(Q^{\sim}P^{\sim}\right)^{\#}Q^{\sim}P^{\sim}\left(Q^{\sim}P^{\sim}\right)^{\#}Q^{\sim} - Q^{\sim}\left(P^{\sim}Q^{\sim}\right)^{\#}\left(P^{\sim}\right)^{2}\left(Q^{\sim}P^{\sim}\right)^{\#}P^{\sim}\left(Q^{\sim}P^{\sim}\right)^{\#}Q^{\sim} \\ &\quad + Q^{\sim}\left(P^{\sim}Q^{\sim}\right)^{\#}P^{\sim}\left(P^{\sim}Q^{\sim}\right)^{\#}\left(P^{\sim}\right)^{2}\left(Q^{\sim}P^{\sim}\right)^{\#}\left(Q^{\sim}P^{\sim}\right)\left(Q^{\sim}P^{\sim}\right)^{\#}Q^{\sim} \\ &\quad - Q^{\sim}\left(P^{\sim}Q^{\sim}\right)^{\#}\left(P^{\sim}\right)^{2}\left(Q^{\sim}P^{\sim}\right)^{\#}Q^{\sim}\left(P^{\sim}Q^{\sim}\right)^{\#}P^{\sim} \\ &\quad + Q^{\sim}\left(P^{\sim}Q^{\sim}\right)^{\#}\left(P^{\sim}\right)^{2}\left(Q^{\sim}P^{\sim}\right)^{\#}Q^{\sim}\left(P^{\sim}Q^{\sim}\right)^{\#}\left(P^{\sim}\right)^{2}\left(Q^{\sim}P^{\sim}\right)^{\#}Q^{\sim} \\ &= \left(Q^{\sim}P^{\sim}\right)^{\#}Q^{\sim} - Q^{\sim}\left(P^{\sim}Q^{\sim}\right)^{\#}\left(P^{\sim}\right)^{2}\left(Q^{\sim}P^{\sim}\right)^{\#} \\ &\quad + Q^{\sim}\left(P^{\sim}Q^{\sim}\right)P^{\sim}\left(P^{\sim}Q^{\sim}\right)^{\#}\left(P^{\sim}\right)^{2}\left(Q^{\sim}P^{\sim}\right)^{\#}Q^{\sim} \end{aligned}$$

$$\begin{aligned} Y_{22} &= -Q^{\sim}\left(P^{\sim}Q^{\sim}\right)^{\#}\left(P^{\sim}\right)^{2}\left(Q^{\sim}P^{\sim}\right)^{\#}Q^{\sim}\left(P^{\sim}Q^{\sim}\right)^{\#}P^{\sim} \\ &= -Q^{\sim}\left(P^{\sim}Q^{\sim}\right)^{\#}\left(P^{\sim}\right)\left(Q^{\sim}P^{\sim}\right)^{\#} \\ &= M_{22} \end{aligned}$$

$$\Rightarrow XMX = \begin{pmatrix} M_{11} & M_{12} \\ M_{21} & M_{22} \end{pmatrix} = X$$

Now

$$\begin{aligned} X_{11} &= \left(P^{\sim}\right)^{2}\left(Q^{\sim}P^{\sim}\right)^{\#}Q^{\sim} + P^{\sim}Q^{\sim}\left(P^{\sim}Q^{\sim}\right)^{\#}P^{\sim} - P^{\sim}Q^{\sim}\left(P^{\sim}Q^{\sim}\right)^{\#}\left(P\right)^{2}\left(Q^{\sim}P^{\sim}\right)^{\#}Q^{\sim} \\ &= \left(P^{\sim}\right)^{2}\left(Q^{\sim}P^{\sim}\right)^{\#}Q^{\sim} + P^{\sim} - \left(P^{\sim}\right)^{2}\left(Q^{\sim}P^{\sim}\right)^{\#}Q^{\sim} \\ &= P^{\sim} \end{aligned}$$

$$X_{12} = P^{\sim}Q^{\sim}\left(P^{\sim}Q^{\sim}\right)^{\#}P^{\sim} = P^{\sim}$$

$$X_{21} = Q^{\sim}P^{\sim}\left(Q^{\sim}P^{\sim}\right)^{\#}Q^{\sim} = Q^{\sim}$$

$$\Rightarrow MXM = \begin{pmatrix} P^{\sim} & P^{\sim} \\ Q^{\sim} & 0 \end{pmatrix} = M$$

$\Rightarrow X = M^{\#}$. □

Theorem 2. Let $M = \begin{pmatrix} P^{\sim} & Q^{\sim} \\ P^{\sim} & 0 \end{pmatrix}$ in*M*, where $P, Q \in F^{n\times n}(\mathcal{C})$,

$$rank\left(Q^{\sim}\right) \geq rank\left(P^{\sim}\right) = r$$

Then,

1) the group inverse of *M* exists in *M* if and only if

$$rank\left(P^{\sim}\right) = rank\left(Q^{\sim}\right) = rank\left(P^{\sim}Q^{\sim}\right) = rank\left(Q^{\sim}P^{\sim}\right).$$

2) if the group inverse of *M* exists *inM*, then $M^{\#} = \begin{pmatrix} Z_{11} & Z_{12} \\ Z_{21} & Z_{22} \end{pmatrix}$, where

$$Z_{11} = \left(P^{\sim}Q^{\sim}\right)^{\#} P^{\sim} - Q^{\sim}\left(P^{\sim}Q^{\sim}\right)^{\#}\left(P^{\sim}\right)^{2}\left(Q^{\sim}P^{\sim}\right)^{\#}$$

$$Z_{12} = Q^{\sim}\left(P^{\sim}Q^{\sim}\right)^{\#} - \left(P^{\sim}Q^{\sim}\right)^{\#}\left(P^{\sim}\right)^{2}\left(Q^{\sim}P^{\sim}\right)Q^{\sim}$$
$$+ Q^{\sim}\left(P^{\sim}Q^{\sim}\right)^{\#}(P^{\sim})^{2}\left(Q^{\sim}P^{\sim}\right)^{\#}P^{\sim}\left(Q^{\sim}P^{\sim}\right)^{\#}Q^{\sim}$$

$$Z_{21} = \left(P^{\sim}Q^{\sim}\right)^{\#} P^{\sim}$$

$$Z_{22} = -\left(P^{\sim}Q^{\sim}\right)^{\#}\left(P^{\sim}\right)^{2}\left(Q^{\sim}P^{\sim}\right)^{\#}Q^{\sim}$$

Proof. 1) Given $rank\left(Q^{\sim}\right) \geq rank\left(P^{\sim}\right) = r$. Suppose $rank\left(P^{\sim}\right) = rank\left(Q^{\sim}\right)$ then,

$$rank\left(P^{\sim}\right)^{2} = rank\left(P^{\sim}Q^{\sim}\right).$$

We know that

$$rank\left(P^{\sim}Q^{\sim}\right) = rank\left(P^{\sim}\right)$$

so,

$$rank\left(P^{\sim}\right)^{2} = rank\left(P^{\sim}\right).$$

Therefore the group inverse of M exists in$M$. Now we show that the condition is necessary,

$$rank(M) = rank\begin{pmatrix} P^{\sim} & P^{\sim} \\ Q^{\sim} & 0 \end{pmatrix} = rank\begin{pmatrix} 0 & P^{\sim} \\ Q^{\sim} & 0 \end{pmatrix} = rank\left(P^{\sim}\right) + rank\left(Q^{\sim}\right)$$

$$M^{2} = \begin{pmatrix} P^{\sim} & Q^{\sim} \\ P^{\sim} & 0 \end{pmatrix} = \begin{pmatrix} \left(P^{\sim}\right)^{2} + Q^{\sim}P^{\sim} & P^{\sim}Q^{\sim} \\ \left(P^{\sim}\right)^{2} & P^{\sim}Q^{\sim} \end{pmatrix}$$

$$\Rightarrow rank\left(M^{2}\right) = rank\begin{pmatrix} \left(P^{\sim}\right)^{2} + Q^{\sim}P^{\sim} & P^{\sim}Q^{\sim} \\ \left(P^{\sim}\right)^{2} & P^{\sim}Q^{\sim} \end{pmatrix} = rank\begin{pmatrix} Q^{\sim}P^{\sim} & 0 \\ \left(P^{\sim}\right)^{2} & P^{\sim}Q^{\sim} \end{pmatrix}$$

Since the group inverse of M exists in M if and only if $rank(M) = rank\left(M^{2}\right)$. We know

$$
\begin{aligned}
rank\left(P^{\sim}\right)+rank\left(Q^{\sim}\right) &= rank\left(M^{2}\right)\\
&\leq rank\begin{pmatrix} Q^{\sim}P^{\sim} \\ \left(P^{\sim}\right)^{2} \end{pmatrix}+rank\left(P^{\sim}Q^{\sim}\right)\\
&\leq rank\left(P^{\sim}\right)+rank\left(P^{\sim}Q^{\sim}\right)
\end{aligned}
$$

Also

$$
\begin{aligned}
rank\left(P^{\sim}\right)+rank\left(Q^{\sim}\right) &= rank\left(M^{2}\right)\\
&\leq rank\left(Q^{\sim}P^{\sim}\right)+rank\left(\left(P^{\sim}\right)^{2} \quad P^{\sim}Q^{\sim}\right)\\
&\leq rank\left(Q^{\sim}P^{\sim}\right)+rank\left(P^{\sim}\right).
\end{aligned}
$$

Then $rank\left(Q^{\sim}\right)\leq rank\left(P^{\sim}Q^{\sim}\right)$ and $rank\left(Q^{\sim}\right)\leq rank\left(Q^{\sim}P^{\sim}\right)$ Therefore $rank\left(Q^{\sim}\right)=rank\left(P^{\sim}Q^{\sim}\right)=rank\left(Q^{\sim}P^{\sim}\right)$

From

$$rank\left(Q^{\sim}\right)=rank\left(P^{\sim}Q^{\sim}\right)\leq rank\left(P^{\sim}\right)$$

and

$$rank\left(P^{\sim}\right)=rank\left(P^{\sim}Q^{\sim}\right)\leq rank\left(Q^{\sim}\right)$$

we have

$$rank\left(P^{\sim}\right)=rank\left(Q^{\sim}\right)$$

Since

$$rank\left(P^{\sim}\right)+rank\left(Q^{\sim}\right)\leq rank\left(P^{\sim}Q^{\sim} \quad \left(P^{\sim}\right)^{2}\right)+rank\left(Q^{\sim}P^{\sim}\right)$$

and

$$rank\left(P^{\sim}Q^{\sim} \quad \left(P^{\sim}\right)^{2}\right)\leq rank\left(P^{\sim}\right)\leq rank\left(P^{\sim}Q^{\sim} \quad \left(P^{\sim}\right)^{2}\right),$$

we get

$$rank\left(P^{\sim}Q^{\sim} \quad \left(P^{\sim}\right)^{2}\right)=rank\left(P^{\sim}\right).$$

Thus

$$rank\left(P^{\sim}Q^{\sim} \quad \left(P^{\sim}\right)^{2}\right)=rank\left(P^{\sim}Q^{\sim}\right).$$

Then there exist a matrix $U^{\sim}\in F^{n\times n}(\mathcal{C})$ such that $P^{\sim}Q^{\sim}U^{\sim}=\left(P^{\sim}\right)^{2}$. Thus

$$rank\left(M^{\sim}\right)^{2}=rank\begin{pmatrix} P^{\sim}Q^{\sim} & 0 \\ 0 & Q^{\sim}P^{\sim}\end{pmatrix}=rank\left(P^{\sim}Q^{\sim}\right)+rank\left(Q^{\sim}P^{\sim}\right)$$

So, we get $rank\left(P^{\sim}\right)=rank\left(Q^{\sim}\right)=rank\left(P^{\sim}Q^{\sim}\right)=rank\left(Q^{\sim}P^{\sim}\right)$.

2) Proof is same as Theorem 1 2).

Theorem 3. Let $Let \;\; P,Q\in F^{n\times n}(\mathcal{C})$ if

$$rank\left(Q^{\sim}\right)=rank\left(P^{\sim}Q^{\sim}\right)=rank\left(Q^{\sim}P^{\sim}\right).$$

Then $P^{\sim}Q^{\sim}$ and $Q^{\sim}P^{\sim}$ are similar.

Proof. Suppose ramk (P) = r, then by using Lemma 1, there are unitary matrices $A,B\in F^{n\times n}(\mathcal{C})$ such

$$P^{\sim}=B^{\sim}\begin{pmatrix} I_r & 0 \\ 0 & 0\end{pmatrix}A^{\sim}, \quad Q^{\sim}=\left(A^{*}\right)^{\sim}\begin{pmatrix} Q_1^{*} & -YQ_1^{*} \\ -Q_1^{*}X & YQ_1^{*}X\end{pmatrix}\left(B^{*}\right)^{\sim}$$

that

where $Q_1^{*}\in F^{r\times r}, X\in F^{r\times(n-r)}, Y\in F^{(n-r)\times r}$. Hence

$$\begin{aligned} P^{\sim}Q^{\sim} &= B^{\sim}\begin{pmatrix} Q_1^{*} & -YQ_1^{*} \\ 0 & 0\end{pmatrix}\left(B^{*}\right)^{\sim} \\ &= B^{\sim}\begin{pmatrix} I_r & Y \\ 0 & I_{n-r}\end{pmatrix}\begin{pmatrix} Q_1^{*} & 0 \\ 0 & 0\end{pmatrix}\begin{pmatrix} I_r & -Y \\ 0 & I_{n-r}\end{pmatrix}\left(B^{*}\right)^{\sim}\end{aligned}$$

$$
\begin{aligned}
Q^{\sim}P^{\sim} &= \left(A^{*}\right)^{\sim}\begin{pmatrix} Q_1^{*} & 0 \\ -Q_1^{*}X & 0 \end{pmatrix}A^{\sim} \\
&= \left(A^{*}\right)^{\sim}\begin{pmatrix} I_r & 0 \\ -X & I_{n-r} \end{pmatrix}\begin{pmatrix} Q_1^{*} & 0 \\ 0 & 0 \end{pmatrix}\begin{pmatrix} I_r & 0 \\ X & I_{n-r} \end{pmatrix}A^{\sim}
\end{aligned}
$$

So $P^{\sim}Q^{\sim}$ and $Q^{\sim}P^{\sim}$ are similar.

REFERENCES

1. Zhuang, W. (1987) Involutory Functions and Generalized Inverses of Matrices over an Arbitrary Skew Fields. Northeast Math, 1, 57-65.
2. Golub, G.H. and Greif, C. (2003) On Solving Blocked-Structured Indefinite Linear Systems. SIAM Journal on Scientific Computing, 24, 2076-2092
3. Ipsen, I.C.F. (2001) A Note on Preconditioning Nonsymmetric Matrices. SIAM Journal on Scientific Computing, 23, 1050-1051
4. Campbell, S.L. and Meyer, C.D. (2013) Generalized Inverses of Linear Transformations. Dover, New York.
5. Bu, C. (2002) On Group Inverses of Block Matrices over Skew Fields. Journal of Mathematics, 35, 49-52
6. Bu, C., Zhao, J. and Zheng, J. (2008) Group inverse for a Class 2 × 2 Block Matrices over Skew Fields. Computers & Mathematics with Applications, 204, 45-49. http://dx.doi.org/10.1016/j.amc.2008.05.145
7. Cao, C. (2001) Some Results of Group Inverses for Partitioned Matrices over Skew Fields. Heilongjiang Daxue Ziran Kexue Xuebao, 18, 5-7.
8. Cao, C. and Tang, X. (2006) Representations of the Group Inverse of Some 2 × 2 Block Matrices. International Mathematical Forum, 31, 1511-1517. http://dx.doi.org/10.12988/imf.2006.06127
9. Chen, X. and Hartwig, R.E. (1996) The Group Inverse of a Triangular Matrix. Linear Algebra and Its Applications, 237/238, 97-108. http://dx.doi.org/10.1016/0024-3795(95)00561-7
10. Catral, M., Olesky, D.D. and van den Driessche, P. (2008) Group Inverses of Matrices with Path Graphs. The Electronic Journal of Linear Algebra, 1, 219-233. http://dx.doi.org/10.13001/1081-3810.1260
11. Cao, C. (2006) Representation of the Group Inverse of Some 2 × 2 Block Matrices. International Mathematical Forum, 31, 1511-1517.
12. Krishnaswamy, D. and Punithavalli, G. (2013) The Anti-Reflexive Solutions of the Matrix Equation A × B=C in Minkowski Space M. International Journal of Research and Reviews in Applied Sciences, 15, 2-9
13. Krishnaswamy, D. and Punithavalli, G. (2013) The Re-nnd Definite Solutions of the Matrix Equation A × B=C in Minkowski Space M. International Journal of Fuzzy Mathematical Archive, 2, 70-77.
14. Krishnaswamy, D. and Punithavalli, G. Positive Semidefinite (and Definite) M-Symmetric Matrices Using Schur Complement in Minkowski Space M. (Preprint)

CHAPTER 12

Computing Approximation GCD of Several Polynomials by Structured Total Least Norm*

Xuefeng Duan[1], Xinjun Zhang[1], Qingwen Wang[2]

[1]College of Mathematics and Computational Science, Guilin University of Electronic Technology, Guilin, China

[2]Department of Mathematics, Shanghai University, Shanghai, China

ABSTRACT

The task of determining the greatest common divisors (GCD) for several polynomials which arises in image compression, computer algebra and speech encoding can be formulated as a low rank approximation problem with Sylvester matrix. This paper demonstrates a method based on structured total least norm (STLN) algorithm for matrices with Sylvester structure. We demonstrate the algorithm to compute an approximate GCD. Both the theoretical analysis and the computational results show that the method is feasible.

Keywords: Sylvester Matrix; Approximate Greatest Common Divisor; Low Rank Approximation; Structured Total Least Norm; Numerical Method

1. INTRODUCTION

Let $\deg(f(x))$ be the degree of $f(x)$ and $C[x]$ be the set of univariate polynomials. $\|A\|_2$ stands for the spectral norm of the matrix *A*. C^n and $C^{m\times n}$ are the vector spaces of complex *n* vectors and *m*×*n* matrices, respectively. Transpose matrices and vectors are denoted by A^T and u^T. *GCD*(*f*,*g*) denotes the greatest common divisor for the polynomials *f* and *g*. We use *rank* (*A*) to stand for the rank of matrix *A*.

In this paper, we consider the following problem. Let $f_1(x),\ f_2(x),\ \cdots,\ f_t(x) \in C[x]\setminus\{0\}$, namely

$$f_1(x) = a_n x^n + a_{n-1}x^{n-1} + \cdots + a_1 x + a_0,$$

$$f_i(x) = b_{ip}x^p + b_{i(p-1)}x^{p-1} + \cdots + b_{i1}x + b_{i0}, i = 2, \cdots, t$$

Problem 1.1. Set *k* be a positive integer with $k \le \min(n, p)$. We wish to compute $\Delta f_1(x),\ \Delta f_2(x), \cdots,\ \Delta f_t(x) \in C[x]\setminus\{0\}$, such that

$$\deg(\Delta f_1(x)) \le n,\ \deg(\Delta f_i(x)) \le p,\ 2 \le i \le t$$

$$\deg(GCD(f_1(x) + \Delta f_1(x), f_2(x) + \Delta f_2(x), \cdots, f_t(x) + \Delta f_t(x))) \ge k$$

and

$$\|\Delta f_1(x)\|_2^2 + \|\Delta f_2(x)\|_2^2 + \cdots + \|\Delta f_t(x)\|_2^2$$

is minimized.

The problem of computing approximate GCD of several polynomials is widely applied in speech encoding and filter design [1], computer algebra [2] and signal processing [3] and has been studied in [4-7] in recent years. Several methods to the problem have been presented. The generally-used computational method is based on the truncated singular decomposition (TSVD) [8] which may not be appropriate when a matrix has a special structure since they do not preserve the special structure (for example, Sylvester matrix). Another common method based on QR decomposition [9,10] may suffer from loss of accuracy when it is applied to ill-conditioned problems and the algorithm derived in [11] can produce a more accurate result for ill-conditioned problems. Cadzow algorithm [12] is also a popular method to solve this problem which has been rediscovered in the literature [13].

Somehow it only finds a structured low rank matrix that is nearby a given target matrix but certainly is not the closet even in the local sense. Another method is based on alternating projection algorithm [14]. Although the algorithm can be applied to any low rank and any linear structure, the speed may be very slow. Some other methods have been proposed such as the ERES method [15], STLS method [16] and the matrix pencil method [17]. An approach to be described is called Structured Total Least Norm (STLN) which has been described for Hankel structure low rank approximation [18,19] and Sylvester structure low rank approximation with two polynomials [20]. STLN is a problem formulation for obtaining an approximate solution $(A+E)X = B+H$ to an overdetermined linear system *AX*=*B* preserving the given structure in *A* or[*A B*].

In this paper, we apply the algorithm to compute the structured preserving rank reduction of Sylvester matrix. We introduce some notations and discuss the relationship between the GCD problems and low rank approximation of Sylvester matrices in Section 2. Based on STLN method, we describe the algorithm to solve Problem 1.1 in Section 3. In Section 4, we use some examples to illustrate the method is feasible.

2. MAIN RESULTS

First of all, we shall prove that Problem 1.1 always has a solution.

Theorem 2.1. Suppose that f_1, f_2, $\cdots$, f_t, $\deg(f_1)$, $\deg(f_2)$, $\cdots$, $\deg(f_t)$ and *k* are defined as those in Problem 1.1. There exist $\hat{f}_1$, $\hat{f}_i \in C[x]$, with, $\deg(\hat{f}_1) \le n$, $\deg(\hat{f}_i) \le p$ and

$\deg\left(GCD\left(\hat{f}_1, \hat{f}_2, \cdots, \hat{f}_t\right)\right) \ge k$ such that for all $\overline{f}_1, \overline{f}_i \in C[x]$ with $\deg(\overline{f}_1) \le n$, $\deg(\overline{f}_i) \le p$, and

$$\deg\left(GCD\left(\overline{f}_1, \overline{f}_2, \cdots, \overline{f}_t\right)\right) \ge k, \quad 2 \le i \le t.$$

We have

$$\left\|\hat{f}_1 - f_1\right\|_2^2 + \left\|\hat{f}_2 - f_2\right\|_2^2 + \cdots + \left\|\hat{f}_t - f_t\right\|_2^2 \le \left\|\overline{f}_1 - f_1\right\|_2^2 + \left\|\overline{f}_2 - f_2\right\|_2^2 + \cdots + \left\|\overline{f}_t - f_t\right\|_2^2$$

Proof. Let $h \in C[x]$ be monic with $\deg(h) = k$ and set $u_i \in C[x]$ with $\deg(u_i) \le \deg(f_i) - k$. For the real and imaginary parts of the coefficients of h and of u_i $(1 \le i \le t)$. We are considered with the continuous objective function

$$F(h, u_1, u_2, \cdots, u_t) = \left\|u_1 h - f_1\right\|_2^2 + \left\|u_2 h - f_2\right\|_2^2 + \cdots + \left\|u_t h - f_t\right\|_2^2$$

We will prove that the function has a value on a closed and bounded set of its real argument vector which is smaller than elsewhere. Consider $\overline{f}_1 = a_n x^n$ and $\overline{f}_i = b_{ip} x^p$ with a GCD of degree $\ge k$ for $2 \le i \le t$. Clearly, any h and u_i with

$$F(h, u_1, u_2, \cdots, u_t) > \left\|\overline{f}_1 - f_1\right\|_2^2 + \left\|\overline{f}_2 - f_2\right\|_2^2 + \cdots + \left\|\overline{f}_t - f_t\right\|_2^2$$

can be discarded. So from above,we know that the coefficients of $u_1 h,\ u_2 h,\ \cdots,\ u_t h$ can be bounded and so can the coefficients of $h,\ u_1,\ u_2,\ \cdots,\ u_t$ by any polynomials factor coefficient bound. Thus the function's domain $F(h, u_1, u_2, \cdots, u_t)$ is restricted to a sufficiently large ball. It remains to exclude $u_1 = u_2 = \cdots = u_t = 0$ as the minimal solution. We have

$$F(h, 0, 0, \cdots, 0) = \left\|f_1\right\|_2^2 + \left\|f_2\right\|_2^2 + \cdots + \left\|f_t\right\|_2^2$$
$$> \left\|\overline{f}_1 - f_1\right\|_2^2 + \left\|\overline{f}_2 - f_2\right\|_2^2 + \cdots + \left\|\overline{f}_t - f_t\right\|_2^2.$$

In conclusion, the theorem is true.

Now we begin to solve Problem 1.1, we first define a $p \times (n+p)$ matrix associated with $f_1(x)$ as follows

$$S_1 = \begin{bmatrix} a_n & a_{n-1} & a_{n-2} & \cdots & a_1 & a_0 & 0 & \cdots & 0 \\ 0 & a_n & a_{n-1} & \cdots & \cdots & a_1 & a_0 & \ddots & \\ \vdots & & \ddots & \ddots & & & \ddots & \ddots & 0 \\ 0 & \cdots & 0 & a_n & a_{n-1} & \cdots & \cdots & a_1 & a_0 \end{bmatrix}$$

and an $n\times(n+p)$ matrix associated with $f_i(x), i = 2,3,\cdots,t$ as

$$S_i = \begin{bmatrix} b_{ip} & b_i(p-1) & b_i(p-2) & \cdots & b_{i1} & b_{i0} & 0 & \cdots & \cdots & 0 \\ 0 & b_{ip} & b_i(p-1) & \cdots & \cdots & b_{i1} & b_{i0} & 0 & \cdots & 0 \\ \vdots & \ddots & \ddots & & & & \ddots & \ddots & & \vdots \\ \vdots & \ddots & \ddots & \ddots & & & & \ddots & \ddots & 0 \\ 0 & \cdots & 0 & b_{ip} & b_i(p-1) & \cdots & \cdots & \cdots & b_{i1} & b_{i0} \end{bmatrix}$$

An extended Sylvester matrix or a generalized resultant is then defined by

$$S = \begin{bmatrix} S_1 \\ S_2 \\ \vdots \\ S_t \end{bmatrix}^{\mathrm{T}} \in C^{(p+n)\times((t-1)n+p)}$$

Deleting the last k-1 rows of s and the last k-1 columns of coefficients of $f_1, f_2, \cdots, f_t$ separately is s, We get the k-th Sylvester matrix $S_k \in C^{(n+p-k+1)\times((t-1)n+p-tk+t)}$

$$S_k = \begin{bmatrix} a_n & 0 & & b_{2p} & & & \cdots & b_{tp} & & \\ a_{n-1} & \ddots & & b_{2(p-1)} & \ddots & & \cdots & b_{t(p-1)} & \ddots & \\ \vdots & \ddots & a_n & \vdots & \ddots & b_{2p} & \vdots & \vdots & \ddots & b_{tp} \\ a_0 & & a_{n-1} & b_{10} & & b_{2(p-1)} & \cdots & b_{t0} & & b_{t(p-1)} \\ & \ddots & \vdots & & \ddots & \vdots & \vdots & & \ddots & \vdots \\ & & a_0 & & & b_{20} & \cdots & & & b_{t0} \end{bmatrix}$$

It is well-known that $(GCD(f_1 \cdots f_t)) \geq k$ if and only if $S_k(f_1 \cdots f_t)$ has rank deficiency at least 1. We have

$$\min_{\deg(GCD(\bar{f}_1 \cdots \bar{f}_t)) \geq k} \left\|\bar{f}_1 - f_1\right\|_2^2 + \left\|\bar{f}_2 - f_2\right\|_2^2 + \cdots + \left\|\bar{f}_t - f_t\right\|_2^2$$

$$\Leftrightarrow \min_{\text{dimNullspace}(\bar{S}_k) \geq 1} \left\|\bar{f}_1 - f_1\right\|_2^2 + \left\|\bar{f}_2 - f_2\right\|_2^2 + \cdots + \left\|\bar{f}_t - f_t\right\|_2^2 \quad (2.1)$$

where $\bar{S}_k$ is the k-th Sylvester matrix generated by $\bar{f}_1, \bar{f}_2, \cdots, \bar{f}_t$.

From above, we know that (2.1) can be transformed to the low rank approximation of a Sylvester matrix.

If we use STLN [16] to solve the following overdetermined system

$A_k X = b_k$

for $S_k = [b_k \quad A_k]$, where b_k is the first column of S_k and A_k are the remainder columns of S_k, then we get a minimal perturbation $[h_k \quad E_k]$ of Sylvester structure satisfying

$$b_k + h_k \in A_k + E_k$$

So the solution with Sylvester structure is $\bar{S}_k = [h_k + b_k \quad E_k + A_k]$ and $\text{dimNullspace}(\bar{S}_k) > 1$.

We will give the following example and theorem to explain why we choose the first column to form the overdetermined system.

Example 2.1. Suppose three polynomials are given

$$f_1(x) = x^2 - 1,$$

$$f_2(x) = x^2 + x - 2,$$

$$f_3(x) = x^2 + 2x - 3.$$

The matrix S is the Sylvester matrix generated by

$f_1(x)$, $f_2(x)$ and $f_3(x)$

$$S = \begin{bmatrix} 1 & 0 & 1 & 0 & 1 & 0 \\ 0 & 1 & 1 & 1 & 2 & 1 \\ -1 & 0 & -2 & 1 & -3 & 2 \\ 0 & -1 & 0 & -2 & 0 & -3 \end{bmatrix}$$

The matrix S is partitioned as $S = \begin{bmatrix} \hat{b}_1 & \hat{A}_1 \end{bmatrix}$ or $S = \begin{bmatrix} \overline{A}_1 & \overline{b}_1 \end{bmatrix}$, where $\hat{b}_1$ is the first column of S, whereas $\overline{b}_1$ is the last column of S.

The overdetermined system

$$\hat{A}_1 X = \hat{b}_1$$

has a solution $X = \begin{bmatrix} 1 & 1 & 1 & -1 & -1 \end{bmatrix}^T$, while the system

$$\overline{A}_1 X = \overline{b}_1$$

has no solution.

Theorem 2.2. Suppose that, $f_1, \ f_2, \ \cdots, \ f_t$, $\deg(f_1), \ \deg(f_2), \ \cdots, \ \deg(f_t)$ and k are defined as those in Problem 1.1 and S_k is the k-th Sylvester matrix of $f_1, \ f_2, \ \cdots, \ f_t$. Then the following statements are equivalent.

1) $\dim \text{Nullspace}(S_k) \geq 1$;

2) $A_k X = b_k$ has a solution, where b_k is the first column of S_k and A_k are the remainder columns of S_k.

Proof. $\Leftarrow$ Suppose $A_k X = b_k$ has a nonzero solution, then $b_k \in \text{Range}(A_k)$. Since b_k is the first column of S_k, obviously, the dimension of the rank deficiency of $S_k = \begin{bmatrix} b_k & A_k \end{bmatrix}$ is at least 1.

$\Rightarrow$ Suppose the rank deficiency of $S_k = \begin{bmatrix} b_k & A_k \end{bmatrix}$ is at least 1 and, $D(x) = GCD(f_1, f_2, \cdots, f_t)$, $f_1^* = f_1/D(x), \quad f_2^* = f_2/D(x), \quad \cdots, \quad f_t^* = f_t/D(x)$. Multiplying the vector $V = \begin{bmatrix} x^{n+p-k}, x^{n+p-k-1}, \cdots x, 1 \end{bmatrix}$ to the matrix S_k, then we obtain

$$VS_k = \left[x^{p-k} f_1, x^{p-k-1} f_1, \cdots, f_1, x^{n-k} f_2, \cdots, f_2, \cdots, x^{n-k} f_t, \cdots, f_t\right] \quad (2.2)$$

Next we will prove that $A_k X = b_k$ has a solution. If we multiply the vector V to two sides of the equation $A_k X = b_k$, it turns out to be

$$\left[x^{p-k-1} f_1, \cdots, f_1, x^{n-k} f_2, \cdots, f_2, \cdots, x^{n-k} f_t, \cdots, f_t\right] X$$
$$= x^{p-k} f_1 \quad (2.3)$$

The solution X of equation (2.3) is equal to the coefficients of polynomials $u_1, \quad u_2, \quad \cdots, \quad u_t$ such that

$$x^{p-k} f_1 = u_1 f_1 + u_2 f_2 + \cdots + u_t f_t$$

We can get $\deg(D(x)) \geq k$ and $\deg(f_1^*) \leq n-k$, $\deg(f_i^*) \leq p-k, \quad (2 \leq i \leq t)$ from $\dim \mathrm{Nullspace}(S_k) \geq 1$.

Dividing x^{p-k} by $f_2^* + \cdots + f_t^*$, we obtain a quotient q and a remainder m satisfy

$$x^{p-k} = q\left(f_2^* + \cdots + f_t^*\right) + m.$$

where $\deg(q) \leq \deg(D(x)) - k, \quad \deg(m) \leq p-k-1$. Now we can get that

$$u_1 = p, u_2 = mf_1^*, \cdots, u_t = mf_1^*$$

are solutions of Equation (2.3), since

$$\deg(u_1) \leq \deg(D(x)) - k,$$

$$\deg(u_i) = \deg(q) + \deg(f_1^*)$$
$$\leq \deg(D(x)) - k + \deg(f_1^*) \leq n-k,$$

and

$$u_1 f_1 + u_2 f_2 + \cdots + u_t f_t = fq\left(f_2^* + \cdots + f_t^*\right) + fp = fx^{p-k}$$

Next, we will illustrate for any given Sylvester matrixas long as all the elements are allowed to be perturbed, we can always find *k*-Sylvester structure matrices $[h_k \;\; E_k]$ satisfy $b_k + h_k \in \mathrm{Range}(A_k + E_k)$, where b_k is the first column of S_k and A_k are the remainder column of S_k.

Theorem 2.3. Given the positive integer *n,p,t,* there exists a Sylvester matrix $\hat{S} \in C^{(n+p)\times(p+(t-1)n)}$ with rank deficiency k.

Proof. We can always find polynomials $f_1, f_2, \cdots, f_t \in C[x]$ with $\deg(f_1) = n$, $\deg(f_i) = p, 2 \le i \le t$ and $\deg(GCD(f_1, f_2, \cdots, f_t)) = k$. Hence *S* is the Sylvester matrix of $f_1, f_2, \cdots, f_t$ and its rank deficiency is k.

Corollary 2.1. Given the positive integer *n,p,t,* and *k*-th Sylvester matrix $S_k = [A_k \;\; b_k]$, where $A_k \in C^{(n+p-k+1)\times(p+(t-1)n-tk+t)}$, $b_k \in C^{(n+p-k+1)\times 1}$, it is always possible to find a *k*-th Sylvester structure perturbation $[h_k \;\; E_k]$ such that $b_k + h_k \in \mathrm{Range}(A_k + E_k)$.

3. STLN FOR OVERDETERMINED SYSTEM WITH SYLVESTER STRUCTURE

In this section, we will use STLN method to solve the overdetermined system

$$A_k X = b_k,$$

According to theorem 2.3 and corollary 2.1, we can always find Sylvester structure $[h_k \;\; E_k]$ with $h_k + b_k \in \mathrm{Range}(A_k + E_k)$. Next we will use STLN method to find the minimum solution.

First, we define the Sylvester structure preserving perturbation $[h_k \;\; E_k]$ of S_k

$$\begin{bmatrix} h_k & E_k \end{bmatrix} = \begin{bmatrix} z_1 & & & z_{n+2} & & & \cdots & z_{n+(t-2)p+t} & & \\ z_2 & \ddots & & z_{n+3} & \ddots & & \cdots & z_{n+(t-2)p+t+1} & \ddots & \\ \vdots & \ddots & z_1 & \vdots & \ddots & z_{n+2} & \vdots & \vdots & \ddots & z_{n+(t-2)p+t} \\ z_{n+1} & & z_2 & z_{n+p+2} & & z_{n+3} & \cdots & z_{n+(t-1)p+t} & & z_{n+(t-2)p+t+1} \\ & \ddots & \vdots & & \ddots & \vdots & \vdots & & \ddots & \vdots \\ & & z_{n+1} & & & z_{n+p+2} & \cdots & & & z_{n+(t-1)p+t} \end{bmatrix}$$

can be represented by a vector

$$Z = \left[z_1, z_2, \cdots, z_{(n+(t-1)p+t)} \right]^{\mathrm{T}}.$$

We can define a matrix P_k such that $h_k = P_k Z$.

$$P_k = \begin{bmatrix} I_{n+1} & 0 \\ 0 & 0 \end{bmatrix} \in C^{(n+p-k+1)\times(n+(t-1)p+t)},$$

where I_{n+1} is a $(n+1)\times(n+1)$ identity matrix.

We will solve the equality-constrained least squares problem

$$\min_{Z,X} \|Z\|_2^2, \text{subject to } r = 0. \tag{3.1}$$

where the structured residual r is

$$r = r(Z, X) = b_k + h_k - (A_k + E_k)X.$$

By using the penalty method, the formulation (3.1) can be transformed into

$$\min_{Z,X} \left\| \begin{matrix} wr(Z,X) \\ Z \end{matrix} \right\|_2^2, w \gg 1 \tag{3.2}$$

where w is a large penalty value.

Let ΔZ and ΔX stand for a small change in Z and X, respectively and ΔE_k be the corresponding change in E_k. Then the first order approximate to $r(Z+\Delta Z, X+\Delta X)$ is

$$r(Z+\Delta Z, X+\Delta X) = b_k + P_k(Z+\Delta Z) - (A_k + E_k + \Delta E_k)(X+\Delta X) \approx b_k + P_k Z - (A_k + E_k)X + P_k \Delta Z - (A_k + E_k)\Delta X - \Delta E_k X$$

Introducing a matrix of Sylvester structure Y_k and

$$X = \left[x_1, x_2, \cdots, x_{p+(t-1)n-tk+t-1}\right]^{\mathrm{T}}$$

(3.2) can be approximated by

$$\min_{\Delta X, \Delta Z} \left\| \begin{bmatrix} w(Y_k - P_k) & w(A_k + E_k) \\ I_{n+(t-1)p+t} & 0 \end{bmatrix} \begin{bmatrix} \Delta Z \\ \Delta X \end{bmatrix} + \begin{bmatrix} -wr \\ Z \end{bmatrix} \right\|_2 \quad (3.3)$$

where $Y_k \in C^{(n+p-k+1)\times(n+(t-1)p+t)}$ satisfies that

$$Y_k Z = E_k X \quad (3.4)$$

In the following, we present a method to obtain the matrix Y_k. Suppose $f_1, \ f_2, \ \cdots, \ f_t, \ E, \ Z$ and X are defined as above. Multiplying the vector

$$V = \left[x^{n+p-k}, x^{n+p-k-1}, \cdots, x, 1\right],$$

to the two sides of equation (3.4), it becomes

$$VY_k Z = VE_k X,$$

Let $\hat{X} = \begin{bmatrix} 0 \\ X \end{bmatrix}$, we obtain

$$VE_k X = V\left[h_k \quad E_k\right]\hat{X} = \hat{g}_1\hat{u}_1 + \hat{g}_2\hat{u}_2 + \cdots + \hat{g}_t\hat{u}_t \quad (3.5)$$

where $\hat{g}_1$ is the polynomial with degree n which is generated by the subvector of Z:

$$\begin{bmatrix} z_1 & z_2 & \cdots & z_{n+1} \end{bmatrix},$$

$\hat{g}_2$ is the polynomial with degree *p* which is generated by the subvector of *Z*:

$$\begin{bmatrix} z_{n+2} & z_{n+3} & \cdots & z_{n+p+2} \end{bmatrix}$$

⋮

$\hat{g}_t$ is the polynomial with degree *P* which is generated by the subvector of *Z*:

$$\begin{bmatrix} z_{n+(t-2)p+t} & z_{n+(t-2)p+t+1} & \cdots & z_{n+(t-1)p+t} \end{bmatrix},$$

$\hat{u}_1$ is the polynomial with degree *p*-*k*-1 which is generated by the subvector of *X*:

$$\begin{bmatrix} 0 & x_1 & \cdots & x_{p-k} \end{bmatrix}$$

$\hat{u}_2$ is the polynomial with degree *n*-*k* which is generated by the subvector of *X*:

$$\begin{bmatrix} x_{p-k+1} & x_{p-k+2} & \cdots & x_{n+p-2k+1} \end{bmatrix},$$

$\hat{u}_t$ is the polynomial with degree *n*-*k* which is generated by the subvector of *X*:

$$\begin{bmatrix} x_{p+(t-2)n-(t-1)k+t-1} & x_{p+(t-2)n-(t-1)k+t} & \cdots & x_{p+(t-1)n-tk+t-1} \end{bmatrix}$$

Here we will present a simple example to illustrate how to find Y_k.

Example 3.1. Suppose $k=1$, $X=\left[x_1,x_2,\cdots,x_7\right]$ and $f_1(x)=a_3x^3+a_2x^2+a_1x+a_0$, $f_2(x)=b_{22}x^2+b_{21}x+b_{20}$, $f_3(x)=b_{32}x^2+b_{31}x+b_{30}$.

$$A=\begin{bmatrix} 0 & b_{22} & 0 & 0 & b_{32} & 0 & 0 \\ a_3 & b_{21} & b_{22} & 0 & b_{31} & b_{32} & 0 \\ a_2 & b_{20} & b_{21} & b_{22} & b_{30} & b_{31} & b_{32} \\ a_1 & 0 & b_{20} & b_{21} & 0 & b_{30} & b_{31} \\ a_0 & 0 & 0 & b_{20} & 0 & 0 & b_{30} \end{bmatrix}, b_1=\begin{bmatrix} a_3 \\ a_2 \\ a_1 \\ a_0 \\ 0 \end{bmatrix},$$

then

$$Y_1=\begin{bmatrix} 0 & 0 & 0 & 0 & x_2 & 0 & 0 & x_5 & 0 & 0 \\ x_1 & 0 & 0 & 0 & x_3 & x_2 & 0 & x_6 & x_5 & 0 \\ 0 & x_1 & 0 & 0 & x_4 & x_3 & x_2 & x_7 & x_6 & x_5 \\ 0 & 0 & x_1 & 0 & 0 & x_4 & x_3 & 0 & x_7 & x_6 \\ 0 & 0 & 0 & x_1 & 0 & 0 & x_4 & 0 & 0 & x_7 \end{bmatrix}.$$

4. APPROXIMATE GCD ALGORITHM AND EXPERIMENTS

The following algorithm is designed to solve Problem 1.1.

Algorithm 4.1.

Input-A Sylvester matrix S generated by $f_1, \quad f_2, \cdots, \quad f_t$, respectively, an integer k and a tolerance *tol.*

Output-Polynomials $\overline{f}_1, \ \overline{f}_2, \ \cdots, \ \overline{f}_t$ and the Euclidean distance $\left\|\overline{f}_1 - f_1\right\|_2^2 + \left\|\overline{f}_2 - f_2\right\|_2^2 + \cdots + \left\|\overline{f}_t - f_t\right\|_2^2$ is to a minimum.

1) Form the k-th Sylvester matrix S_k as the above section, set the first column of S_k as b_k and the remainder columns of S_k as A_k. Let $E_k = 0, \ h_k = 0$.

2) Calculate X from $\min\left\|A_k X - b_k\right\|_2$ and $r = b_k - A_k X$. Compute P_k and Y_k as the above section.

3) Repeat

(1) $$\min_{\Delta X,\Delta Z}\left\|\begin{bmatrix} w(Y_k - P_k) & w(A_k + E_k) \\ I_{n+(t-1)p+t} & 0 \end{bmatrix}\begin{bmatrix}\Delta Z \\ \Delta X\end{bmatrix}+\begin{bmatrix}-wr \\ Z\end{bmatrix}\right\|_2$$

(2) Let $X = X + \Delta X,\ Z = Z + \Delta Z$

(3) Form the matrix E_k and h_k from Z, and Y_k from X. Let $A_k = A_k + E_k$

$b_k = b_k + h_k,\quad r = b_k - A_k X$ until $\|\Delta X\|_2 \le tol$ and $\|\Delta Z\|_2 \le tol$

4) Output the polynomials $\bar{f}_1,\ \bar{f}_2,\ \cdots,\ \bar{f}_t$ constructed from $[b_k\ \ A_k]$.

Given a tolerance ε, we can use the Algorithm 4.1 to compute an approximate GCD of $f_1, f_2, \cdots, f_t$. The method begin with $k \le \min\{n, p\}$ using Algorithm 4.1 to compute the minimum perturbation $N = \sum_i \|\bar{f}_i - f_i\|_2^2$

with $\dim\mathrm{Nullspace}(S) \ge 1$. If $N < \varepsilon$, then we can compute the approximate GCD form matrix S_k. Otherwise, we reduce k by one and repeat the Algorithm 4.1.

Example 4.1. We wish to find the minimal polynomial perturbations Δf and Δg of

$f = x^2 - 6x + 5$,

$g = x^2 - 6.3x + 5.72$,

satisfy that the polynomials $\Delta f + f$ and $\Delta g + g$ have a common root. We take two cases into account.

Case 1: The leading coefficients can be perturbed. Let $k = 1$ and $tol = 10^{-3}$, after 3 iterations, we get the polynomials $\bar{f}$ and $\bar{g}$

$\bar{f} = 0.9850x^2 - 6.0029x + 4.9994$,

$\bar{g} = 1.0150x^2 - 6.2971x + 5.7206$,

with a minimum distance

$N = \|\bar{f} - f\|_2^2 + \|\bar{g} - g\|_2^2 = 0.0004663$.

Case 2: The leading coefficients can be perturbed. Let K=1 and $tol = 10^{-3}$, after 3 iterations, we have the polynomials $\bar{f}$ and $\bar{g}$:

$$\bar{f} = x^2 - 6.0750x + 4.9853, \quad \bar{g} = x^2 - 6.222x + 5.7353,$$

with a minimum distance

$$N = \left\|\bar{f} - f\right\|_2^2 + \left\|\bar{g} - g\right\|_2^2 = 0.01213604583.$$

Example 4.2. Let $k = 1, tol = 10^{-3}$ and

$$f_1 = 1.98x^3 + 5x^2 + 5x + 2.96,$$

$$f_2 = 1.99x^2 - 1.01 + 3.01,$$

$$f_3 = 2x^2 + 4.99x + 2.99,$$

after 8 iterations, we have the polynomials

$$\bar{f}_1 = 1.9800x^3 + 5.0000x^2 + 5.0000x + 2.9600,$$

$$\bar{f}_2 = 1.9963x^2 + 1.0100x - 3.0099,$$

$$\bar{f}_3 = 2.0071x^2 + 4.9900x + 2.9902,$$

with a minimum distance,

$$N = \left\|\bar{f}_1 - f_1\right\|_2^2 + \left\|\bar{f}_2 - f_2\right\|_2^2 + \left\|\bar{f}_3 - f_3\right\|_2^2 = 9.0763 \times 10^{-5},$$

and the CPU time

$$t = 0.974920(s)$$

Example 4.3. Let $k = 1, tol = 10^{-3}$ and

$$f_1 = 1.85x^4 - 2x^3 - 2.69x^2 - 1.42x,$$

$$f_2 = 1.47x^3 - 2.94x^2 + 1.18x - 2.36,$$

$$f_3 = 0.52x^3 - 4.01x^2 + 5.94x,$$

$$f_4 = 0.52x^3 - 0.13x^2 - 1.05x - 2.58.$$

after 11 iterations, we have the polynomials

$$\overline{f}_1 = 1.85x^4 - 2x^3 - 2.69x^2 - 1.42x,$$
$$\overline{f}_2 = 1.47x^3 - 2.94x^2 + 1.18x - 2.36,$$
$$\overline{f}_3 = 0.5242x^3 - 4.01x^2 + 5.9405x,$$
$$\overline{f}_4 = 0.52x^3 - 0.134x^2 + -1.05x - 2.58.$$

with a minimum distance

$$N = \left\|\overline{f}_1 - f_1\right\|_2^2 + \left\|\overline{f}_2 - f_2\right\|_2^2 + \left\|\overline{f}_3 - f_3\right\|_2^2 + \left\|\overline{f}_4 - f_4\right\|_2^2$$
$$= 3.4387 \times 10^{-5},$$

and the CPU time

$$t = 0.642582(s).$$

Example 4.4. Let $k = 2, tol = 10^{-3}$ and

$$f_1 = 0.144x^4 - 0.761x^3 + 1.316x^2 - 0.74x,$$
$$f_2 = 0.393x^3 - 2.212x^2 + 4.132x - 2.56$$
$$f_3 = 0.182x^3 - 0.358x^2 - 0.752x + 1.48,$$
$$f_4 = 0.173x^3 - 0.544x^2 + 0.01x + 0.592.$$

after 1 iteration, we have the polynomials

$$f_1 = 0.144x^4 - 0.761x^3 + 1.316x^2 - 0.74x,$$
$$f_2 = 0.393x^3 - 2.212x^2 + 4.132x - 2.56,$$
$$f_3 = 0.182x^3 - 0.3548x^2 - 0.752x + 1.48,$$
$$f_4 = 0.173x^3 - 0.4701x^2 + 0.01x + 0.6257.$$

with a minimum distance

$$N = \left\|\overline{f}_1 - f_1\right\|_2^2 + \left\|\overline{f}_2 - f_2\right\|_2^2 + \left\|\overline{f}_3 - f_3\right\|_2^2 + \left\|\overline{f}_4 - f_4\right\|_2^2 = 0.0066,$$

and the CPU time

$$t = 0.001335(s).$$

Example 4.5. Let $k = 2, tol = 10^{-3}$ and

$$f_1 = 0.64x^6 - 2.56x^5 + 2.56x^4$$
$$+ 0.62x^3 - 1.03x^2 - 3.32x + 5.8,$$
$$f_2 = 1.85x^5 - 5.7x^4 + 1.31x^3 + 3.96x^2 + 2.84,$$
$$f_3 = 0.44x^5 - 3.23x^4 + 7.64x^3 - 4.7x^2 - 4.72x + 4.72,$$
$$f_4 = 1.23x^5 - 4.92x^4 + 2.65x^3 + 9.645x^2 - 11.12x + 2.04,$$
$$f_5 = 1.31x^5 - 5.76x^4 + 8.49x^3 - 5.47x^2 - 0.48x + 5.16.$$

after 1 iteration, we have the polynomials

$$f_1 = 0.6401x^6 - 2.56x^5 + 2.5603x^4$$
$$+ 0.62x^3 - 1.03x^2 - 3.32x + 5.8,$$
$$f_2 = 1.85099x^5 - 5.7x^4 + 1.3102x^3 + 3.96x^2 + 2.84,$$
$$f_3 = 0.4389x^5 - 3.23x^4 + 7.64x^3$$
$$- 4.7003x^2 - 4.72x + 4.72,$$
$$f_4 = 1.23x^5 - 4.92x^4 + 2.65x^3 + 9.645x^2 - 11.12x + 2.04,$$
$$f_5 = 1.307x^5 - 5.76x^4 + 8.4902x^3$$
$$- 5.47x^2 - 0.48x + 5.16.$$

with a minimum distance

$$N = \left\|\overline{f}_1 - f_1\right\|_2^2 + \left\|\overline{f}_2 - f_2\right\|_2^2 + \left\|\overline{f}_3 - f_3\right\|_2^2$$
$$+ \left\|\overline{f}_4 - f_4\right\|_2^2 + \left\|\overline{f}_5 - f_5\right\|_2^2 = 1.1460 \times 10^{-5},$$

and the CPU time

$$t = 0.0923583(s).$$

Examples 4.1, 4.2, 4.3, 4.4 and **4.5** show that Algorithm **4.1** is feasible to solve Problem **1.1**.

In **Table 1**, we present the performance of Algorithm **4.1** and compare the accuracy of the new fast algorithm with the algorithms in [9,21]. Denote n be the total degree of polynomials f_1 and p be the total degree of polynomials $f_i,\ 2 \le i \le t$. It (Chu) stands for the number of iterations by the method in [14] whereas it (STLN) denotes the number of iterations by Algorithm 4.1. Denoted by error(Zeng) and error (STLN) are the perturbations $\sum_i \|\bar{f}_i - f_i\|_2^2$ computed by the method in

[21] and Algorithm 4.1, respectively. The last two columns denote the CPU time in seconds costed by AFMP algorithm and our algorithm, respectively.

As shown in the above table, we show that our method based on STLN algorithm converges quickly to the minimal approximate solutions, needing no more than 2 iterations whereas the method in [14] requires more iteration steps. We also note that our algorithm still converges very quickly when the degrees of polynomials become large while the algorithm in [14] needs more iteration steps. Besides, our algorithm needs less CPU time than the AFMP algorithm. So the convergence speed of our method is faster. From the errors, we demonstrate that our method has smaller magnitudes compared with the method in [21]. So our algorithm can generate much more accurate solutions.

Table 1. Algorithm performance on benchmarks.

Ex	n, p	k	it (Chu)	it (STLN)	error (Zeng)	error (STLN)	time (s) (AFMP)	time (s) (STLN)
1	2, 2	1	5	2	1.89e-4	2.87e-5	7.76	2.5
2	3, 3	2	8	2	1.36e-3	1.05e-4	19.51	7.81
3	5, 4	3	11	2	1e-3	1.56e-6	6.81	2.44
4	6, 6	4	23	2	1.46e-3	1.96e-10	31.829	12.08
5	8, 7	5	33	2	6.53e-4	1.98e-16	50.222	16.95
6	10, 10	6	43	2	1.61e-3	1.51e-12	157.09	61.40
7	14, 13	7	58	2	1.23e-3	2.61e-4	273.7	122.31
8	28, 28	10	634	2	2.6e-3	3.54e-4	559.3	210.65

5. CONCLUSION

In this paper, we present that approximation GCD of several polynomials can be solved by a practical and reliable way based on STLN method and transformed to the approximation of Sylvester structure problem. For the matrices related to the minimization problems are all structured matrix with low displacement rank, applying the algorithm to solve these minimization problems would be possible. The complexity of the algorithm is reduced with respect to the degrees of the given polynomials. Although the problem of structured low rank approximation has been studied in many literatures and obtained many accomplishments, there is still much work to be done, for example, low rank approximation of finite dimensional matrix has not been fully resolved.

REFERENCES

1. B. DeMoor, "Total Least Squares for Affinely Structured Matrices and the Noisy Realization Problem," IEEE Transactions on Signal Process, Vol. 42, No. 11, 1994, pp. 3104-3113. http://dx.doi.org/10.1109/78.330370
2. R. M. Corless, P. M. Gianni, B. M. Tragerm and S. M. Watt, "The Singular Value Decomposition for Polynomial System," Proceedings of International Symposium on Symbolic and Algebraic Computation, Montreal, 1995, pp. 195-207
3. S. R. Khare, H. K. Pillai and M. N. Belur, "Numerical Algorithm for Structured Low Rank Approximation Problem," Proceeding of the 19th International Symposium on Mathematical Theory of Networks and Systems, Budapest, Hungary, 2010.
4. E. Kaltofen, Z. F. Yang and L. H. Zhi, "Approximate Greatest Common Divisors of Several Polynomials with Linearly Constrained Coecients and Simgular Polynomials," Proceedings of International Symposium on Symbolic and Algebraic Computations, Genova, 2006.
5. N. Karkanias, S. Fatouros, M. Mitrouli and G. H. Halikias, "Approximate Greatest Common Divisor of Many Polynomials, Generalised Resultants, and Strength of Approximation," Computers & Mathematics with Applications, Vol. 51, No. 12, 2006, pp. 1817-1830.http://dx.doi.org/10.1016/j.camwa.2006.01.010
6. Markovsky, "Structured Low-Rank Approximation and Its Applications," Automatica, Vol. 44, No. 4, 2007, pp. 891-909. http://dx. doi.org/ 10.1016 /j.automatica.2007.09.011
7. D. Rupprecht, "An Algorithm for Computing Certied Approximate GCD of Univariate Polynomials," Journal of Pure and Applied Algebra, Vol. 139, No. 1-3, 1999, pp. 255-284.http://dx.doi.org/10.1016/S0022-4049(99)00014-6
8. J. A. Cadzow, "Signal Enhancement: A Composite Property Mapping Algorithm," IEEE Transactions on Acoustic Speech Signal Process, Vol. 36, No. 1, 1988, pp. 49-62. http://dx.doi.org/10.1109/29.1488
9. G. Cybenko, "A General Orthogonalization Technique with Applications to Time Series Analysis and Signal Processing," Mathematics of Computation, Vol. 40, 1983, pp. 323-336. http://dx.doi.org/10.1090/S0025-5718-1983-0679449-6
10. J. R. Winkler and J. D. Allan, "Structured Total Least Norm and Approximate GCDs of Inexact Polynomials," Journal of Computational and Applied Mathematics, Vol. 215, No. 1, 2008, pp. 1-13. http://dx.doi.org/10.1016/j.cam.2007.03.018
11. Frieze, R. Kannaa and S. Vempala, "Fast Monte-Carlo Algorithm for Finding Low Rank Approximations," Journal of ACM, Vol. 51, No. 6, 2004, pp. 1025-1041.http://dx.doi.org/10.1145/1039488.1039494
12. R. Beer, "Quantitative in Vivo NMR (Nuclear Magnetic Resonance on Living Objects)," University of Technology Delft, 1995.
13. B. Paola, "Structured Matrix-Based Methods for Approximate Polynomial GCD," Edizioni della Normale, 2011.
14. M. T. Chu, R. E. Funderlic and R. J. Plemmons, "Structured Low Rank Approximation," Linear Algebra Applications, Vol. 366, No. 1, 2003, pp. 157-172.http://dx.doi.org/10.1016/S0024-3795(02)00505-0
15. B. Beckermann and G. Labahn, "A Fast and Numerically Stable Euclidean-Like Algorithm for Detecting Relative Prime Numerical Polynomials," IEEE Journal of Symbolic Computation, Vol. 26, No. 6, 1998, pp. 691-714.http://dx.doi.org/10.1006/jsco.1998.0235

16. B. Y. Li, Z. F. Yang and L. H. Zhi, "Fast Low Rank Approximation of a Sylvester Matrix by Structured Total Least Norm," Journal of JSSAC (Japan Society for Symbolic and Algebraic Computation), Vol. 11, No. 34, 2005, pp. 165-174.
17. B. Botting, M. Giesbrecht and J. May, "Using Riemannian SVD for Problems in Approximate Algebra," Proceedings of the 2005 International Workshop on Symbolic-Numeric, Xi'an, 2005.
18. E. Kaltofen, Z. F. Yang and L. H. Zhi, "Structured Low Rank Approximation of a Sylvester Matrix," International Workshop on Symbolic-Numeric Computation, Xi'an, 2005, pp. 19-21.
19. H. Park, L. Zhang and J. B. Rosen, "Low Rank Approximation of a Hankel Matrix by Structured Total Least Norm," BIT Numerical Mathematics, Vol. 39, No. 4, 1999, pp. 757-779. http://dx.doi.org/10.1023/A:1022347425533
20. L. H. Zhi and Z. F. Yang, "Computing Approximate GCD of Univariate Polynomials by Structure Total Least Norm," Mathematics-Mechanization Research Preprints, No. 24, 004, pp. 375-387.
21. Z. Zeng and B. H. Dayton, "The Approximate GCD of Inexact Polynomials Part 2: A Multivariate Algorithm," Proceedings of the 2004 International Symposium on Symbolic and Algebraic Computation, Santander, 2004.

CHAPTER 13

Cognitive Mapping: Testing Application of Causal Algebra

Nassreddine Garoui, Anis Jarboui

Faculty of Economics and Management, Higher Institute of Business Administration, Sfax, Tunisia

ABSTRACT

This paper takes the basic mathematical operations required to manipulate the cognitive maps. This paper start by presenting all the values that a causal relationship can take. By the using of causal algebra, cognitive map (CC) become not only a graphical representation of a person's beliefs, an agent or a particular area but also can capture the causal relationships existing between the concepts of a given system in a simple manner. Cognitive maps do not use a conventional algebra; algebra causal is needed to treat.

Keywords: Cognitive Mapping; Causal Algebra; Causal Relationships

1. INTRODUCTION

Causal algebra in cognitive maps plays a critical role in the analysis. Improving causal algebra in cognitive maps to model complicated situations has been one of the major research topics in this area. In this paper we propose a causal algebra which can improve representation of mental models. We use the causal algebra to manipulate the cognitive map. Generally, the basic elements of a CM are simple. The concepts an individual uses are represented as points, and the causal relationships between these concepts are represented as arrows between these points. This representation gives a graph of points and arrows, called a cognitive map. The strategic alternatives, all of the various causes and effects, goals and the ultimate utility of the decision-making agent can all be considered as concept variables, and represented as points in the causal map. Causal relationships can take on basic values +, – and 0 ([1]and taken from [2]). With this representation, it is relatively easy to see how concepts and causal relationships are related to each other and to see the overall causal relationships of one concept with another.

2. LITERATURE REVIEW

In this section we will establish the basic mathematical operations required to manipulate the cognitive maps. We will start by presenting all the values that a causal relationship can take.

2.1. Values of the Causal Relationships

The description of classical cognitive maps comes mostly from [3] and [4]. Generally, causal links (causal relations) between two concepts i and j have one of the eight values indicated in **Table 1**.

A causal map is a directed graph that represents an individual's (an agent, a group of agents or an organization) assertions about its beliefs with respect to its environment. The components of this graph are a set of points (the vertices) and a set of arrows A (the edges) between these points. A point represents a concept (also called a concept variable in the sequel), which may be a goal or an action option of any agent. It can also represent the utility of any agent or the utility of a group or an organization, or any other concept appropriate to multiagent reasoning. An arrow represents a causal relation between concepts, that is, it represents a causal assertion of how one concept variable affects another. The concept variable at the origin of an arrow is called a cause variable and that at the end point of the arrow is called the effect variable.

We have already seen that there are three basic types of causal relationships: $(+,-,0)$. But for a more comprehensive set of causal relationships is an overview of the eight resulting values of all possible logical combina tions of positive, negative and zero (+, - ,0 , ⊕,θ , ±, ?, a) $(+,-,0,\oplus,\theta,\pm,?,a)$, where

Table 1. Adjacency matrix.

	Concept 1	Concept 2	……	Concept n
Concept 1	L11			
Concept 2	L21	L22		L2n
……			……	
Concept n	Ln1	Ln2		Lnn

+ : *positive*;

- : *negative*;

0 : *zero*;

⊕ : *notnegative*(+,0) ;

θ : *notpositive*(0,-) ;

± : *notzero*(+, -) ;

? : *universal*(+ , -,) ;

a : *ambivment.*

If for any two concepts there are more of a relationship to describe the effect that A to B, then we must make the intersection of all values for a single relationship. For example, the intersection of + and – is represented by a, ambivalence. That is, if a person or an agent that "A promotes B" and also said that "A is harmful to B", while in the cognitive map is represented the relation of A to B as ambivalent. This type of reasoning is based on the rules for operations, in this case: the addition.

2.2. Operations

There are four types of operations defined for causal relationships: addition (⁝),

Multiplication (*), union (∩) and intersection (∪).

The addition is used when two or more arcs with the same starting point and end with the same point. This will add up all the direct effects for only one outcome total. Here are the rules for addition:

$A \in$ for all x, y

$0 \vdots y = y$

$a \vdots y = y$

$y \vdots y = y$

$+ \vdots - = ?$

$\vdots$- *do* (addition distributes over union)

$x \vdots y = y \vdots x$

The multiplication is used to calculate the indirect effects of a map [1,2]. The rules of multiplication are used:

$$A \in \text{for all } x, y$$

$$+ \otimes y = y$$

$$0 \otimes y = yify \neq a$$

$$a \otimes y = a$$

$$- \otimes - = +$$

$$\otimes do - \text{(multiplication distributes over union)}$$

$$x \otimes x \otimes y = y$$

Laws for the operations of union and intersection are obtained from the set of values

$$(+, -, 0, \oplus, \theta, \pm, ?, a)$$

$$\oplus = 0 \cup +$$

$$? = + \cup - \cup 0$$

$$\theta = +0 \cup -$$

$$a = + \cap 0 = + \cap - = 0 \cap -$$

$$\pm = + \cup -$$

3. RESEARCH METHODOLOGY

3.1. Description of the Empirical Investigation

To meet the research objectives mentioned above, a survey was conducted among players in the company of Tunisia. I have chosen as exploratory approach using multiple case studies. The multiple case studies seek a better understanding of the phenomenon. They are to study a phenomenon in its natural setting by working with a limited number of cases. They are particularly interesting in the case of exploration of little-known phenomena. The case studies thus allow multiple accounts the specificities and characteristics of corporate governance. The data is from 10 firms. The decision to base my study on a sample of firms from various sectors is based on the assumption that a variety of issues will be addressed as well. The output is a cognitive map for actors reflecting their perceptions.

The method used to create cognitive maps is the questionnaire.

3.1.1. Presentation of the Questionnaire

The questionnaire is divided into two parts: the first identifies the company and the second deals with corporate governance. For the second part, relating to corporate governance, we interview actors from the firm on stakeholder approach of corporate governance by providing a list of concepts for each approach with systematic exploration grids and matrices cross. Systematic exploration of the grid **Figure 1** is a technique for collecting materials. Each player is encouraged to explore their own ideas or cognitive representations in relation to its strategic vision. The subject is asked to identify important factors that he said will have an impact on the key concept related to an approach to corporate governance.

Regarding the cross-matrix (**Table 1**), it is also a technique of data collection and the basis for the construction of the cognitive map. The matrix is presented in the form of a table with n rows and n columns. Box of index (i,j) indicates the relationship between concept i and concept j. The actors manipulate the key concepts and assign pairs of concepts depending on the nature and degree of proximity sensed between these concepts. Causal relationships can take on basic values +, – and 0.

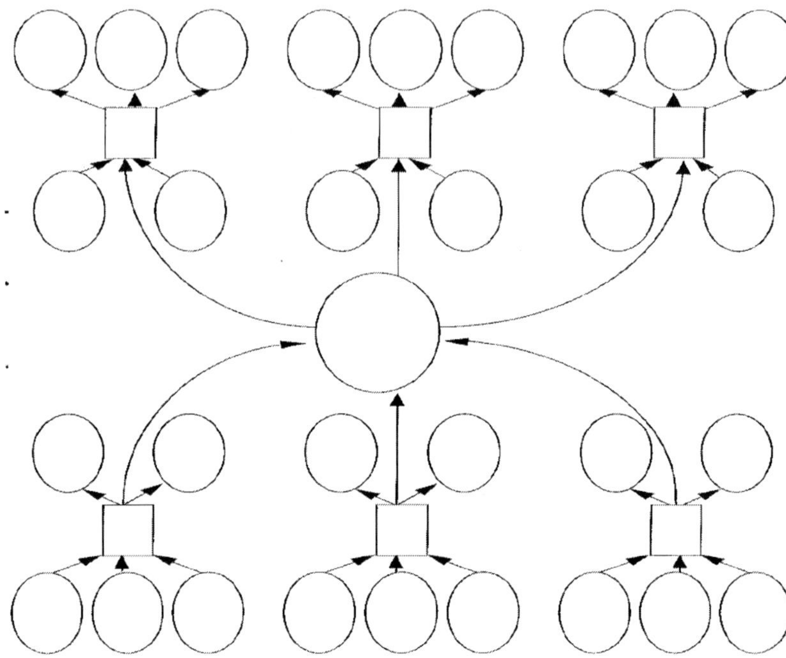

Figure 1. Grid systematic exploration.

3.1.2. Proposal for Modeling Cognitive Maps

When it is difficult to identify the goals, an integrated approach of performance provides a holistic view in which the performance is analyzed by the processes that lead, through the performances of the actors. These representation processes are two problems of implementation: the sharing of representations of actors and the identification of dominant representations in the organization in order to act upon them. The construction of this representation necessarily requires a model that allows understanding to act is "an action of intentional design and construction, for composition of symbols, patterns that would make a complex phenomenon intelligible perceived.

In this context, the use of cognitive maps seems relevant, because they can take into account the complexity and comprehensiveness of the system in which [the behavior] is embedded, while maintaining access to the analysis" [5]. The value of the tool is instrumental [6], it allows both improving their actions and making sense. Cognitive mapping is used as a tool for representation of an idiosyncratic schema [7], a pattern is "a cognitive structure that guides the cutting of reality, the interpretation of events, and action individuals", pattern unique to each individual, causing it to have its own behavior.

3.1.3. The Construction of Cognitive Maps

We will see at first step that allowed the construction of concepts, methodological approach that we discuss. Then we will examine how the cards were dealt.

3.1.4. Concepts

We addressed this issue by the representations constructed by players using the method of cognitive maps, a method that can be applied to poorly structured situations. An analysis based on cognitive maps can understand this process of structuring, as this model is to build or rebuild the mental simultaneously modeling. This construction takes the form of a structure, carrier for clarification. It helps to identify ways to implement to achieve a given goal, the same way it helps to identify the goals justifying the use of such means. Finally, it facilitates communication and negotiation. There are two major trends in the construction method of the cards: the determination of the concepts can be ex ante, or subsequent interviews with respondents for whom the cards are built. The emphasis is on describing the world from the experiences of people who experience it. Nodes and links are determined directly by the participants that advocate [8], not to deprive the subject of representations: the questions should be invitations for the respondent verbalizes his thoughts on what he considers important subject of research. In addition, the researcher cannot force the subject to consider every possible link because the links must be made spontaneously or in response to open questions, so that the subject constructs its reality. In the normative paradigm, the universe is more or less determined. The focus is on operational definitions and research plans reproducible. Observers, different participants, may determine the relationship between variables and nodes that can be. **Table 2** gives 19 concepts selected for the governance approachs to their ability to describe the field of governance. We were guided in this by a literature review and an exploratory study based on a questionnaire made up of grids of systematic exploration and crossmatrices.

4. STAKEHOLDER APPROACH: APPLICATION OF CAUSAL ALGEBRA

Tables 3 and 4 define cross matrix, used as a source base by trying to apply the causal algebra on two case studies. All eight resulting values of all possible logical combinations of positive, negative and zero: $(+,-,0,\oplus,\theta,\pm,?,a)$ where

$+: positive$;

$-: negative$;

$0: zero$;

$\oplus: notnegative(+,0)$;

$\theta: notpositive(0,-)$;

$\pm: notzero(+,-)$;

$?: universal(+,-,)$;

$a: ambivment$.

4.1. Addition of Cross Matrices

The addition of the matrices is defined for two matrices of the same type. The sum of two matrices of type $(m,n), A = a_{ij}$ and $B = b_{ij}$, denoted by $A+B$, is again a matrix c_{ij} type (m,n) obtained by adding corresponding elements. For all $i, j, c_{ij} = a_{ij} + b_{ij}$

$$A = \begin{pmatrix} a_{11} & a_{12} \\ a_{21} & a_{22} \end{pmatrix}$$

$$B = \begin{pmatrix} b_{11} & b_{12} \\ b_{21} & b_{22} \end{pmatrix}$$

$$A + B = \begin{pmatrix} a_{11} + b_{11} & a_{12} + b_{12} \\ a_{21} + b_{21} & a_{22} + b_{22} \end{pmatrix}$$

Table 2. Key concepts.

Stakeholder approach	Cognitive approach	Behavioral approach
1) Creating Value (CV)	1) Knowledge (KN)	1) Cognitive dissonance(Dis cogn)
2) Opportunity (Opp)	2) Creation of Value (C V)	2) Dependence (Dep)
3) Contract node (Nc)	3) Competence (COMP)	3) Optimism (Op)
4) Specific Investment (Inv sp)	4) Organizational Learning (LORG)	4) Overconfidence (Over c)
5) Specific human capital (C H S)	5) Control (CON)	5) Mimicry (Mim)
6) Responsibility multiple (Res Mul)	6) Communication (COMM)	
7) Power (Pou)	7) Training (TR)	
8) Legitimacy (Leg)	8) Cognitive Resource (RES COG)	
9) Profit (Pro)	9) Growth Opportunity (GR OPP)	
10) Residual Claim (CR)	10) Innovation (INN)	
11) Annuity (R)	11) Specific Capacitance (SP C)	
12) Distribution (Rep)	12) Rationality (RAT)	
13) Conflict (C)	13) Patterns of Creation and Ownership of the Annuity (PCOA)	
14) Asymmetric Information (AI)	14) Repertoire of knowledge (REP KN)	
15) Property (Prop)		
16) Decision (D)		
17) Dispersion of Property (Dis Pro)		
18) Business strategy (S Ese)		
19) Business Assets (A Ese)		

Table 3. Matrix influences taking into account three relations: +, – and 0 (case 1 stakeholder approach of corporate governance).

	CV	Opp	Nc	Inv sp	C H S	Res Mul	Pou	Leg	Pro	CR	R	Rep	C	AI	Prop	D	Dis Pro	S Ese	A Ese
CV	0	–	0	+	0	–	0	+	0	+	0	+	0	0	+	0	–	0	0
Opp	+	0	0	0	0	0	0	0	0	0	0	–	0	0	0	+	0	0	0
Nc	0	0	0	0	0	0	0	0	0	0	0	0	0	0	–	0	+	0	0
Inv sp	+	0	0	0	0	0	+	0	0	0	0	0	0	–	0	0	0	–	0
C H S	0	0	0	0	0	0	0	0	0	+	0	0	0	0	0	0	–	0	0
Res Mul	–	0	+	0	0	0	0	0	0	0	0	0	+	0	0	0	–	0	0
Pou	0	0	0	+	0	0	0	0	0	0	0	0	+	0	0	0	–	0	+
Leg	+	0	–	0	0	0	0	0	–	0	0	0	0	0	0	+	0	0	0
Pro	0	0	0	0	+	0	0	0	0	0	0	+	0	0	0	0	0	0	0
CR	–	0	0	0	–	0	0	–	0	0	0	0	0	+	0	0	+	0	0
R	0	0	0	0	+	0	0	0	+	0	0	0	0	0	0	+	0	0	+
Rep	+	0	0	–	0	0	0	0	0	0	0	0	+	+	0	0	0	0	0
C	0	0	+	0	0	+	0	0	0	+	0	0	0	0	0	0	–	0	+
AI	0	0	0	+	0	0	0	–	0	0	0	0	0	0	+	0	0	0	0
Prop	0	0	+	0	0	0	0	0	0	0	0	+	0	+	0	0	0	0	0
D	0	0	0	0	0	0	0	0	+	0	–	0	0	0	–	0	0	0	0
Dis Pro	–	0	–	0	+	0	0	0	0	0	0	0	0	0	0	0	0	0	+
S Ese	0	0	0	+	0	0	0	–	0	0	0	0	0	0	0	+	0	0	0
A Ese	0	0	0	0	0	+	0	0	0	–	0	0	0	0	+	0	0	0	0

Table 4. Matrix influences taking into account three relations: +, – and 0 (Case 2 stakeholder approach of corporate governance).

	CV	Opp	Nc	Inv sp	C H S	Res Mul	Pou	Leg	Pro	CR	R	Rep	C	AI	Prop	D	Dis Pro	S Ese	A Ese
CV	0	–	0	+	0	–	0	0	+	0	+	0	–	0	+	0	0	0	+
Opp	+	0	+	0	0	+	0	0	–	0	0	0	0	0	+	0	0	0	0
Nc	0	0	0	0	0	0	0	+	0	0	0	+	0	0	–	0	+	0	0
Inv sp	+	+	0	0	0	0	0	–	+	0	0	0	0	0	0	0	0	0	0
C H S	0	0	0	0	0	0	0	0	0	0	0	0	0	0	0	0	0	0	–
Res Mul	–	–	0	0	+	0	0	–	0	0	+	0	–	0	–	0	0	0	0
Pou	0	0	0	+	0	+	0	0	0	0	0	0	0	+	0	0	0	0	+
Leg	0	+	0	0	+	0	0	0	0	0	0	0	0	–	0	0	0	0	0
Pro	+	0	+	0	0	0	+	0	0	0	0	+	0	0	0	0	0	0	–
CR	0	0	0	0	–	0	0	–	0	0	0	0	–	0	0	0	0	0	0
R	+	0	0	0	+	0	0	0	+	0	0	–	0	0	+	0	–	0	+
Rep	0	0	0	–	0	+	0	+	0	0	0	0	0	0	0	+	0	0	0
C	–	0	0	0	0	0	0	0	–	0	0	0	0	0	–	0	0	0	0
AI	0	0	–	0	+	0	0	0	0	0	–	0	0	0	+	0	0	0	+
Prop	+	0	0	0	0	0	0	0	0	0	0	0	0	0	0	0	0	0	0
D	0	0	+	0	–	0	0	0	+	0	0	+	0	0	+	0	0	0	–
Dis Pro	0	0	0	0	0	0	+	0	–	0	0	0	+	0	0	0	0	0	0
S Ese	0	0	+	+	0	0	0	–	0	+	0	–	0	0	–	0	0	0	+
A Ese	+	0	0	–	0	+	0	0	+	0	0	–	0	+	0	0	–	0	0

We will try to add the two matrices respecting the different logical combinations of possible causal relationships. **Table 5** gives value based on this operation.

Table 6 gives the following notations.

Table 7 is the result of application of algebra and the causal summation rule.

4.2. Cross-Multiplication of Matrices

For two matrices of the same type, the product of two matrices $A=(a_{ij})$ and $B=(b_{ij})$ type (m,n) noted $A*B=c_{ij}$, is a matrix type (m,n) given by $c_{ij}=a_{ij}*b_{ij}$.

Condition: the product of two matrices is only possible if the number of lines of the first equals the number of columns in the second and the number of columns of the first number is equal to the second line. The number of rows of the result is equal to that of the first matrix; the number of columns is equal to that of the second.

We will try to add the two matrices respecting the different combinations possible of logical causal relations. **Table 8** gives values with respecting multiplication operation.

Table 5. Addition values.

	+	–	0	Nz	Np	Nn	?	c
+	+	?	+	?	?	+	?	c
–	?	–	–	?	–	?	?	c
0	+	–	0	Nz	Np	Nn	?	c
Nz	?	?	Nz	Nz	?	?	?	c
Np	?	–	Np	?	Np	?	?	c
Nn	+	?	Nn	?	?	Nn	?	c
?	?	?	?	?	?	?	?	c
c	C	C	c	c	C	c	c	c

Table 6. Notations.

The sign	The name of causal relatioship	The set
+	Positive	{+}
–	Negative	{–}
0	Zero	{0}
Nn	Not negative	{0, +}
Np	Not positive	{0, –}
Nz	Not zero	{–, +}
?	undetermined	{–, 0, +}
C	Conflict (ambivalent)	Ø empty set

Table 7. Matrix crossed with the application of algebra and the causal summation rule.

	CV	Opp	Nc	Inv sp	C H S	Res Mul	Pou	Leg	Pro	CR	R	Rep	C	AI	Prop	D	Dis Pro	S Ese	A Ese
CV	0	–	0	+	0	–	0	+	+	+	+	+	–	0	+	0	–	0	+
Opp	+	0	+	0	0	+	0	0	–	0	0	–	0	0	+	+	0	0	0
Nc	0	0	0	0	0	0	0	+	0	0	0	0	0	0	–	0	+	0	0
Inv sp	+	+	0	0	0	0	+	–	+	0	0	0	0	–	0	0	0	–	0
C H S	0	0	0	0	0	0	0	0	0	+	0	0	0	0	0	0	–	0	–
Res Mul	–	–	+	0	+	0	0	–	0	0	+	0	+	0	–	0	–	0	0
Pou	0	0	0	+	0	+	0	0	0	0	0	0	+	–	0	0	–	0	+
Leg	+	+	–	0	+	0	0	0	–	0	0	0	0	–	0	+	0	0	0
Pro	+	0	+	0	+	0	+	0	0	0	0	+	0	0	0	0	0	0	–
CR	–	0	0	0	–	0	0	–	–	0	0	0	–	+	0	+	+	0	0
R	+	0	0	0	+	0	0	0	+	0	0	–	0	0	+	+	–	0	+
Rep	+	0	0	–	0	0	0	+	+	+	0	0	+	+	0	0	0	0	0
C	–	0	+	0	0	+	0	+	0	0	0	0	0	0	–	0	–	0	+
AI	0	0	–	+	+	0	0	–	0	0	–	0	0	0	+	0	0	0	+
Prop	+	0	+	0	0	0	0	0	0	0	0	+	0	+	0	0	0	0	0
D	0	0	+	0	–	0	0	0	+	0	–	+	0	0	?	0	0	0	–
Dis Pro	–	0	–	0	+	0	+	0	–	0	0	0	+	0	0	0	0	0	+
S Ese	0	0	+	+	0	0	0	–	0	+	0	–	0	0	–	+	0	0	+
A Ese	+	0	0	–	0	+	0	0	0	–	0	–	0	+	+	0	–	0	0

Table 8. Multiplication values.

	+	–	0	Nz	Np	Nn	?	c
+	+	–	0	Nz	Np	Nn	?	c
–	–	+	0	Nz	Nn	Np	?	c
0	0	0	0	0	0	0	0	c
Nz	Nz	Nz	0	Nz	?	?	?	c
Np	Np	Nn	0	?	Nn	Np	?	c
Nn	Nn	Np	0	?	Np	Nn	?	c
?	?	?	0	?	?	?	?	c
c	C	C	c	c	c	c	c	c

Table 9 is the result of the application of algebra causal and the multiplication.

Table 9. Matrix crossed with the application of algebra causal and the multiplication.

	CV	Opp	Nc	Inv sp	C H S	Res Mul	Pou	Leg	Pro	CR	R	Rep	C	AI	Prop	D	Dis Pro	S Ese	A Ese
CV	0	+	0	+	0	+	0	0	0	0	0	0	0	0	+	0	0	0	0
Opp	+	0	0	0	0	0	0	0	0	0	0	0	0	0	0	0	0	0	0
Nc	0	0	0	0	0	0	0	0	0	0	0	0	0	0	+	0	+	0	0
Inv sp	+	0	0	0	0	0	0	0	0	0	0	0	0	0	0	0	0	0	0
C H S	0	0	0	0	0	0	0	0	0	0	0	0	0	0	0	0	0	0	0
Res Mul	+	0	0	0	0	0	0	0	0	0	0	0	+	0	0	0	0	0	0
Pou	0	0	0	+	0	0	0	0	0	0	0	0	0	0	0	0	0	0	+
Leg	+	0	0	0	0	0	0	0	0	0	0	0	0	0	0	0	0	0	0
Pro	0	0	+	0	0	0	0	0	0	0	0	+	0	0	0	0	0	0	0
CR	0	0	0	0	+	0	0	+	0	0	0	0	0	0	0	+	0	0	0
R	0	0	0	0	+	0	0	0	+	0	0	0	0	0	0	0	0	0	+
Rep	0	0	0	+	0	0	0	0	0	0	0	0	0	0	0	0	0	0	0
C	0	0	0	0	0	0	0	0	0	0	0	0	0	0	0	0	0	0	0
AI	0	0	-	0	0	0	0	0	0	0	0	0	0	0	0	0	0	0	0
Prop	0	0	0	0	0	0	0	0	0	0	0	0	0	0	0	0	0	0	0
D	0	0	0	0	0	0	0	0	+	0	0	0	0	0	0	0	0	0	0
Dis Pro	0	0	0	0	0	0	0	0	0	0	0	0	0	0	0	0	0	0	0
S Ese	0	0	0	+	0	0	0	+	0	0	0	0	0	0	0	0	0	0	0
A Ese	0	0	0	0	0	+	0	0	0	0	0	0	0	0	+	0	0	0	0

5. COGNITIVE APPROACH: APPLICATION OF CAUSAL ALGEBRA

All eight resulting values of all possible logical combinations of positive, negative and zero: $(+,-,0,\oplus,\theta,\pm,?,a)$, where

$+: positive$;

$-: negative$;

$0: zero$;

$\oplus: notnegative(+,0)$;

$\theta : notpositive(0,-)$;

$\pm : notzero(+,-)$;

$? : universal(+,-,)$;

$a : ambivment$.

Table 10 define values matrix taking into account the influences of three relations: +, – and 0.

Table 10. Matrix taking into account the influences of three relations: +, – and 0 (Case 1 of the cognitive approach).

	KN	CV	COMP	LORG	CON	COMM	TR	RES COG	GR OPP	INN	SP C	RAT	PCOA	REP KN
KN	0	0	0	0	0	0	0	0	0	+	0	–	0	+
CV	0	0	0	+	0	0	0	+	0	0	0	0	0	0
COMP	0	0	0	0	0	0	0	0	0	0	0	0	0	0
LORG	0	+	–	0	+	0	0	0	0	–	0	+	0	+
CON	0	+	0	0	0	+	0	0	0	+	0	0	0	0
COMM	0	0	0	0	–	0	0	+	0	0	0	–	0	+
TR	0	+	0	0	0	0	0	0	0	–	0	0	0	0
RES COG	0	0	+	0	0	0	+	0	0	0	+	0	0	0
GR OPP	0	–	0	0	+	0	0	0	0	0	0	0	0	+
INN	+	0	0	+	0	0	+	0	–	0	0	+	0	0
SP C	0	+	0	+	0	0	0	+	0	0	0	0	0	0
RAT	–	0	0	–	0	0	0	–	0	0	0	0	0	+
PCOA	0	0	–	0	–	0	+	0	0	+	0	–	0	0
REP KN	+	0	0	+	0	0	+	0	0	0	+	0	0	0

Table 11 define values matrix taking into account the influences of three relations: +, – and 0.

Table 11. Matrix of direct influences taking into account three relations: +, – and 0 (Case 2 of the cognitive approach).

	KN	CV	COMP	LORG	CON	COMM	TR	RES COG	GR OPP	INN	SP C	RAT	PCOA	REP KN
KN	0	+	–	+	+	–	0	0	0	0	0	0	0	+
CV	+	0	0	+	0	0	0	+	0	+	0	0	+	0
COMP	–	0	0	0	0	0	+	0	0	–	0	+	0	+
LORG	–	0	0	0	+	0	–	0	0	0	–	0	–	0
CON	+	0	0	0	0	0	0	0	0	0	0	0	–	–
COMM	–	0	0	0	–	0	0	0	+	0	0	+	0	0
TR	0	0	+	0	0	+	0	0	0	0	0	0	+	+
RES COG	0	0	+	0	0	+	0	0	–	0	0	–	0	0
GR OPP	0	–	0	+	0	0	–	0	0	+	0	0	–	0
INN	0	0	0	+	+	0	+	0	0	0	0	0	0	0
SP C	0	0	–	0	0	–	–	0	0	0	0	0	0	+
RAT	0	+	0	0	–	0	0	–	0	–	0	0	+	0
PCOA	0	0	–	0	0	–	0	0	+	0	0	–	0	0
REP KN	+	0	0	0	+	0	+	0	0	+	0	0	–	0

Table 12 gives the results by the application of algebra and the causal rule of addition.

Table 12. Matrix crossed with the application of algebra and the causal rule of addition.

	KN	CV	COMP	LORG	CON	COMM	TR	RES COG	GR OPP	INN	SP C	RAT	PCOA	REP KN
KN	0	+	–	+	+	–	0	0	0	+	0	–	0	+
CV	+	0	0	+	0	0	0	+	0	+	0	0	+	0
COMP	–	0	0	0	0	0	+	0	0	–	0	+	0	+
LORG	–	+	–	0	+	0	–	0	0	–	–	+	–	+
CON	+	0	0	0	+	0	0	0	+	0	0	0	–	–
COMM	–	0	0	0	–	0	+	0	+	0	0	?	0	+
TR	0	+	+	0	0	+	0	0	0	–	0	0	+	+
RES COG	0	0	+	0	0	0	+	0	–	0	+	–	0	0
GR OPP	0		+	0	+	0	+	0	0	+	0	0	–	0
INN	+	0	0	+	0	0	+	0	–	0	+	+	0	0
SP C	0	+	–	+	0	0	–	?	0	0	0	0	0	+
RAT	–	+	0	–	0	–	0	–	0	–	0	0	+	+
PCOA	0	0	–	0	–	–	+	0	+	+	0	–	0	0
REP KN	+	0	0	+	+	0	+	+	0	0	+	0	–	0

Table 13 gives the results by the application of algebra and the multiplication rule.

Table 13. Matrix crossed with the application of algebra causal and the multiplication rule.

	KN	CV	COMP	LORG	CON	COMM	TR	RES COG	GR OPP	INN	SP C	RAT	PCOA	REP KN
KN	0	0	0	0	0	0	0	0	0	0	0	0	0	+
CV	0	0	0	+	0	0	0	0	0	0	0	0	0	0
COMP	0	0	0	0	0	0	0	0	0	0	0	0	0	0
LORG	0	0	0	0	+	0	0	0	0	0	0	0	0	0
CON	0	0	0	0	0	0	0	0	0	0	0	0	0	0
COMM	0	0	0	0	+	0	0	0	0	0	0	–	0	0
TR	0	0	0	0	0	0	0	0	0	0	0	0	0	0
RES COG	0	0	+	0	0	0	0	0	0	0	0	0	0	0
GR OPP	0	+	0	0	0	0	0	0	0	0	0	0	0	0
INN	0	0	0	+	0	0	+	0	0	0	0	0	0	0
SP C	0	0	0	0	0	0	0	0	0	0	0	0	0	0
RAT	0	0	0	0	0	0	0	+	0	0	0	0	0	0
PCOA	0	0	+	0	0	0	0	0	0	0	0	+	0	0
REP KN	+	0	0	0	0	0	+	0	0	0	0	0	0	0

6. BEHAVIORAL APPROACH: APPLICATION OF CAUSAL ALGEBRA

Table 14 gives the values of direct influences taking into account three relations: +, – and 0 in the case 1.

Table 14. Matrix of direct influences, taking into account three relations: +, – and 0 (case1de behavioral approach).

	Dis cogn	Dep	Op	Over c	Mim
Dis cogn	0	+	–	0	+
Dep	+	0	0	–	0
Op	+	+	0	–	+
Over c	0	0	+	0	–
Mim	–	0	0	+	0

Table 15 gives the values of direct influences taking

Table 15. Matrix of direct influences taking into account three relations: +, – and 0 (Case 2 of the behavioral approach).

	Dis cogn	Dep	Op	Over c	Mim
Dis cogn	0	+	–	+	+
Dep	+	0	+	0	–
Op	+	0	0	0	0
Over c	–	0	+	0	–
Mim	–	0	+	0	0

Table 16 gives the matrix results crossed with the application of causal algebra and the rule of addition.

Table 16. Matrix crossed with the application of algebra and the causal rule of addition.

	Dis cogn	Dep	Op	Over c	Mim
Dis cogn	0	+	–	+	+
Dep	+	0	+	?	–
Op	+	+	0	–	+
Over c	–	0	+	0	–
Mim	–	0	+	+	0

Table 17 gives the matrix results crossed with the application of causal algebra and the multiplication rule.

Table 17. Matrix crossed with the application of algebra causal and the multiplication rule.

	Dis cogn	Dep	Op	Over c	Mim
Dis cogn	0	+	+	0	+
Dep	+	0	0	0	0
Op	+	0	0	0	0
Over c	0	0	+	0	+
Mim	+	0	0	0	0

7. CONCLUSIONS AND IMPLICATIONS OF THE RESEARCH

The benefit of cognitive maps is that they are based on a mathematical basis, which makes their results more reliable and faithful.

The use of causal algebra is advantageous in terms of cognitive maps. Indeed one can enumerate the following benefits:

1) The causal assertions contain the best way to understand how the world is organized.

2) Causality is the first idea to assign an explanation of events.

3) To choose among several alternative actions imply a causal assessment.

In this application we tried to apply algebra on two causal cognitive maps of two companies using the rules of addition and multiplication which aims to bring together and thought patterns and find a convergence in understanding the concepts Government's key businesses. Through the use of algebra one can find a causal mental pattern common among actors from different companies.

Finally, we can say that a cognitive map (CC) is a graphical representation of a person's beliefs, an agent or a particular area. The aim is to capture the causal relationships existing between the concepts of a given system in a simple manner.

Nassreddine Garoui, Anis Jarboui Cognitive maps not use a conventional algebra; algebra causal is needed to treat.

REFERENCES

1. M. Wellman, "Inference in Cognitive Maps," Mathematics and Computers in Simulation, Vol. 36, No. 2, 1994, pp. 137-148. doi:10.1016/0378-4754(94)90028-0
2. Levi and P. E. Tetlock, "A Cognitive Analysis of Japan's 1941 Decision for War," Journal of Conflict Resolution, Vol. 24, No. 2, 1980, pp. 195-211. doi:10. 1177/ 002200278002400201
3. K. Nakumara, S. Iwai and T. Sawaragi, "Decision Support Using Causation Knowledge Base," IEEE Transactions on Systems, Man and Cybernetics, Vol. 12, No. 6, 1982, pp. 765-777.
4. R. Axelrod, Ed., "Structure of Decision: The Cognitive Maps of Political Elites," Princeton University Press, Princeton, 1976.
5. J. M. Komocar, "Cartes Causales d'un Milieu de Travail," In: P. Cossette, Ed., Cartes Cognitives et Organisations, éd. Eska, 1994, pp. 155-184.
6. M. Audet, "Plasticity, Instrumentality and Reflexivity," In: P. Cossette, Ed., Cognitive Maps and Organizations, Collection "Business Administration", Quebec/Paris: The Laval University Press/Editions ESKA, 1994, pp. 187-198.
7. P. Cossette, "The Cognitive Maps of Service Study Organizations," In: P. Cossette, Ed., Cognitive Maps Organizations and Collection "Administrative Sciences", The University Press Laval/ESKA Publishing, Quebec/Paris,1994, pp. 3-12.
8. P. Cossette and M. Audet, "What is a Cognitive Map?" In: P. Cossette (Eds.), Cognitive Maps and organizations, Collection "Business administration", Presses de l'Université Laval/Publishing ESKA, Quebec/Paris, 1994, pp. 13- 33.

CHAPTER 14

A QMR-Type Algorithm for Drazin-Inverse Solution of Singular Nonsymmetric Linear Systems

Alireza Ataei

Department of Mathematics, Faculty of Science, Persian Gulf University, Bushehr, Iran

ABSTRACT

In this paper, we propose DQMR algorithm for the Drazin-inverse solution of consistent or inconsistent linear systems of the form Ax=b where $A \in \mathbb{C}^{N \times N}$ is a singular and in general non-hermitian matrix that has an arbitrary index. DQMR algorithm for singular systems is analogous to QMR algorithm for non-singular systems. We compare this algorithm with DGMRES by numerical experiments.

Keywords: Singular Linear Systems, DGMRES Method, Quasi-Minimal Residual Methods, Drazin-Inverse Solution, Krylov Subspace Methods

1. INTRODUCTION

Consider the linear system

$$Ax=b, \quad (1)$$

where $A \in \mathbb{C}^{N \times N}$ is a singular matrix and ind(A) is arbitrary. Here ind(A), the index of A is the size of the largest Jordan block corresponding to the zero eigenvalue of A. We recall that the Drazin-inverse solution of (1) is the vector $A^D b$, where A^D is the Drazin-inverse of the singular matrix A. For the Drazin-inverse and its properties, we can refer to [1] or [2]. In the important special case nd(A), this matrix is called the group inverse of A and denoted by A^g. The Drazin-inverse has various applications in the theory of finite Markov chains [2] , the study of singular differential and difference equations [2] , the investigation of Cesaro-Neumann iterations [3] , cryptography [4] , iterative methods in numerical analysis, [5] [6] , multibody system dynamics [7] and so forth. The problem of finding the solution of the form $A^D b$ for (1) is very common in the literature and many different techniques have been developed in order to solve it.

In [6] [8] [9] [10] [11] , authors presented some Krylov subspace methods [9] to solve singular linear system with some restriction. However, the treatment of singular linear inconsistent system by Krylov subspace has been proved extremely hard. In [12] , Sidi had not put any restrictions on the matrix A and the system (1). In his paper, the spectrum of A can have any shape and no restrictions are put on the linear system (1). The only assumption is that index(A) is known. Although the index(A) of A is overestimated, the method is valid.

In [12] , Sidi proposed a general approach to Krylov subspace methods to compute Drazin-inverse solution. In addition, he presented several Krylov subspace methods of Arnoldi, DGCR and Lancoze types. Furthermore, in [13] [14] , Sidi has continued to drive two Krylov subspace methods to compute $A^D b$. One is DGMRES method, which is the implementation of the DGCR method for singular systems which is analogues to GMRES for non-singular systems. Other is DBI-CG method which is Lanczos type algorithm [13]. DGMRES, like, GMRES method, is a stable numerically and economical computationally, which is a storage wise method. DBI-CG method, also like BI-CG for non-singular systems, is a fast algorithm, but when we need a high accuracy, the algo- rithm is invalid. DFOM algorithm is another implementation of the projection method for singular linear systems is analogues to Arnoldi for non-singular systems. DFOM algorithm may be less accurate but faster than DGMRES, and more precise and slower than DBI-CG [15].

In the present paper, the Drazin-Quasi-minimal residual algorithm (DQMR here- after) is another implementation of the projection method for singular linear systems is analogues to Lanczos algorithm for non-singular systems. DGMRES algorithm, in prac- tice, cannot afford to run the full algorithm and it is necessary to use restart. For dif- ficult problems, in most cases, this results in extremely slow convergence, While DQMR algorithm can be implemented using only short recurrences and hence it can be com- puted with little work and low storage requirements per iteration.

The outline of this paper is as follows. In Section 2, we will provide a brief of sum- mary of the review of the theorem and projection method in [12] which is relevant to us. We shall discuss the projection methods approach to solve (1) in general and DQMR particular. In Section 3, we will drive the DQMR method. We design DQMR when we set ind(A)=0 throughout, DQMR reduces to QMR. In this sense, DQMR is an extension of QMR that archives the Drazin-inverse solution of singular systems. In Sec- tion 4, by numerical examples, we show that the computation time and iteration number of DQMR algorithm is substantially less than that of DGMRES algorithm. Section 5 is devoted to concluding remarks.

2. SOME BASIC THEOREM AND PROJECTION METHODS FOR A^DB

The method we are interested in starts with an arbitrary initial vector x_0 and generate sequences of vector x_1, x_2,... according to

$$x_m = x_0 + q_{m-1}(A) r_0, \quad r_0 = b - Ax_0, \tag{2}$$

where $q_{m-1}(\lambda)$ is a polynomial in λ of degree at most m-1, given by

$$q_{m-1}(\lambda) = \sum_{i=1}^{m-a} c_i \lambda^{a+i-1}, \quad a = \text{ind}(A) \tag{3}$$

Let us define

$$p_{m-1}(\lambda) = 1 - \lambda q_{m-1}(\lambda) = 1 - \sum_{i=1}^{m-a} c_i \lambda^{a+i} \tag{4}$$

We call $p_m(\lambda)$ the mth residual polynomial since

$$r_m = b - Ax_m = \left[I - q_{m-1}(A)\right] r_0 = p_m(A) r_0 \tag{5}$$

Note that

$$p_m(0) = 1 \quad \text{and} \quad p_m^{(i)}(0) = 0, \quad i = 1, 2, \cdots, a \tag{6}$$

The condition (6) is due to Eiermann et al. [5].

For convenience we denote by II_m the class of polynomials of degree at most m and define

$$\Pi_m^0 = \left\{ p \in \Pi_m : p_m(0) = 1 \text{ and } p_m^{(i)}(0) = 0, i = 1, 2, \cdots, a \right\} \tag{7}$$

Thus, the polynomial p_m described above is in Π^0_m.

The projection methods of [12] are now defined by demanding that the vector $A^0 r_m$ to be orthogonal to a given W subspace of dimension m-a. In addition, If we denote by W the $N\times$(m-a) matrix whose columns span the subspace W, then this ortho- gonality demand is equivalent to $W^* A^0 r_m$.=0 As we have $r_m = r_0 + \sum_{i=1}^{m-a} c_i A^{i+a} r_0$ from (4), $W^* A^0 r_m$.=0 amounts to the requirement that $c_1, c_2, \ldots, c_{m-a}$ satisfy the linear system

$$W^* A^{a+1} Vc = W^* A^a r_0 \tag{8}$$

where $V \in \mathbb{C}^{N\times(m-a)}$ and $c \in \mathbb{C}^{m-a}$ are given by

$$V = \left[A^a r_0 \mid A^{a+1} r_0 \mid \cdots \mid A^{m-a} r_0 \right] \quad \text{and} \quad c = \left[c_1, c_2, \cdots, c_{m-a} \right]^T \tag{9}$$

We see that unique solution for c exists provided det $\left(W^* A^{a+1} V\right) \neq 0$, and when it does we have $x_m = x_0 + V\left(W^* A^{a+1} V\right)^{-1} W^* A^a r_0$.

As we choose different W, we have a different algorithm: for DGMRES, we choose $W = A^{(a+1)} V$ for DBI-CG, we choose $W = \text{span}\left\{A^* \tilde{r}_0, \left(A^*\right)^2 \tilde{r}_0, \cdots, \left(A^*\right)^{m-a} \tilde{r}_0\right\}$

In this paper, for DQMR, we choose

$$W = \text{span}\left\{A^a r_0, A^*\left(A^a r_0\right), \cdots, \left(A^*\right)^{m-a} A^a r_0\right\}$$

We will mention several definitions and theorems, which have projection method converge below.

We will denote by $\hat{\mathcal{L}}$ the direct sum of the variant subspaces of A corresponding to its non-zeros eigenvalues, and by $\hat{\mathcal{L}}$, its invariant subspaces corresponding to its zeros eigenvalue. Thus, $\hat{\mathcal{L}}$ is $R(A^a)$, the range of A^a, and $\hat{\mathcal{L}}$ is $N(A^a)$, the null space of A^a. So every vector in $\mathbb{C}^N$ can be expressed as the sum of two vector, one in $\hat{\mathcal{L}}$ and other in $\hat{\mathcal{L}}$.

Definition 1 [12]. Let A be singular and ind $(A)=a$, and let $\hat{u} \in \hat{\mathcal{L}}$ be given. Then a polynomial $p(\lambda)$ will be called the minimal a—incomplete polynomial of A with respect to the vector $\hat{u}$ if $p(\lambda) \in \Pi^0_m$ and m is the smallest possible one so that $p(A)\hat{u} = 0$

The following theorems will ensure the success of projection method.

Theorem 1 [12] $p(\lambda)$ exists and is unique. Furthermore, its degree m satisfies $q \le m \le q+a$, where q is the degree of the minimal polynomial of A with respect to $\hat{u}$.

The following result that is the justification of the above-mentioned projection approach is Theorem 4.2 in [12].

Theorem 2 Let $x_0 = \hat{x}_0 + \tilde{x}_0$, where $\hat{x}_0 \in \hat{\mathcal{L}}$ and $\hat{x}_0 \in \hat{\mathcal{L}}$, are the initial vector in the projection method to compute $A^D b$. Moreover, Let also $p(\lambda)$ the minimal a—incomplete polynomial of A with respect to $\hat{x}_0 + A^D b$, and let m be its degree. Finally, let x_m be the vector generated by the projection method through (2)-(8) with $(W^* A^{a+1} V) \neq 0$. Then $x_m = A^D b + \tilde{x}_0$.

Obviously, of Theorem 2 if $(W^* A^{a+1} V) \neq 0$, the projection method will terminate successfully in a finite number of steps, this number being at most N. If we pick $x_0 = 0$, which also forces $\tilde{x}_0 = 0$, they produce the Drazine-inverse solution $A^D b$ for (1) upon termination.

Theorem 3 [14]. The vector x_m exists uniquely and unconditionally for all

$m \le m_0$, being the degree of the minimal a-incomplete polynomial of A with respect to $\hat{x}_0 - A^D b \in \hat{\mathcal{L}}$. Furthermore, $x_m = \hat{x}_m + \tilde{x}_m$ with $\hat{x}_m \in \hat{\mathcal{L}}$ and $\tilde{x}_m = \tilde{x}_0$ for all m.

3. DQMR ALGORITHM

In this section, we will introduce a different implementation of projection method. The algorithm is analogous to QMR algorithm. We must note that in spite of the analogy, DQMR seems to be quite different from QMR, which is for non-singular systems.

As $x_m = x_0 + \sum_{i=1}^{m-a} c_i A^{a+i-1} r_0$, we start with $v_1 = w_1 = A^a r_0 / \|A^a r_0\|_2$, the lanczos algorithm [16] generates two sequences of vectors $v_1, v_2, \cdots, v_n$ and $w_1, w_2, \cdots, w_n, n = 1, 2, \cdots$, that satisfy

$$\operatorname{span}\{v_1, v_2, \cdots, v_k\} = \operatorname{span}\{A^a r_0, A^{a+1} r_0, \cdots, A^{a+k-1} r_0\} = K_k(A; A^a r_0)$$

$$\operatorname{span}\{w_1, w_2, \cdots, w_k\} = \operatorname{span}\left\{A^a r_0, A^*(A^a r_0), \cdots, (A^*)^{a+k-1} A^a r_0\right\} = K_k(A^*; A^a r_0) \tag{10}$$

where they are clear that $K_k\left(A;A^a r_0\right)$ and $K_k\left(A^*;A^a r_0\right)$ denote the Krylov subspaces

$\text{span}\left\{A^a r_0, A^{a+1} r_0, \cdots, A^{a+k-1} r_0\right\}$ and $\text{span}\left\{A^a r_0, A^* A^a r_0, \cdots, \left(A^*\right)^{a+k-1} A^a r_0\right\}$

respectively.

If we define that $N\times K$ matrix $\hat{V}_k$ by

$$\hat{V}_k = \begin{bmatrix} v_1 & v_2 & \cdots & v_k \end{bmatrix}, \quad k = 1, 2, \cdots$$

then, we can write for $m \le m_0$

$x_m = x_0 + \hat{V}_{m-a}\zeta_m$ for some $\zeta_m \in C^{m-a}$

Therefore, it is obvious that we need to determine ζ_m. Since $r_m = r_0 - A\hat{V}_{m-a}\zeta_m$ we have

$$A^a r_m = A^a r_0 - A^{a+1}\hat{V}_{m-a}\zeta_m = \gamma_1 v_1 - A^{a+1}\hat{V}_{m-a}\zeta_m \quad (11)$$

Where $\gamma_1 = \left\|A^a r_0\right\|_2$.

Moreover, provided that $k \le q-1$, from (11) we can write

$$A\hat{V}_k = \hat{V}_{k+1}\overline{T}_k, \quad \hat{T}_m = \begin{bmatrix}
\hat{t}_{11} & \hat{t}_{12} & \cdots & \hat{t}_{1,a+2} & 0 & \cdots & & 0 \\
\hat{t}_{21} & \hat{t}_{22} & \cdots & \hat{t}_{2,a+2} & \hat{t}_{2,a+3} & \ddots & & \vdots \\
\vdots & \hat{t}_{32} & \ddots & \vdots & \ddots & \ddots & & \vdots \\
\hat{t}_{a+2,1} & \vdots & \ddots & \hat{t}_{a+2,a+2} & \vdots & \ddots & & 0 \\
0 & \hat{t}_{a+3,2} & \ddots & \hat{t}_{a+3,a+2} & \ddots & \vdots & & \hat{t}_{m-2a-1,m-a} \\
\vdots & 0 & \ddots & \vdots & \ddots & \ddots & & \hat{t}_{m-2a,m-a} \\
\vdots & & \ddots & \hat{t}_{2a+3,a+2} & \vdots & \ddots & & \vdots \\
\vdots & & & \ddots & \ddots & \vdots & & \hat{t}_{m-a,m-a} \\
\vdots & & & & \ddots & \ddots & & \hat{t}_{m-a+1,m-a} \\
\vdots & & & & \ddots & \ddots & & \vdots \\
\vdots & & & & & \ddots & & \vdots \\
\vdots & & & & & & & \vdots \\
0 & \cdots & \cdots & \cdots & \cdots & \cdots & 0 & \hat{t}_{m+1,m-a}
\end{bmatrix} \quad (12)$$

Note that $\hat{T}_m \in C^{(m+1)\times(m-a)}$. Since the vectors Av_1 Av_2 ... Av_k are linearly independent when $k \le q-1$, we have rank $\left(A\hat{V}_k\right) = k$. Since rank $\left(\hat{V}_{k+1}\right) = k+1$, and rank $\left(A\hat{V}_k\right) \le \min\left\{\text{rank}\left(\hat{V}_k\right), \text{rank}\left(\overline{T}_k\right)\right\}$, we also have that $\text{rank}\left(\overline{T}_k\right) = k$. In other words,

$\overline{T}_k$ has full rank. In additon, if we apply (12) to $A^{a+1}\hat{V}_{m-a}$. Provided that $m \le q$-1, we have:

$$A^{a+1}\hat{V}_{m-a} = A^{a}\hat{V}_{m-a+1}\overline{T}_{m-a} = A^{a-1}\hat{V}_{m-a+2}\overline{T}_{m-a+1}\overline{T}_{m-a}$$
$$= \cdots = \hat{V}_{m+1}\hat{V}_{m}, \quad \hat{V}_{m} \equiv \overline{T}_{m}\overline{T}_{m-1}\cdots\overline{T}_{m-a}.$$

Consequently, provided $m \le q$-1, from (11) we can write:

$$A^{a}r_{m} = \gamma_1 v_1 - \hat{V}_{m+1}\hat{T}_{m}\zeta_{m}$$

and since $v_1 = \hat{V}_{m+1}e_1$, where $e_1 = [1,0,\cdots,0]^{\mathrm{T}} \in C^{m+1}$, hence

$$\left\|A^{a}r_{m}\right\|_2 = \left\|\hat{V}_{m+1}\left(\gamma_1 e_1 - \hat{T}_{m}\zeta_{m}\right)\right\|_2$$

If the column vectors of $\hat{V}_{m+1}$ were orthonormal, then we would have:

$$\left\|A^{a}r_{m}\right\|_2 = \left\|\gamma_1 e_1 - \hat{T}_{m}\zeta_{m}\right\|_2$$

as in GMRES. Therefore, a least, squares solution could be obtained from the krylove space $K_m\left(A; A^a r_0\right)$ by minimizing $\left\|\gamma_1 e_1 - \hat{T}_m\zeta\right\|_2$ over ζ. In the Lanczos algorithm, the v_i's are not orthonormal. However, it is still a reasonable idea to minimize the function

$$T(\zeta) = \left\|\beta e_1 - \hat{T}_m\zeta\right\|_2$$

over ζ and compute the corresponding approximate solution $x_m = x_0 + \hat{V}_{m-a}\zeta_m$. The resulting solution is called the Drazin-Quasi-minimal residual (hereafter DQMR) approximation. Since $\overline{T}_m \in C^{(m+1)\times m}$ is a tridiagonal matrix, Therefore, the $\hat{T}_m \in C^{(m+1)\times(m-a)}$ is a matrix with 2a+3 diagonal to form (12).

Similar to [14] , $\hat{T}_{m+1}$ can be obtained as a simple update of $\hat{T}_m$ by first appending a row of zeros at the bottom of $\hat{T}_m$ and following that by appending the (m+1)-

vector $\left[\hat{t}_{m-a,m+1-a}, \hat{t}_{m+1-2a,m+1-a}, \cdots, \hat{t}_{m+2,m+1-a}\right]^{\mathrm{T}}$ as the $(m+1-a)^{th}$ column as follows.

Let us define

$$G_k^{(0)} = \overline{T}_k,\ k = 1, 2, \cdots \tag{13}$$

$$G_k^{(j)} = \overline{T}_k G_{k-1}^{(j-1)},\ k = j+1, j+2, \cdots;\ j = 1, 2, \cdots \tag{14}$$

where $G_k^{(j)} \in C^{(k+1)\times(k-j)}$ also $G_m^{(a)} = \hat{T}_m$ for each $m \geq a+1$.

Since $\hat{T}_m$ is tridiagonal matrix we have:

$$\overline{T}_{m+1} = \begin{bmatrix} \alpha_1 & \beta_2 & & \\ \delta_2 & \ddots & \ddots & \\ & \ddots & \ddots & \beta_{m+1} \\ & & \delta_{m+1} & \alpha_{m+1} \\ & & & \delta_{m+2} \end{bmatrix} = \left[\begin{array}{c|c} \overline{T}_m & g_{m+1}^{(0)} \\ \hline 0_m^{\mathrm{T}} & \alpha_{m+1}^{(0)} \end{array}\right],$$

where, certainly, $g_{m+1}^{(0)} = \begin{bmatrix} 0 & 0 & \cdots & \beta_{m+1} & \alpha_{m+1} \end{bmatrix}^{\mathrm{T}} \in C^{m+1}$, 0_k denotes the k-dimensional (column) zeros vector, and $\alpha_{m+1}^{(0)} = \delta_{m+2}$ that is scalar.

Supposed that

$$T_{m+1}^{(s+1)} = \left[\begin{array}{c|c} T_m^{(s+1)} & g_{m+1}^{(s+1)} \\ \hline O_{m-s-1}^{\mathrm{T}} & \alpha_{m+1}^{(s+1)} \end{array}\right],\ m \geq s+2. \tag{15}$$

$$g_{m+1}^{(s+1)} = \overline{T}_m g_m^{(s)} + \alpha_m^{(s)} g_{m+1}^{(0)}, \tag{16}$$

From [14] , we have: $\alpha_{m+1}^{(s+1)} = \alpha_{m+1}^{(0)} \alpha_m^{(s)}$. (17)

Equation (16) can be simplified as follows:

$$\overline{T}_m g_m^{(s)} = \begin{bmatrix} \alpha_1 & \beta_2 & \cdots & \cdots & 0 \\ \delta_2 & \alpha_2 & \ddots & & \vdots \\ 0 & \delta_3 & \ddots & \ddots & \vdots \\ \vdots & & \ddots & \ddots & \beta_m \\ \vdots & & \ddots & \ddots & \alpha_m \\ 0 & \cdots & \cdots & 0 & \delta_m \end{bmatrix} \begin{bmatrix} g_{1,m}^{(s)} \\ g_{2,m}^{(s)} \\ \vdots \\ g_{m,m}^{(s)} \end{bmatrix} = \begin{bmatrix} \alpha_1 g_{1,m}^{(s)} + \beta_2 g_{2,m}^{(s)} \\ \delta_2 g_{1,m}^{(s)} + \alpha_2 g_{2,m}^{(s)} + \beta_3 g_{3,m}^{(s)} \\ \delta_3 g_{2,m}^{(s)} + \alpha_3 g_{3,m}^{(s)} + \beta_4 g_{4,m}^{(s)} \\ \vdots \\ \delta_m g_{m-1,m}^{(s)} + \alpha_m g_{m,m}^{(s)} \\ \delta_{m+1} \cdot g_{m,m}^{(s)} \end{bmatrix}$$

$$= \begin{bmatrix} 0 \\ \delta_2 \cdot g_{1,m}^{(s)} \\ \delta_3 \cdot g_{2,m}^{(s)} \\ \vdots \\ \delta_m \cdot g_{m-1,m}^{(s)} \\ \delta_{m+1} \cdot g_{m,m}^{(s)} \end{bmatrix} + \begin{bmatrix} \alpha_1 \cdot g_{1,m}^{(s)} \\ \alpha_2 \cdot g_{2,m}^{(s)} \\ \alpha_3 \cdot g_{3,m}^{(s)} \\ \vdots \\ \alpha_m \cdot g_{m,m}^{(s)} \\ 0 \end{bmatrix} + \begin{bmatrix} \beta_2 \cdot g_{2,m}^{(s)} \\ \beta_3 \cdot g_{3,m}^{(s)} \\ \vdots \\ \beta_m \cdot g_{m,m}^{(s)} \\ \alpha_m \cdot g_{m,m}^{(s)} \\ 0 \\ 0 \end{bmatrix} \tag{18}$$

By using the Hadamard product Equation (18) is reduced. For this purpose, we first introduce the concepts of Hadamard matrix product.

Definition 2 Let A and B be m×n matrices with entries in *C*. The Hadamard product of *A* and *B* is defined by as follows

$$A \odot B = \begin{pmatrix} a_{11}b_{11} & a_{12}b_{12} & \cdots & a_{1n}b_{1n} \\ \vdots & \vdots & & \vdots \\ a_{m1}b_{m1} & a_{m2}b_{m2} & \cdots & a_{mn}b_{mn} \end{pmatrix}.$$

Let us denote

$$\Delta_m = \begin{bmatrix} \delta_2 & \delta_3 & \cdots & \delta_{m+1} \end{bmatrix}^{\mathrm{T}}, \quad \overline{\alpha}_m = \begin{bmatrix} \alpha_1 & \alpha_2 & \cdots & \alpha_m \end{bmatrix}^{\mathrm{T}}, \quad \overline{\beta}_m = \begin{bmatrix} \beta_2 & \beta_3 & \cdots & \beta_m \end{bmatrix}^{\mathrm{T}},$$

$$g_m^{(s)} = \begin{bmatrix} g_{1,m}^{(s)} & g_{2,m}^{(s)} & \cdots & g_{m,m}^{(s)} \end{bmatrix}^{\mathrm{T}}, \quad \underline{g}_m^{(s)} = \begin{bmatrix} g_{2,m}^{(s)} & g_{3,m}^{(s)} & \cdots & g_{m,m}^{(s)} \end{bmatrix}^{\mathrm{T}}.$$

Now, we can be simplified (18) as follows

$$\overline{T}_m g_m^{(s)} = \begin{bmatrix} 0 \\ \Delta_m \odot g_m^{(s)} \end{bmatrix} + \begin{bmatrix} \overline{\alpha}_m \odot g_m^{(s)} \\ 0 \end{bmatrix} + \begin{bmatrix} \overline{\beta}_m \odot \underline{g}_m^{(s)} \\ 0 \\ 0 \end{bmatrix} \tag{19}$$

For solution system

$$\hat{T}_m \zeta = \beta e_1 \tag{20}$$

We must reduce the band matrix, $\hat{T}_m$, into upper triangular by using Givens rotation. $\hat{T}_m$ matrix has bandwidth $2a+3$. To reduce the matrix $\hat{T}_m$ to a upper triangular matrix we need to $(m\text{-}a)(a+1)$ Givens rotations matrix. We denote with $\overline{g}_0 = \beta e_1$ right-hand side (20), and we multiply both sides of (20) from left by Givens rotations. To update the mth column of matrix $\hat{T}_m$, we must first multiply the previous Givens rotations by this column and then we annihilate the main subdiagonal elements with appropriate rotations. It should be noted that number of the previous rotations is min $(m\text{-}1, 2a+2)\times(a+1)$, and the number of the rotations to annihilate the main subdiagonal elements is)×$(a+1)$. Finally, the mth end step we have an upper triangular matrix as follows

$$\hat{R}_m = \begin{bmatrix} \hat{r}_{1,1} & \hat{r}_{1,2} & \cdots & \hat{r}_{1,2a+3} & 0 & \cdots & \cdots & 0 \\ 0 & \hat{r}_{2,2} & \cdots & \hat{r}_{2,2a+3} & \hat{r}_{2,2a+4} & \cdots & \cdots & 0 \\ \vdots & \ddots & \ddots & \vdots & & \ddots & & \vdots \\ \vdots & & \ddots & \hat{r}_{2a+3,2a+3} & & & \ddots & 0 \\ & & & \ddots & \hat{r}_{2a+4,2a+4} & & & \hat{r}_{m-3a-2,m-a} \\ & & & & \ddots & \ddots & & \vdots \\ & & & & & \ddots & \ddots & \hat{r}_{m-a-1,m-a} \\ & & & & & & \ddots & \hat{r}_{m-a,m-a} \\ & & & & & & & 0 \\ & & & & & & & \vdots \\ 0 & \cdots & & & & & & 0 \end{bmatrix} \tag{21}$$

Generally, if we define $\Omega^{(m-a)}$ the product of matrices $\Omega^{(m)}_{i+1,i}$, then

$$\Omega^{(m-a)} = \Omega^{(m-a)}_{m-a+1,m-a} \cdots \Omega^{(m-a)}_{m,m-1}\, \Omega^{(m-a)}_{m+1,m}, \quad i = m+1, m, \cdots, m-a$$

where $\Omega^{(m)}_{i+1,i}$ be the Givens matrices use to transform $\hat{T}_m$ into an upper triangular $\hat{R}_m$ and the vector of $\overline{g}_m = \Omega^{(m-a)}(\gamma e_1)$. Finally , the approximate solution is given by

$$x_m = x_0 + \hat{V}_{m-a} R_{m-a}^{-1} g_{m-a}$$

where R_{m-a} and g_{m-a} are obtained by removing the a+1 row of the matrix $\hat{R}_m$ and right-hand side $\overline{g}_m$. The approximation solution x_m can be updated at each step by the relation,

$$x_m = x_{m-1} + p_{m-a}\gamma_{m-a} \tag{22}$$

Since if we assume $P_{m-a} = [p_1 \cdots p_{m-a}]$, then we have:

$$P_{m-a} = \hat{V}_{m-a} R_{m-a}^{-1}$$

Consequently,

$$[p_1 \cdots p_{m-a}] R_{m-a} = [v_1 \cdots v_{m-a}]$$

and

$$p_{m-a} = \left(v_{m-a} - \sum_{i=m-3a-2}^{m-a-1} p_i \hat{r}_{i,m-a} \right) \Big/ \hat{r}_{m-a,m-a}$$

where p_{m-a} is the last column of p_{m-a}. Therefore, it can be written

$$x_m = x_0 + P_{m-a} g_{m-a} = x_0 + [P_{m-a-1}, p_{m-a}] \begin{bmatrix} g_{m-a-1} \\ \gamma_{m-a} \end{bmatrix}$$

or

$$x_m = x_{m-1} + p_{m-a}\gamma_{m-a} \tag{23}$$

Thus, x_m can be updated easily at each step, via the relation (23) using x_{m-1}.

This gives the following algorithm, which we call the Drazin-QMR for Drazin- inverse solution of singular nonsymmetric linear equations.

Algorithm 4.1 DQMR Algorithm

1. *Pick* x_0 *and compute* $r_0 = b - Ax_0$ *and* $A^a r_0$.
2. *Compute* $\gamma_1 = \|A^a r_0\|_2$ *and set* $v_1 = (A^a r_0)/\gamma$.
3. *For* $i = 1,2,3,\ldots,m$.
4. *Compute* α_i, δ_{i+1}, β_{i+1} *and* v_{i+1}, w_{i+1} as in ([17] , Algorithm 7.1).
5. *For* k=1, 2, ... *m form the matrices* $\widehat{V}_k \in R^{n\times k}$ *and* $\overline{T}_k \in R^{(k+1)\times k}$:

$$\widehat{V}_k = [v_1\ v_2 \cdots v_k] \text{ and } \overline{T}_k = \begin{bmatrix} \alpha_1 & \beta_2 & 0 & \cdots & \cdots & 0 \\ \delta_2 & \alpha_2 & \beta_3 & \ddots & & \vdots \\ 0 & \delta_3 & \ddots & \ddots & & 0 \\ \vdots & \ddots & & & \ddots & \beta_m \\ \vdots & & & \ddots & & \alpha_m \\ 0 & \cdots & 0 & & & \delta_{m+1} \end{bmatrix}$$

6. *Compute the matrix* $\widehat{T}_m = \overline{T}_m \overline{T}_{m-1} \cdots \overline{T}_{m-a}$ *by using* (15).

7. *Update the QR factorization of* $\widehat{T}_m$, *i.e,*

8. *Set* $i_1 = \min(m-1, 2a+2)$.

9. *Apply* $\Omega^{(i)}$,$i=i_1{:}m-a+1$ to the (m-a)th column of $\widehat{T}_m$

10. For i = 1: a +1.

11. *Compute the rotation coefficients* s_i, c_i *of rotation matrix* $\Omega^{m-a}_{m+2-i,m+1-i}$

12. *Apply rotation* $\Omega^{m-a}_{m+2-i,m+1-i}$ *to the* (m-a)th *column of* $\widehat{T}_m$ *and* $\overline{g}_m$.

13. EndFor.

14. *Compute* $$p_{m-a} = \left(v_{m-a} - \sum_{i=m-3a-2}^{m-a-1} p_i \hat{t}_{i,m-a}\right) / \hat{t}_{m-a,m-a}$$

15. *Compute* $x_m = x_{m-1} - p_{m-a}\gamma_{m-a}$.

4. NUMERICAL EXAMPLES

In this section, we will compute the linear system $Ax=b$ by discretization Poisson equation with Neumann boundary conditions:

$$\begin{cases} \left(\dfrac{\partial^2}{\partial x^2} + \dfrac{\partial^2}{\partial y^2}\right) u(x,y) = f(x,y), & (x,y) \in \Omega = [0,1] \times [0,1] \\ \dfrac{\partial}{\partial n} u(x,y) = \varphi(x,y) & x,y \in \partial\Omega. \end{cases}$$

This linear system has also been computed by Sidi [14] for testing DGMRES algorithm. The problem has also been considered by Hank and Hochbruck [18] for testing the Chebyshev-type semi-iterative method. The numerical computations are performed in MATLAB (R213a) with double precision. The results were obtains by running the code on an Intel (R) Core (TM) i7-2600 CPU Processor running 3.40 GHz with 8 GB of RAM memory using Windows 7 professional 64-bit operating system. The initial vector x_0 is the zero vector. All the tests were stopped as soon as

$$\text{Relative Error} = \frac{\left\| x_n - A^D b \right\|_\infty}{\left\| A^D b \right\|_\infty} \le 10^{-8}.$$

Let M be an odd integer, we discretize the Poisson equation on a uniform grid of mesh size $h=1/M$ via central differences, and then by taking the unknowns in the red-black order we obtain the system $Ax=b$, where the $(M+1)^2 \times (M+1)^2$ non- symmetric matrix A is as follows

$$A = \begin{bmatrix}
4I & o & \cdots & \cdots & \cdots & \cdots & \cdots & o & T_1 & -2I & o & \cdots & \cdots & \cdots & \cdots & o \\
o & 4I & \ddots & & & & & \vdots & -I & T_2 & -I & o & & & & \vdots \\
\vdots & \ddots & \ddots & \ddots & & & & \vdots & o & -I & T_1 & -I & o & & & \vdots \\
\vdots & & \ddots & \ddots & \ddots & & & \vdots & \vdots & o & -I & T_2 & -I & o & & \vdots \\
\vdots & & & \ddots & \ddots & \ddots & & \vdots & \vdots & & \ddots & \ddots & \ddots & \ddots & \ddots & \vdots \\
\vdots & & & & \ddots & \ddots & \ddots & \vdots & \vdots & & & \ddots & \ddots & \ddots & \ddots & o \\
\vdots & & & & & \ddots & 4I & o & \vdots & & & & o & -I & T_1 & -I \\
o & \cdots & \cdots & \cdots & \cdots & \cdots & o & 4I & o & \cdots & \cdots & \cdots & \cdots & o & -2I & T_2 \\
T_2 & -2I & o & \cdots & \cdots & \cdots & \cdots & o & 4I & o & \cdots & \cdots & \cdots & \cdots & \cdots & o \\
-I & T_1 & -I & o & & & & \vdots & o & 4I & \ddots & & & & & \vdots \\
o & -I & T_2 & -I & o & & & \vdots & \vdots & \ddots & \ddots & \ddots & & & & \vdots \\
\vdots & o & -I & T_1 & -I & o & & \vdots & \vdots & & \ddots & \ddots & \ddots & & & \vdots \\
\vdots & & \ddots & \ddots & \ddots & \ddots & \ddots & \vdots & \vdots & & & \ddots & \ddots & \ddots & & \vdots \\
\vdots & & & \ddots & \ddots & \ddots & \ddots & o & \vdots & & & & \ddots & \ddots & \ddots & \vdots \\
\vdots & & & & o & -I & T_2 & -I & \vdots & & & & & \ddots & 4I & o \\
o & \cdots & \cdots & \cdots & \cdots & o & -2I & T_1 & o & \cdots & \cdots & \cdots & \cdots & \cdots & o & 4I
\end{bmatrix} \tag{24}$$

Here, I and 0 denote, respectively, the $(M+1)/2 \times (M+1)/2$ identity and zero matrices and the $(M+1)/2 \times (M+1)/2$ matrices T_1 and T_2 are given by

$$T_1 = \begin{bmatrix} -2 & o & \cdots & \cdots & o \\ -1 & -1 & \ddots & & \vdots \\ o & \ddots & \ddots & \ddots & \vdots \\ \vdots & \ddots & & -1 & o \\ o & & o & -1 & -1 \end{bmatrix}, \qquad T_2 = \begin{bmatrix} -1 & -1 & o & \cdots & o \\ o & -1 & \ddots & \ddots & \vdots \\ \vdots & \ddots & \ddots & \ddots & o \\ \vdots & & & -1 & -1 \\ o & \cdots & \cdots & o & -2 \end{bmatrix}$$

The numerical experiment was performed for M=31, 63.

It should be noted A is singular with 1D null space spanned by the vector $e=[1, \ldots, 1]^T$. Furthermore, (A)=1, as mentioned in [13] [14] [15] [18]. Even if continues problem has a solution, the discretized problem need not be consistent. In the sequel we consider only the Drazin-inverse solution of the system for arbitrary right side b, not necessarily related to f and φ.

We first construct a consistent system with the known solution $\hat{s} \in R(A)$ via $\hat{s} = Ay$, where $y = [0, \cdots, 0, 1]^T$. Then we perturb $A\hat{s}$, the right-hand side of $Ax = A\hat{s} = \hat{b}$, with a constant multiple of the null space vector e and we obtain the right-hand side

$$b = \hat{b} + \delta \frac{e}{\|e\|_2}$$

Consequently the system $Ax = \hat{b} + \delta \frac{e}{\|e\|_2}$ is solved for x. The perturbation parameter δ is selected as 10^{-2} in our experiments. The solution we intend to obtain is the vector $\hat{S}$, whose components are zeros except

$$\hat{s}_{2\hat{M}^2-\hat{M}} = -1,\ \hat{s}_{2\hat{M}^2-1} = -1,\ \hat{s}_{2\hat{M}^2} = -2,\ \hat{s}_{4\hat{M}^2} = 4, \ \text{where } \hat{M} = (M+1)/2$$

In Table 1, Table 2, we give the number of iterations (Its), the CPU time (Time) required for convergence, the relative error (RE), for the DGMRES and DQMR methods. As shown in Table 1, Table 2 the DQMR algorithm is effective and less expensive than the DGMRES algorithm.

Table 1. Application of DQMR implementation to the consistent singular system.

Size of A		1024 × 1024		4096 × 4096		
Method	Its	Time	RE	Its	Time	RE
DQMR	155	0.61	5.9674e−09	267	7.39	8.8234e−09
DGMRES	165	0.88	3.0294e−09	307	10.89	8.1578e−09

Table 2. Application of DQMR implementation to the inconsistent singular system.

Size of A		4096×4096			16384×16384	
Method	Its	Time	RE	Its	Time	RE
DQMR	155	0.53	5.9674e–09	267	6.88	8.8234e–09
DGMRES	165	0.81	3.0294e–09	307	10.34	8.1578e–09

5. CONCLUSION

In this paper, we presented a new method, called DQMR, for Drazin-inverse solution of singular nonsymmetric linear systems. The DQMR algorithm for singular systems is analogous to the QMR algorithm for non-singular systems. Numerical experiments indi- cate that the Drazin-inverse solution obtained by this method is reasonably accurate, and its computation time is less than that of solution obtained by the DGMRES method. Thus, we can conclude that the DQMR algorithm is a robust and efficient tool to compute the Drazin-inverse solution of singular linear systems.

ACKNOWLEDGEMENTS

First, we would like to thank professor F. Toutounian for her comments that improved our results. Second, we thank the editor and the referees for their carefully reading and useful comments.

REFERENCES

1. Ben-Israel, A. and Grevile, T.N.E. (2003) Generalized Inverses: Theory and Applications. 2nd Edition, Springer-Verlag, New York.
2. Campell, S.L. and Meyer, C.D. (1979) Generalized Inverses of Linear Transformations. Pitman (Advanced Publishing Program), Boston.
3. Hartwig, R.E. and Hall, F. (1982) Applications of the Drazin Inverse to Cesaro-Neumann Iterations. In: Campbell, S.L., Ed., Recent Applications of Generalized Inverses, 66, 145-195.
4. Hartwig, R.E. and Levine, J. (1981) Applications of the Drazin Inverse to the Hill Cryptographic system. Part III, Cryptologia, 5, 67-77. https://doi.org/10.1080/0161-118191855850
5. Eiermann, M., Marek, I. and Niethammer, W. (1988) On the Solution of Singular Linear Systems of Algebric Equations by Semiiterative Methods. Numerische Mathematik, 53, 265-283. https://doi.org/10.1007/BF01404464
6. Freund, R.W. and Hochbruck, M. (1994) On the Use of Two QMR Algorithms for Solving Singular Systems and Applications in Markov Chain Modeling. Numerical

Linear Algebra with Applications, 1, 403-420. https://doi. org/10. 1002/nla.1680010406
7. Simeon, B., Fuhrer, C. and Rentrop, P. (1993) The Drazin Inverse in Multibody System Dynamics. Numerische Mathematik, 64, 521-539. https://doi.org/ 10.1007/BF01388703
8. Brown, P.N. and Walker, H.F. (1997) GMRES on (Nearly) Singular Systems. SIAM Journal on Matrix Analysis and Applications, 18, 37-51. https://doi.org/ 10.1137/S0895479894262339
9. Ipsen, I.F. and Meyer, C.D. (1998) The Idea behind Krylov Methods. The American Mathematical Monthly, 105, 889-899. https://doi.org/10.2307/2589281
10. Smoch, L. (1999) Some Result about GMRES in the Singular Case. Numerical Algorithms, 22, 193-212. https://doi.org/10.1023/A:1019162908926
11. Wei, Y. and Wu, H. (2000) Convergence Properties of Krylov Subspace Methods for Singular System with Arbitrary Index. Journal of Computational and Applied Mathematics, 114, 305-318. https://doi.org/10.1016/S0377-0427(99)90237-6
12. Sidi, A. (1999) A Unified Approach to Krylov Subspace Methods for the Drazin-Inverse Solution of Singular Non-Symmetric Linear Systems. Linear Algebra and Its Applications, 298, 99-113. https://doi.org/10.1016/S0024-3795(99)00153-6
13. Sidi, A. and Kluzner, V. (1999) A Bi-CG Type Iterative Method for Drazin Inverse Solution of Singular Inconsistent Non-Symmetric Linear Systems of Arbitrary Index. The Electronic Journal of Linear Algebra, 6, 72-94.
14. Sidi, A. (2001) DGMRES: A GMRES-Type Algorithm for Drazin-Inverse Solution of Singular Non-Symmetric Linear Systems. Linear Algebra and Its Applications, 335, 189-204. https://doi.org/10.1016/S0024-3795(01)00289-0
15. Zhou, J. and Wei, Y. (2004) DFOM Algorithm and Error Analysis for Projection Methods for Solving Singular Linear System. Applied Mathematics and Computation, 157, 313-329. https://doi.org/10.1016/j.amc.2003.08.036
16. Lanczos, C. (1950) An Iteration Method for the Solution of the Eigenvalue Problem of Linear Differential and Integral Operators. Journal of Research of the National Bureau of Standards, 45, 255-282. https://doi.org/10.6028/jres.045.026
17. Saad, Y. (2003) Iterative Methods for Sparse Linear Systems. 2nd Edition, SIAM, Philadelphia. https://doi.org/10.1137/1.9780898718003
18. Hank, M. and Hochbruck, M. (1993) A Chebyshev-Like Semiiteration for Inconsistent Linear Systems. Electronic Transactions on Numerical Analysis, 1, 315-339.

CHAPTER 15

Minimum Covering Randić Energy of a Graph

M. R. Rajesh Kanna[1,2], R. Jagadeesh[3]

[1]*Department of Mathematics, Government First Grade College, Puttur, India*

[2]*Department of Mathematics, Maharani's Science College for Women, Mysore, India*

[3]*Research and Development Centre, Bharathiar University, Coimbatore, India*

ABSTRACT

Randić energy was first defined in the paper [1]. Using minimum covering set, we have introduced the minimum covering Randić energy $RE_C(G)$ of a graph *G* in this paper. This paper contains computation of minimum covering Randić energies for some standard graphs like star graph, complete graph, thorn graph of complete graph, crown graph, complete bipartite graph, cocktail graph and friendship graphs. At the end of this paper, upper and lower bounds for minimum covering Randić energy are also presented.

Keywords: Minimum Covering Set, Minimum Covering Randić Matrix, Minimum Covering Randić Eigenvalues, Minimum Covering Randić Energy

1. INTRODUCTION

Study on energy of graphs goes back to the year 1978, when I. Gutman [2] defined this while working with energies of conjugated hydrocarbon containing carbon atoms. All graphs considered in this paper are assumed to be simple without loops and multiple edges. Let $A=(a_{ij})$ be the adjacency matrix of the graph *G* with its eigenvalues $\lambda_1, \lambda_2, \cdots, \lambda_n$ assumed in decreasing order. Since A is real symmetric, the eigenvalues of G are real numbers whose sum equal to zero. The sum of the absolute eigenvalues values of G is called the energy *E(G)* of *G*. i.e.,

$$E(G)=\sum_{i=1}^{n}|\lambda_i|.$$

Theories on the mathematical concepts of graph energy can be seen in the reviews [3] , papers [4] [5] [6] and the references cited there in. For various upper and lower bounds for energy of a graph can be found in papers [7] [8] and it was observed that graph energy has chemical applications in the molecular orbital theory of conjugated mo- lecules [9] [10].

1.1. Randić Energy

It was in the year 1975, Milan Randić invented a molecular structure descriptor called Randić index which is defined as [11]

$$R(G)=\sum_{v_iv_j\in E(G)}\frac{1}{\sqrt{d_id_j}}$$

Motivated by this S.B. Bozkurt *et al.* [1] defined Randić matrix and Randić energy as follows. Let *G* be graph of order n with vertex set $V=\{v_1,v_2 \dots v_n\}$ and edge set *E*. Randić matrix of *G* is a $n\times n$ symmetric matrix defined by$R(G):=(r_{ij})$, where

$$r_{ij}=\begin{cases}\dfrac{1}{\sqrt{d_id_j}} & \text{if } v_iv_j\in E(G)\\ 0 & \text{otherwise}\end{cases}.$$

The characteristic equation of *R(G)* is defined by $f_n(G,\rho)=\det(\rho I-R(G))=0$. The roots of this equation is called Randić eigenvalues of *G*. Since *R(G)* is real and symmetric, its eigenvalues are real numbers and we label them in decreasing order$p_1\geq p_2\geq\cdots p_n$.Randić energy of *G* is defined as

$$RE(G)=\sum_{i=1}^{n}|\rho_i|$$

Further studies on Randić energy can be seen in the papers [12] [13] [14] and the references cited there in.

1.2. Minimum Covering Energy

In the year 2012 *C* Adiga *et al.* [15] introduced minimum covering energy of a graph, which depends on its particular minimum cover. *A* subset *C* of vertex set *V* is called a covering set of *G* if every edge of *G* is incident to at least one vertex of *C*. Any covering set with minimum cardinality is called a minimum covering set. If *C* is a minimum covering set of a graph *G* then the minimum covering matrix of G is the *n*×*n* matrix defined by $A_C(G) := (a_{ij})$, where

$$a_{ij} = \begin{cases} 1 & \text{if } v_i v_j \in E(G) \\ 1 & \text{if } i = j \text{ and } v_i \in C \\ 0 & \text{otherwise} \end{cases}$$

The minimum covering eigenvalues of the graph *G* are roots of the characteristic equation $f_n(G,\lambda) = \det(\lambda I - A_C(G)) = 0$, obtained from the matrix $A_c(G)$. Since $A_c(G)$ is real and symmetric, its eigenvalues are real numbers and we label them in the order $\lambda_1 \geq \lambda_2 \geq \cdots \geq \lambda_n$. The minimum covering energy of *G* is defined as $E_C(G) = \sum_{i=1}^{n} |\lambda|_i$

1.3. Minimum Covering Randić Energy

Results on Randić energy and minimum covering energy of graph *G* motivates us to define minimum covering Randić energy. Consider a graph *G* with vertex set $V=\{v_1,v_2 \ldots v_n\}$and edge set *E*. If *C* is a minimum covering set of a graph G then the minimum covering Randić matrix of G is the *n*×*n* matrix defined by $R_C(G) := (r_{ij})$,

where

$$r_{ij} = \begin{cases} \frac{1}{\sqrt{d_i d_j}} & \text{if } v_i v_j \in E(G) \\ 1 & \text{if } i = j \text{ and } \upsilon_i \in C \\ 0 & \text{otherwise} \end{cases}.$$

The characteristic polynomial of $R_c(G)$ is defined by $f_n(G,\rho) = \det(\rho I - R_C(G))$. The minimum covering Randić eigenvalues of the graph G are the eigenvalues of *Rc(G)*. Since *Rc(G)* is real and symmetric matrix so its eigenvalues are real numbers. We label the eigenvalues in order $p_1 \geq p_2 \geq \cdots p_n$. The minimum covering Randić energy of G is defined as $RE_C(G) = \sum_{i=1}^{n} |\rho_i|$

Example 1: i) $C_1 = \{v_1, v_2, v_5\}$ ii) $C_2 = \{v_2, v_4, v_5\}$ are the possible minimum cove- ring sets for the Figure 1 as shown below.

$$R_{C_1}(G) = \begin{pmatrix} 1 & \frac{1}{3} & 0 & \frac{1}{\sqrt{6}} & \frac{1}{\sqrt{15}} & 0 \\ \frac{1}{3} & 1 & \frac{1}{\sqrt{6}} & 0 & \frac{1}{\sqrt{15}} & 0 \\ 0 & \frac{1}{\sqrt{6}} & 0 & 0 & \frac{1}{\sqrt{10}} & 0 \\ \frac{1}{\sqrt{6}} & 0 & 0 & 0 & \frac{1}{\sqrt{10}} & 0 \\ \frac{1}{\sqrt{15}} & \frac{1}{\sqrt{15}} & \frac{1}{\sqrt{10}} & \frac{1}{\sqrt{10}} & 1 & \frac{1}{\sqrt{5}} \\ 0 & 0 & 0 & 0 & \frac{1}{\sqrt{5}} & 0 \end{pmatrix}$$

i)

Minimum covering Randić eigenvalues are

$\rho_1 \approx -0.3143, \rho_2 \approx -0.1937, \rho_3 \approx -0.0646, \rho_4 \approx 0.8604, \rho_5 \approx 0.9108, \rho_6 \approx 1.8014$ Minimum covering Randić energy, $RE_{C_1}(G) \approx 4.1453.$

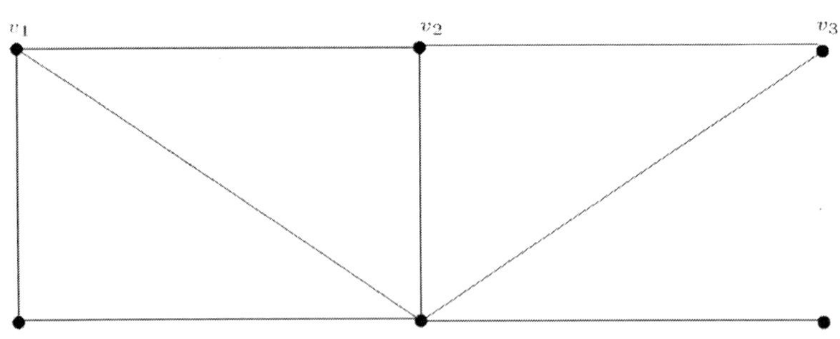

Figure 1. Minimum covering Randić energy depends on the covering set.

$$R_{C_2}(G)=\begin{pmatrix} 0 & \frac{1}{3} & 0 & \frac{1}{\sqrt{6}} & \frac{1}{\sqrt{15}} & 0 \\ \frac{1}{3} & 1 & \frac{1}{\sqrt{6}} & 0 & \frac{1}{\sqrt{15}} & 0 \\ 0 & \frac{1}{\sqrt{6}} & 0 & 0 & \frac{1}{\sqrt{10}} & 0 \\ \frac{1}{\sqrt{6}} & 0 & 0 & 1 & \frac{1}{\sqrt{10}} & 0 \\ \frac{1}{\sqrt{15}} & \frac{1}{\sqrt{15}} & \frac{1}{\sqrt{10}} & \frac{1}{\sqrt{10}} & 1 & \frac{1}{\sqrt{5}} \\ 0 & 0 & 0 & 0 & \frac{1}{\sqrt{5}} & 0 \end{pmatrix}$$

ii)

Minimum covering Randić eigenvalues are

$\rho_1 \approx -0.3515, \rho_2 \approx -0.1788, \rho_3 \approx -0.0611, \rho_4 \approx 0.7699, \rho_5 \approx 1.0767, \rho_6 \approx 1.7488$

.

Minimum covering Randić energy, $RE_{C_2}(G) \approx 4.1828.$

Therefore minimum covering Randić energy depends on the covering set.

2. MAIN RESULTS AND DISCUSSION

2.1. Minimum Covering Randić Energy of Some Standard Graphs

Theorem 2.1 *For n≥2, the minimum covering Randić energy, $R_c(G)$of complete graph K_n is* $\dfrac{(n-2)^2+\sqrt{4n^2-8n+5}}{n-1}$.

Proof. Let K_n be a complete graph with vertex set$V=\{v_1,v_2...v_n\}$. The minimum covering set for K_n is$C=\{v_1,v_2...v_n\}$. Then

$$R_C(K_n)=\begin{pmatrix} 1 & \frac{1}{n-1} & \frac{1}{n-1} & \cdots & \frac{1}{n-1} & \frac{1}{n-1} & \frac{1}{n-1} \\ \frac{1}{n-1} & 1 & \frac{1}{n-1} & \cdots & \frac{1}{n-1} & \frac{1}{n-1} & \frac{1}{n-1} \\ \frac{1}{n-1} & \frac{1}{n-1} & 1 & \cdots & \frac{1}{n-1} & \frac{1}{n-1} & \frac{1}{n-1} \\ \vdots & \vdots & \vdots & \ddots & \vdots & \vdots & \vdots \\ \frac{1}{n-1} & \frac{1}{n-1} & \frac{1}{n-1} & \cdots & 1 & \frac{1}{n-1} & \frac{1}{n-1} \\ \frac{1}{n-1} & \frac{1}{n-1} & \frac{1}{n-1} & \cdots & \frac{1}{n-1} & 1 & \frac{1}{n-1} \\ \frac{1}{n-1} & \frac{1}{n-1} & \frac{1}{n-1} & \cdots & \frac{1}{n-1} & \frac{1}{n-1} & 0 \end{pmatrix}_{n\times n}$$

Characteristic polynomial is $\dfrac{(-1)^n\left[(n-1)\rho-(n-2)\right]^{n-2}\left[(n-1)\rho^2-(2n-3)\rho-1\right]}{(n-1)^{n-1}}$

Characteristic equation is $(-1)^n\left[(n-1)\rho-(n-2)\right]^{n-2}\left[(n-1)\rho^2-(2n-3)\rho-1\right]=0$

Minimum covering Randić Spec

$$(K_n)=\begin{pmatrix} \frac{n-2}{n-1} & \frac{(2n-3)+\sqrt{4n^2-8n+5}}{2(n-1)} & \frac{(2n-3)-\sqrt{4n^2-8n+5}}{2(n-1)} \\ n-2 & 1 & 1 \end{pmatrix}$$

Minimum covering Randić energy,

$$RE_C(K_n)=\left|\frac{n-2}{n-1}\right|(n-2)+\left|\frac{(2n-3)+\sqrt{4n^2-8n+5}}{2(n-1)}\right|(1)+\left|\frac{(2n-3)-\sqrt{4n^2-8n+5}}{2(n-1)}\right|$$

$$=\frac{(n-2)^2}{n-1}+\frac{(2n-3)+\sqrt{4n^2-8n+5}}{2(n-1)}+\frac{\sqrt{4n^2-8n+5}-(2n-3)}{2(n-1)}$$

$$=\frac{(n-2)^2}{n-1}+\frac{\sqrt{4n^2-8n+5}}{(n-1)}=\frac{(n-2)^2+\sqrt{4n^2-8n+5}}{n-1}. \tag{1}$$

Definition 2.1 Thorn graph of k_n is denoted by $(k_n)^{+1}$ and it is obtained by attaching one edge to each vertex of k_n.

Theorem 2.2 For n≥2 , the minimum covering Randić energy, $R_c(G)$of thorn graph$(k_n)^{+1}$ is $\frac{(n^2-2)+\sqrt{4n^2+1}}{n}$.

Proof. $(k_n)^{+1}$ is a thorn graph of complete graph k_n with vertex set $V=\{v_1,v_2,\cdots,v_n,v_1',v_2',\cdots,v_n'\}$. The minimum covering set for thorn graph $(k_n)^{+1}$ is $C=\{v_1,v_2,v_3,\cdots,v_n\}$. Then

$$R_C\left((K_n)^{+1}\right)=\left(\begin{array}{c|ccccc|ccccc}
 & v_1 & v_2 & v_3 & \cdots & v_n & v_1' & v_2' & v_3' & \cdots & v_n' \\
\hline
v_1 & 1 & \frac{1}{n} & \frac{1}{n} & \cdots & \frac{1}{n} & \frac{1}{\sqrt{n}} & 0 & 0 & 0 & 0 \\
v_2 & \frac{1}{n} & 1 & \frac{1}{n} & \cdots & \frac{1}{n} & 0 & \frac{1}{\sqrt{n}} & 0 & \cdots & 0 \\
v_3 & \frac{1}{n} & \frac{1}{n} & 1 & \cdots & \frac{1}{n} & 0 & 0 & \frac{1}{\sqrt{n}} & \cdots & 0 \\
\vdots & \vdots & \vdots & \vdots & \ddots & \vdots & \vdots & \vdots & \vdots & \ddots & \vdots \\
v_n & \frac{1}{n} & \frac{1}{n} & \frac{1}{n} & \frac{1}{n} & 1 & 0 & 0 & 0 & \cdots & \frac{1}{\sqrt{n}} \\
\hline
v_1' & \frac{1}{\sqrt{n}} & 0 & 0 & \cdots & 0 & 0 & 0 & 0 & \cdots & 0 \\
v_2' & 0 & \frac{1}{\sqrt{n}} & 0 & \cdots & 0 & 0 & 0 & 0 & \cdots & 0 \\
v_3' & 0 & 0 & \frac{1}{\sqrt{n}} & \cdots & 0 & 0 & 0 & 0 & \cdots & 0 \\
\vdots & \vdots & \vdots & \vdots & \ddots & \vdots & \vdots & \vdots & \vdots & \ddots & \vdots \\
v_n' & 0 & 0 & 0 & \cdots & \frac{1}{\sqrt{n}} & 0 & 0 & 0 & \cdots & 0
\end{array}\right)$$

Characteristic polynomial is

$$\frac{[\rho-1]^{n-1}[n\rho+1]^{n-2}\left[n\rho^2-(2n-1)\rho-1\right]}{n^n}.$$

Characteristic equation is $[\rho-1]^{n-1}[n\rho+1]^{n-2}\left[n\rho^2-(2n-1)\rho-1\right]=0$

Minimum covering Randić Spec

$$\left((K_n)^{+1}\right)=\begin{pmatrix} 1 & -\frac{1}{n} & \frac{(2n-1)+\sqrt{4n^2+1}}{2n} & \frac{(2n-1)-\sqrt{4n^2+1}}{2n} \\ n-1 & n-2 & 1 & 1 \end{pmatrix}$$

Minimum covering Randić energy is,

$$RE_C\left((K_n)^{+1}\right)=|1|(n-1)+\left|-\frac{1}{n}\right|(n-2)+\left|\frac{(2n-1)+\sqrt{4n^2+1}}{2n}\right|(1)+\left|\frac{(2n-1)-\sqrt{4n^2+1}}{2n}\right|$$

$$=(n-1)+\frac{n-2}{n}+\frac{(2n-1)+\sqrt{4n^2+1}}{2n}+\frac{\sqrt{4n^2+1}-(2n-1)}{2n}$$

$$=(n-1)+\frac{n-2}{n}+\frac{\sqrt{4n^2+1}}{n}=\frac{(n^2-2)+\sqrt{4n^2+1}}{n}. \tag{1}$$

Definition 2.2 Cocktail party graph is denoted by $K_{n\times 2}$, is a graph having the vertex set $V=\bigcup_{i=1}^{n}\{u_i,v_i\}$ and the edge set $E=\{u_iu_j, v_iv_j : i\neq j\}\bigcup\{u_iv_j, v_iu_j : 1\leq i<j\leq n\}$.

Theorem 2.3 The minimum covering Randić energy, $R_c(G)$ of cocktail party graph is $K_{n\times 2}$ is $\dfrac{(2n^2-6n+5)+\sqrt{4n^2-8n+5}}{n-1}$

Proof. Consider cocktail party graph $K_{n\times 2}$ with vertex set $V=\bigcup_{i=1}^{n}\{u_i,v_i\}$.

The mi- nimum covering set of cocktail party graph $K_{n\times 2}$ is $C=\bigcup_{i=1}^{n-1}\{u_i,v_i\}$. Then

$$R_C\left(K_{n\times 2}\right)=\left(\begin{array}{c|ccccc|ccccc}
 & u_1 & u_2 & u_3 & \cdots & u_n & v_1 & v_2 & v_3 & \cdots & v_n \\
\hline
u_1 & 1 & \frac{1}{2n-2} & \frac{1}{2n-2} & \cdots & \frac{1}{2n-2} & 0 & \frac{1}{2n-2} & \frac{1}{2n-2} & \cdots & \frac{1}{2n-2} \\
u_2 & \frac{1}{2n-2} & 1 & \frac{1}{2n-2} & \cdots & \frac{1}{2n-2} & \frac{1}{2n-2} & 0 & \frac{1}{2n-2} & \cdots & \frac{1}{2n-2} \\
u_3 & \frac{1}{2n-2} & \frac{1}{2n-2} & 1 & \cdots & \frac{1}{2n-2} & \frac{1}{2n-2} & \frac{1}{2n-2} & 0 & \cdots & \frac{1}{2n-2} \\
\vdots & \vdots & \vdots & \vdots & \ddots & \vdots & \vdots & \vdots & \vdots & \ddots & \vdots \\
u_n & \frac{1}{2n-2} & \frac{1}{2n-2} & \frac{1}{2n-2} & \cdots & 0 & \frac{1}{2n-2} & \frac{1}{2n-2} & \frac{1}{2n-2} & \cdots & 0 \\
\hline
v_1 & 0 & \frac{1}{2n-2} & \frac{1}{2n-2} & \cdots & \frac{1}{2n-2} & 1 & \frac{1}{2n-2} & \frac{1}{2n-2} & \cdots & \frac{1}{2n-2} \\
v_2 & \frac{1}{2n-2} & 0 & \frac{1}{2n-2} & \cdots & \frac{1}{2n-2} & \frac{1}{2n-2} & 1 & \frac{1}{2n-2} & \cdots & \frac{1}{2n-2} \\
v_3 & \frac{1}{2n-2} & \frac{1}{2n-2} & 0 & \cdots & \frac{1}{2n-2} & \frac{1}{2n-2} & \frac{1}{2n-2} & 1 & \cdots & \frac{1}{2n-2} \\
\vdots & \vdots & \vdots & \vdots & \ddots & \vdots & \vdots & \vdots & \vdots & \ddots & \vdots \\
v_n & \frac{1}{2n-2} & \frac{1}{2n-2} & \frac{1}{2n-2} & \cdots & 0 & \frac{1}{2n-2} & \frac{1}{2n-2} & \frac{1}{2n-2} & \cdots & 0
\end{array}\right)$$

Characteristic polynomial is,

$$\frac{\rho(\rho-1)^{n-1}\left[(n-1)\rho-(n-2)\right]^{n-2}\left[(n-1)\rho^2-(2n-3)\rho-1\right]}{(n-1)^{n-1}}.$$

Characteristic equation is,

$$\rho(\rho-1)^{n-1}\left[(n-1)\rho-(n-2)\right]^{n-2}\left[(n-1)\rho^2-(2n-3)\rho-1\right]=0.$$

Minimum covering Randić Spec

$$\left(K_{n\times 2}\right)=\begin{pmatrix} 1 & \frac{n-2}{n-1} & \frac{(2n-3)+\sqrt{4n^2-8n+5}}{2(n-1)} & \frac{(2n-3)-\sqrt{4n^2-8n+5}}{2(n-1)} \\ n-1 & n-2 & 1 & 1 \end{pmatrix}$$

Minimum covering Randić energy,

$$RE_C(K_{n\times 2}) = |1|(n-1) + \left|\frac{n-2}{n-1}\right|(n-2) +$$

$$\left|\frac{(2n-3)+\sqrt{4n^2-8n+5}}{2(n-1)}\right|(1) + \left|\frac{(2n-3)-\sqrt{4n^2-8n+5}}{2(n-1)}\right|(1)$$

$$=(n-1)+\frac{(n-2)^2}{n-1}+\frac{(2n-3)+\sqrt{4n^2-8n+5}}{2(n-1)}+\frac{\sqrt{4n^2-8n+5}-(2n-3)}{2(n-1)}$$

$$=(n-1)+\frac{(n-2)^2}{n-1}+\frac{\sqrt{4n^2-8n+5}}{(n-1)}$$

$$=\frac{(n-1)^2+(n-2)^2+\sqrt{4n^2-8n+5}}{n-1}$$

$$=\frac{(2n^2-6n+5)+\sqrt{4n^2-8n+5}}{n-1}.$$

Theorem 2.4 For *n≥2*, minimum covering Randić energy, $R_c(G)$of star graph k_{1n-1} is equal to $\sqrt{5}$.

Proof. Let k_{1n-1} be a star graph with vertex set $V=\{\upsilon_0,\upsilon_1,\upsilon_2,\cdots,\upsilon_{n-1}\}$.Then its Minimum covering set is $C=\{v_0\}$.

$$R_C(K_{1,n-1}) = \begin{pmatrix} 1 & \frac{1}{\sqrt{n-1}} & \frac{1}{\sqrt{n-1}} & \frac{1}{\sqrt{n-1}} & \cdots & \frac{1}{\sqrt{n-1}} \\ \frac{1}{\sqrt{n-1}} & 0 & 0 & 0 & \cdots & 0 \\ \frac{1}{\sqrt{n-1}} & 0 & 0 & 0 & \cdots & 0 \\ \frac{1}{\sqrt{n-1}} & 0 & 0 & 0 & \cdots & 0 \\ \vdots & \vdots & \vdots & \vdots & \ddots & \vdots \\ \frac{1}{\sqrt{n-1}} & 0 & 0 & 0 & \cdots & 0 \end{pmatrix}_{n\times n}$$ Characteristi c equation is $(-1)^n \rho^{n-2}\left[\rho^2-\rho-1\right]=0$.

Minimum covering Randić Spec

$$(K_{1,n-1}) = \begin{pmatrix} 0 & \frac{1+\sqrt{5}}{2} & \frac{1-\sqrt{5}}{2} \\ n-2 & 1 & 1 \end{pmatrix}$$

Minimum covering Randić energy,

$$RE_C\left(K_{1,n-1}\right)=|0|(n-2)+\left|\frac{1+\sqrt{5}}{2}\right|(1)+\left|\frac{1-\sqrt{5}}{2}\right|(1)=0+\frac{1+\sqrt{5}}{2}+\frac{\sqrt{5}-1}{2}=\sqrt{5}.$$

Definition 2.3 Crown graph S_n^0 for an integer $V=\{u_1,u_2,\cdots,u_n,v_1,v_2,\cdots,v_n\}$ is the graph with vertex set and edge set $\{u_1,u_2,\cdots,u_n,v_1,v_2,\cdots,v_n\}$ and *edge set* $\{u_iv_j : 1\le i,j\le n, i\ne j\}$.

Theorem 2.5 For *n*≥2, minimum covering Randić energy, $R_c(G)$ of the crown graph S_n^0 is equal to $\sqrt{5}+\sqrt{n^2-2n+5}$.

Proof. For the crown graph S_n^0 with vertex set $V=\{u_1,u_2,\cdots,u_n,v_1,v_2,\cdots,v_n\}$, mi- nimum covering set of crown graph S_n^0 is. Then

$$R_C\left(S_n^0\right)=\left(\begin{array}{c|ccccc|ccccc}
 & u_1 & u_2 & u_3 & \cdots & u_n & v_1 & v_2 & v_3 & \cdots & v_n \\
\hline
u_1 & 1 & 0 & 0 & \cdots & 0 & 0 & \frac{1}{n-1} & \frac{1}{n-1} & \cdots & \frac{1}{n-1} \\
u_2 & 0 & 1 & 0 & \cdots & 0 & \frac{1}{n-1} & 0 & \frac{1}{n-1} & \cdots & \frac{1}{n-1} \\
u_3 & 0 & 0 & 1 & \cdots & 0 & \frac{1}{n-1} & \frac{1}{n-1} & 0 & \cdots & \frac{1}{n-1} \\
\vdots & \vdots & \vdots & \vdots & \ddots & \vdots & \vdots & \vdots & \vdots & \ddots & \vdots \\
u_n & 0 & 0 & 0 & \cdots & 1 & \frac{1}{n-1} & \frac{1}{n-1} & \frac{1}{n-1} & \cdots & 0 \\
\hline
v_1 & 0 & \frac{1}{n-1} & \frac{1}{n-1} & \cdots & \frac{1}{n-1} & 0 & 0 & 0 & \cdots & 0 \\
v_2 & \frac{1}{n-1} & 0 & \frac{1}{n-1} & \cdots & \frac{1}{n-1} & 0 & 0 & 0 & \cdots & 0 \\
v_3 & \frac{1}{n-1} & \frac{1}{n-1} & 0 & \cdots & \frac{1}{n-1} & 0 & 0 & 0 & \cdots & 0 \\
\vdots & \vdots & \vdots & \vdots & \ddots & \vdots & \vdots & \vdots & \vdots & \ddots & \vdots \\
v_n & \frac{1}{n-1} & \frac{1}{n-1} & \frac{1}{n-1} & \cdots & 0 & 0 & 0 & 0 & \cdots & 0
\end{array}\right)$$

Characteristic polynomial is

$$\frac{(\rho^2-\rho-1)\left[(n-1)^2\rho^2-(n-1)^2\rho-1\right]^{n-1}}{(n-1)^{2(n-1)}}.$$

Characteristic equation is

$$(\rho^2-\rho-1)\left[(n-1)^2\rho^2-(n-1)^2\rho-1\right]^{n-1}=0$$

Minimum covering Randić Spec

$$S_n^0=\begin{pmatrix}\frac{1+\sqrt{5}}{2} & \frac{1-\sqrt{5}}{2} & \frac{(n-1)+\sqrt{n^2-2n+5}}{2(n-1)} & \frac{(n-1)-\sqrt{n^2-2n+5}}{2(n-1)}\\ 1 & 1 & n-1 & n-1\end{pmatrix}$$

Minimum covering Randić energy,

$$RE_C(S_n^0)=\left|\frac{1+\sqrt{5}}{2}\right|(1)+\left|\frac{1-\sqrt{5}}{2}\right|(1)+\left|\frac{(n-1)+\sqrt{n^2-2n+5}}{2(n-1)}\right|(n-1)+\left|\frac{(n-1)-\sqrt{n^2-2n+5}}{2(n-1)}\right|(n-1)$$
$$=\frac{\sqrt{5}+1}{2}+\frac{\sqrt{5}-1}{2}+\frac{\left[(n-1)+\sqrt{n^2-2n+5}\right]}{2(n-1)}(n-1)+\frac{\left[\sqrt{n^2-2n+5}-(n-1)\right]}{2(n-1)}(n-1)$$
$$=\sqrt{5}+\sqrt{n^2-2n+5}.$$

Theorem 2.6 The minimum covering Randić energy, $R_c(G)$ of the complete bipartite graph $RE_C(K_{m,n})$ is equal to $m+\sqrt{5}-1$

Proof. For the complete bipartite graph $K_{m,n}\,(m\le n)$ with vertex set $V=\{u_1,u_2,\cdots,u_m,v_1,v_2,\cdots,v_n\}$, minimum covering set is $C=\{u_1,u_2,\cdots,u_m\}$. Then

$$R_C\left(K_{m,n}\right)=\left(\begin{array}{c|ccccc|ccccc} & u_1 & u_2 & u_3 & \cdots & u_m & v_1 & v_2 & v_3 & \cdots & v_n \\ \hline u_1 & 1 & 0 & 0 & \cdots & 0 & \frac{1}{\sqrt{mn}} & \frac{1}{\sqrt{mn}} & \frac{1}{\sqrt{mn}} & \cdots & \frac{1}{\sqrt{mn}} \\ u_2 & 0 & 1 & 0 & \cdots & 0 & \frac{1}{\sqrt{mn}} & \frac{1}{\sqrt{mn}} & \frac{1}{\sqrt{mn}} & \cdots & \frac{1}{\sqrt{mn}} \\ u_3 & 0 & 0 & 1 & \cdots & 0 & \frac{1}{\sqrt{mn}} & \frac{1}{\sqrt{mn}} & \frac{1}{\sqrt{mn}} & \cdots & \frac{1}{\sqrt{mn}} \\ \vdots & \vdots & \vdots & \vdots & \ddots & \vdots & \vdots & \vdots & \vdots & \ddots & \vdots \\ u_m & 0 & 0 & 0 & \cdots & 1 & \frac{1}{\sqrt{mn}} & \frac{1}{\sqrt{mn}} & \frac{1}{\sqrt{mn}} & \cdots & \frac{1}{\sqrt{mn}} \\ \hline v_1 & \frac{1}{\sqrt{mn}} & \frac{1}{\sqrt{mn}} & \frac{1}{\sqrt{mn}} & \cdots & \frac{1}{\sqrt{mn}} & 0 & 0 & 0 & \cdots & 0 \\ v_2 & \frac{1}{\sqrt{mn}} & \frac{1}{\sqrt{mn}} & \frac{1}{\sqrt{mn}} & \cdots & \frac{1}{\sqrt{mn}} & 0 & 0 & 0 & \cdots & 0 \\ v_3 & \frac{1}{\sqrt{mn}} & \frac{1}{\sqrt{mn}} & \frac{1}{\sqrt{mn}} & \cdots & \frac{1}{\sqrt{mn}} & 0 & 0 & 0 & \cdots & 0 \\ \vdots & \vdots & \vdots & \vdots & \ddots & \vdots & \vdots & \vdots & \vdots & \ddots & \vdots \\ v_n & \frac{1}{\sqrt{mn}} & \frac{1}{\sqrt{mn}} & \frac{1}{\sqrt{mn}} & \cdots & \frac{1}{\sqrt{mn}} & 0 & 0 & 0 & \cdots & 0 \end{array}\right)$$

Characteristic equation is

$$(-1)^{m+n}\rho^{n-1}(\rho-1)^{m-1}\left[\rho^2-\rho-1\right]=0$$

Minimum covering Randić Spec

$$\left(K_{m,n}\right)=\begin{pmatrix} 0 & 1 & \frac{1+\sqrt{5}}{2} & \frac{1-\sqrt{5}}{2} \\ n-1 & m-1 & 1 & 1 \end{pmatrix}$$

Minimum covering Randić energy,

$$RE_C\left(K_{m,n}\right)=|0|(n-1)+|1|(m-1)+\left|\frac{1+\sqrt{5}}{2}\right|(1)+\left|\frac{1-\sqrt{5}}{2}\right|(1)$$

$$=(m-1)+\frac{1+\sqrt{5}}{2}+\frac{\sqrt{5}-1}{2}=m+\sqrt{5}-1.$$

Definition 2.4 Friendship graph is the graph obtained by taking n copies of the cycle graph c_3 with a vertex in common. It is denoted by F_3^n .Friendship graph F_3^n contains *2*n+1 vertices and 3*n* edges.

Theorem 2.7 The minimum covering Randić energy, $R_c(G)$ of friendship graph

$RE_C\left(F_3^n\right)$ is equal to $\frac{\left(1+\sqrt{17}-2\sqrt{2}\right)+2\sqrt{2}n}{2}$.

Proof. For a friendship graph F_3^n with vertex set $V=\{v_0,v_1,v_2,\cdots,v_n,v_{n+1},v_{n+2},\cdots,v_{2n}\}$, minimum covering set is $C=\{v_0,v_1,v_3,v_5,\cdots,v_{2n-1}\}$. Then

$R_C\left(F_3^n\right)=$

$$\left(\begin{array}{c|cccccccccc}
 & v_0 & v_1 & v_2 & v_3 & v_4 & v_5 & v_6 & v_7 & \cdots & v_{2n} \\
\hline
v_0 & 1 & \frac{1}{\sqrt{4n}} & \frac{1}{\sqrt{4n}} & \frac{1}{\sqrt{4n}} & \frac{1}{\sqrt{4n}} & \frac{1}{\sqrt{4n}} & \frac{1}{\sqrt{4n}} & \frac{1}{\sqrt{4n}} & \cdots & \frac{1}{\sqrt{4n}} \\
v_1 & \frac{1}{\sqrt{4n}} & 1 & \frac{1}{2} & 0 & 0 & 0 & 0 & 0 & \cdots & 0 \\
v_2 & \frac{1}{\sqrt{4n}} & \frac{1}{2} & 0 & 0 & 0 & 0 & 0 & 0 & \cdots & 0 \\
v_3 & \frac{1}{\sqrt{4n}} & 0 & 0 & 1 & \frac{1}{2} & 0 & 0 & 0 & \cdots & 0 \\
v_4 & \frac{1}{\sqrt{4n}} & 0 & 0 & \frac{1}{2} & 0 & 0 & 0 & 0 & \cdots & 0 \\
v_5 & \frac{1}{\sqrt{4n}} & 0 & 0 & 0 & 0 & 1 & \frac{1}{2} & 0 & \cdots & 0 \\
v_6 & \frac{1}{\sqrt{4n}} & 0 & 0 & 0 & 0 & \frac{1}{2} & 0 & 0 & \cdots & 0 \\
v_7 & \frac{1}{\sqrt{4n}} & 0 & 0 & 0 & 0 & 0 & 0 & 1 & \cdots & \frac{1}{2} \\
\vdots & \vdots & \vdots & \vdots & \vdots & \vdots & \vdots & \vdots & \vdots & \ddots & \vdots \\
v_{2n} & \frac{1}{\sqrt{4n}} & 0 & 0 & 0 & 0 & 0 & 0 & \frac{1}{2} & \cdots & 0
\end{array}\right)_{(2n+1)\times(2n+1)}$$

Characteristic equation is

$$\frac{(2\rho-1)\left(2\rho^2-3\rho-1\right)\left(4\rho^2-4\rho-1\right)^{n-1}}{4^n}=0$$

Minimum covering Randić Spec

$$RE_C\left(F_3^n\right)=\begin{pmatrix} \frac{1}{2} & \frac{3+\sqrt{17}}{4} & \frac{3-\sqrt{17}}{4} & \frac{1+\sqrt{2}}{2} & \frac{1-\sqrt{2}}{2} \\ 1 & 1 & 1 & n-1 & n-1 \end{pmatrix}$$

Minimum covering Randić energy,

$$RE_C\left(F_3^n\right)=\left|\frac{1}{2}\right|(1)+\left|\frac{3+\sqrt{17}}{4}\right|(1)+\left|\frac{3-\sqrt{17}}{4}\right|(1)+\left|\frac{1+\sqrt{2}}{2}\right|(n-1)+\left|\frac{1-\sqrt{2}}{2}\right|(n-1)$$

$$=\frac{1}{2}+\frac{3+\sqrt{17}}{4}+\frac{\sqrt{17}-3}{4}+\frac{1+\sqrt{2}}{2}(n-1)+\frac{\sqrt{2}-1}{2}(n-1)$$

$$=\frac{1}{2}+\frac{\sqrt{17}}{2}+\frac{n-1}{2}\left(2\sqrt{2}\right)$$

$$=\frac{\left(1+\sqrt{17}-2\sqrt{2}\right)+2\sqrt{2}n}{2}.$$

2.2. Properties of Minimum Covering Randić Eigenvalues

Theorem 2.8 Let G be a graph with vertex set $V=\{v_1,v_2,\cdots,v_n\}$, edge set E and $D=\{u_1,u_2,\cdots,u_k\}$ be a minimum covering set. If $p_1,p_2\ldots p_n$ are the eigenvalues of minimum covering Randić matrix R_c *(G)* then (i) $\sum_{i=1}^{n}\rho_i=|C|$ (ii) $\sum_{i=1}^{n}\rho_i^2=|C|+2\sum_{i<j}\frac{1}{d_id_j}$.

Proof. i) We know that the sum of the eigenvalues of $R_c(G)$ is the trace of *Rc(G)*

$$\therefore\sum_{i=1}^{n}\rho_i=\sum_{i=1}^{n}r_{ii}=|C|$$

ii) Similarly the sum of squares of the eigenvalues of $R_c(G)$ is trace of $\left[R_C(G)\right]^2$

$$\therefore\sum_{i=1}^{n}\rho_i^2=\sum_{i=1}^{n}\sum_{j=1}^{n}r_{ij}r_{ji}=\sum_{i=1}^{n}(r_{ii})^2+\sum_{i\neq j}r_{ij}r_{ji}=\sum_{i=1}^{n}(r_{ii})^2+2\sum_{i<j}(r_{ij})^2=\sum_{i=1}^{n}(r_{ii})^2+2\sum_{i<j}\left(\frac{1}{\sqrt{d_id_j}}\right)^2$$

$$=|C|+2\sum_{i<j}\frac{1}{d_id_j}=|C|+2M$$ where $M=\sum_{i<j}\frac{1}{d_id_j}$

2.3. Bounds for Minimum Covering Randić Energy

Mclelland's [8] gave upper and lower bounds for ordinary energy of a graph. Similar bounds for $RE_c(G)$ are given in the following theorem.

Theorem 2.9 Let G be a simple graph with n vertices and m edges . If C is the minimum covering set and $P = \left|\det R_C(G)\right|$ then

$$\sqrt{|C| + 2\sum_{i<j}\frac{1}{d_i d_j} + n(n-1)P^{\frac{2}{n}}} \le RE_C(G) \le \sqrt{n\left(|C| + 2\sum_{i<j}\frac{1}{d_i d_j}\right)}.$$

Proof.

Canchy Schwarz inequality is $\left(\sum_{i=1}^{n} a_i b_i\right)^2 \le \left(\sum_{i=1}^{n} a_i^2\right)\left(\sum_{i=1}^{n} b_i^2\right)$

If $a_i = 1, b_i = |\rho_i|$ then $\left(\sum_{i=1}^{n} |\rho_i|\right)^2 \le \left(\sum_{i=1}^{n} 1\right)\left(\sum_{i=1}^{n} \rho_i^2\right)$

$$\left[RE_C(G)\right]^2 \le n\left(|C| + 2\sum_{i<j}\frac{1}{d_i d_j}\right)$$ [From theorem 2.8]

$$\Rightarrow RE_C(G) \le \sqrt{n\left(|C| + 2\sum_{i<j}\frac{1}{d_i d_j}\right)}.$$

Since arithmetic mean is greater than or equal to geometric mean we have

$$\frac{1}{n(n-1)}\sum_{i \ne j} |\rho_i||\rho_j| \ge \left[\prod_{i \ne j} |\rho_i||\rho_j|\right]^{\frac{1}{n(n-1)}} = \left[\prod_{i=1}^{n} |\rho_i|^{2(n-1)}\right]^{\frac{1}{n(n-1)}}$$

$$= \left[\prod_{i=1}^{n} |\rho_i|\right]^{\frac{2}{n}} = \left|\prod_{i=1}^{n} \rho_i\right|^{\frac{2}{n}} = \left|\det R_C(G)\right|^{\frac{2}{n}} = P^{\frac{2}{n}}$$

$$\therefore \sum_{i \ne j} |\rho_i||\rho_j| \ge n(n-1)P^{\frac{2}{n}}. \qquad (2.1)$$

Now consider, $\left[RE_C(G)\right]^2 = \left(\sum_{i=1}^{n}|\rho_i|\right)^2 = \sum_{i=1}^{n}|\rho_i|^2 + \sum_{i\neq j}|\rho_i||\rho_j|$

$$\therefore \left[RE_C(G)\right]^2 \geq |C| + 2\sum_{i<j}\frac{1}{d_i d_j} + n(n-1)P^{\frac{2}{n}}$$

$$i.e., RE_C(G) \geq \sqrt{|C| + 2\sum_{i<j}\frac{1}{d_i d_j} + n(n-1)P^{\frac{2}{n}}}.$$

[From (2.1)]

Theorem 2.10 If $p_1(g)$ is the largest minimum covering Randić eigenvalue of $Rc(G)$, then $\rho_1(G) \geq \frac{|C| + R(G)}{n}$.

Proof. For any nonzero vector X, we have by [16] $\rho_1(A) = \max_{X\neq 0}\left\{\frac{X'AX}{X'X}\right\}$,

$$\therefore \rho_1(G) \geq \frac{J'AJ}{J'J} = \frac{|C| + 2\sum_{i<j}\frac{1}{\sqrt{d_i d_j}}}{n} = \frac{|C| + R(G)}{n}$$

where J is a unit column matrix.

Just like Koolen and Moulton's [17] upper bound for energy of a graph, an upper bound for $RE_c(G)$ is given in the following theorem.

Theorem 2.11 If G is a graph with n vertices and m edges and $\left(|C| + 2\sum_{i<j}\frac{1}{d_i d_j}\right) \geq n$ then

$$RE_C(G) \leq \frac{|C| + 2\sum_{i<j}\frac{1}{d_i d_j}}{n} + \sqrt{(n-1)\left[|C| + 2\sum_{i<j}\frac{1}{d_i d_j} - \left(\frac{|C| + 2\sum_{i<j}\frac{1}{d_i d_j}}{n}\right)^2\right]}.$$

Proof.

Cauchy-Schwartzin equality is $\left[\sum_{i=2}^{n} a_i b_i\right]^2 \le \left(\sum_{i=2}^{n} a_i^2\right)\left(\sum_{i=2}^{n} b_i^2\right)$

Put $a_i = 1, b_i = |\rho_i|$ then $\left(\sum_{i=2}^{n} |\rho_i|\right)^2 = \sum_{i=2}^{n} 1 \sum_{i=2}^{n} |\rho_i|^2$

Let

$$f(x) = x + \sqrt{(n-1)\left(|C| + 2\sum_{i<j} \frac{1}{d_i d_j} - x^2\right)}$$

For decreasing function

$$f'(x) \le 0 \Rightarrow 1 - \frac{x(n-1)}{\sqrt{(n-1)\left(|C| + 2\sum_{i<j} \frac{1}{d_i d_j} - x^2\right)}} \le 0$$

$$\Rightarrow x \ge \sqrt{\frac{|C| + 2\sum_{i<j} \frac{1}{d_i d_j}}{n}}$$

Since $\left(|C| + 2\sum_{i<j} \frac{1}{d_i d_j}\right) \ge n$, we have $\sqrt{\frac{|C| + 2\sum_{i<j} \frac{1}{d_i d_j}}{n}} \le \frac{|C| + 2\sum_{i<j} \frac{1}{d_i d_j}}{n} \le \rho_1$

$$\therefore f(\rho_1) \le f\left(\frac{|C| + 2\sum_{i<j}\frac{1}{d_i d_j}}{n}\right)$$

$$i.e., RE_C(G) \le f(\rho_1) \le f\left(\frac{|C| + 2\sum_{i<j}\frac{1}{d_i d_j}}{n}\right)$$

$$i.e., RE_C(G) \le f\left(\frac{|C| + 2\sum_{i<j}\frac{1}{d_i d_j}}{n}\right)$$

$$i.e., RE_C(G) \le f(\rho_1) \le f\left(\frac{|C| + 2\sum_{i<j}\frac{1}{d_i d_j}}{n}\right)$$

$$i.e., RE_C(G) \le f\left(\frac{|C| + 2\sum_{i<j}\frac{1}{d_i d_j}}{n}\right)$$

$$i.e., RE_C(G) \le \frac{|C| + 2\sum_{i<j}\frac{1}{d_i d_j}}{n} + \sqrt{(n-1)\left[|C| + 2\sum_{i<j}\frac{1}{d_i d_j} - \left(\frac{|C| + 2\sum_{i<j}\frac{1}{d_i d_j}}{n}\right)^2\right]}.$$

Milovanović [18] bounds for minimum covering Randić energy of a graph are given in the following theorem.

Theorem 2.12 Let G be a graph with n vertices and m edges. Let $|\rho_1| \ge |\rho_2| \ge \cdots \ge |\rho_n|$ be a non-increasing order of minimum covering Randić eigen- values of *Rc(G)* and C is minimum covering set then

$$RE_C(G) \geq \sqrt{n\left(|C| + 2\sum_{i<j}\frac{1}{d_i d_j}\right) - \alpha(n)\left(|\rho_1| - |\rho_n|\right)^2}$$ where

$$\alpha(n) = n\left[\frac{n}{2}\right]\left(1 - \frac{1}{n}\left[\frac{n}{2}\right]\right)$$ and [x] denotes the integral part of a real number.

Proof. For real numbers $a, a_1, a_2, \cdots, a_n, A$ and $b, b_1, b_2, \cdots, b_n, B$ .with $a \leq a_i \leq A$ and $b \leq b_i \leq B \; \forall i = 1, 2, \cdots n$ the following inequality is proved in [19]. $\left|n\sum_{i=1}^{n} a_i b_i - \sum_{i=1}^{n} a_i \sum_{i=1}^{n} b_i\right| \leq \alpha(n)(A - a)(B - b)$ where $\alpha(n) = n\left[\frac{n}{2}\right]\left(1 - \frac{1}{n}\left[\frac{n}{2}\right]\right)$ and equality holds if and only if $a_1 = a_2 = \cdots = a_n$ and $b_1 = b_2 = \cdots = b_n$.

If $a_i = |\rho_i|, b_i = |\rho_i|, a = b = |\rho_n|$ and $A = B = |\rho_1|$, then

$$\left|n\sum_{i=1}^{n}|\rho_i|^2 - \left(\sum_{i=1}^{n}|\rho_i|\right)^2\right| \leq \alpha(n)\left(|\rho_1| - |\rho_n|\right)^2$$

But $\sum_{i=1}^{n}|\rho_i|^2 = |C| + 2\sum_{i<j}\frac{1}{d_i d_j}$

and $RE_C(G) \leq \sqrt{n\left(|C| + 2\sum_{i<j}\frac{1}{d_i d_j}\right)}$ then the above inequality becomes

$$n\left(|C|+2\sum_{i<j}\frac{1}{d_i d_j}\right)-\left(RE_C(G)\right)^2 \leq \alpha(n)\left(|\rho_1|-|\rho_n|\right)^2$$

$$i,e.,\ RE_C(G) \geq \sqrt{n\left(|C|+2\sum_{i<j}\frac{1}{d_i d_j}\right)-\alpha(n)\left(|\rho_1|-|\rho_n|\right)^2}.$$

Theorem 2.13 Let *G* be a graph with n vertices and m edges. Let $|\rho_1| \geq |\rho_2| \geq \cdots \geq |\rho_n| > 0$ be a non-increasing order of minimum covering eigenvalues of $R_c(G)$ then

$$RE_C(G) \geq \frac{|C|+2\sum_{i<j}\frac{1}{d_i d_j}+n|\rho_1||\rho_n|}{\left(|\rho_1|+|\rho_n|\right)}.$$

Proof. Let $a_i \neq 0, b_i, r$ and *R* be real numbers satisfying $ra_i \leq b_i \leq Ra_i$, , then the following inequality is proved in [20].

$$\sum_{i=1}^{n} b_i^2 + rR\sum_{i=1}^{n} a_i \leq (r+R)\sum_{i=1}^{n} a_i b_i$$

Put $b_i = |\rho_i|,\ a_i = 1,\ r = |\rho_n|$ and $R = |\rho_1|$ then

$$\sum_{i=1}^{n}|\rho_i|^2 + |\rho_1||\rho_n|\sum_{i=1}^{n}1 \leq \left(|\rho_1|+|\rho_n|\right)\sum_{i=1}^{n}|\rho_i|$$

$$i.e.,|C|+2\sum_{i<j}\frac{1}{d_i d_j}+|\rho_1||\rho_n|n \leq \left(|\rho_1|+|\rho_n|\right)RE_C(G)$$

$$\therefore RE_C(G) \geq \frac{|C|+2\sum_{i<j}\frac{1}{d_i d_j}+n|\rho_1||\rho_n|}{\left(|\rho_1|+|\rho_n|\right)}.$$

The question of when does the graph energy becomes a rational number was answered by Bapat and S. pati in their paper [21]. Similar result for minimum covering Randić energy is obtained in the following theorem.

Theorem 2.14 *Let G* be a graph with a minimum covering set *C*. If the minimum covering Randić energy $RE_C(G)$ is a rational number, then $RE_C(G) \equiv |C| \ \ (mod\, 2)$ (mod 2).

Proof. Proof is similar to theorem 3.7 of [15]

3. CONCLUSION

It was proved in this paper that the minimum covering Randić energy of a graph G depends on the covering set that we take for consideration. Upper and lower bounds for minimum covering Randić energy are established. A generalized expression for minimum covering Randić energies for star graph, complete graph, thorn graph of com- plete graph, crown graph, complete bipartite graph, cocktail party graph and friendship graphs are also computed.

ACKNOWLEDGEMENTS

The authors are thankful to anonymous referees for their valuable comments and useful suggestions.

REFERENCE

1. Bozkurt, S.B., Güngör, A.D. and Gutman, I. (2010) Randić Spectral Radius and Randić Energy. Communications in Mathematical and in Computer Chemistry, 64, 239-250
2. Gutman, I. (1978) The Energy of a Graph. Ber. Math-Statist. Sekt. Forschungsz. Graz, 103, 1-22
3. Gutman, I., Li, X. and Zhang, J. (2009) Graph Energy. In: Dehmer, M. and Emmert-Streib, F., Eds., Analysis of Complex Networks. From Biology to Linguistics, Wiley-VCH, Weinheim, 145-174
4. Cvetković, D. and Gutman, I., Eds. (2009) Applications of Graph Spectra. Mathematical Institution, Belgrade
5. Cvetković, D. and Gutman, I., Eds. (2011) Selected Topics on Applications of Graph Spectra. Mathematical Institute Belgrade
6. Gutman, I. (2001) The Energy of a Graph: Old and New Results. In: Betten, A., Kohnert, A., Laue, R. and Wassermann, A., Eds., Algebraic Combinatorics and Applications, Springer, Berlin, 196-211

7. Liu, H.Q., Lu, M. and Tian, F. (2007) Some Upper Bounds for the Energy of Graphs. Journal of Mathematical Chemistry, 41, 45-57. https://doi.org/ 10.1007/s10910-006-9183-9
8. McClelland, B.J. (1971) Properties of the Latent Roots of a Matrix: The Estimation of π - Electron Energies. The Journal of Chemical Physics, 54, 640-643. https://doi.org/10.1063/1.1674889
9. Gutman, I. and Polansky, O.E. (1986) Mathematical Concepts in Organic Chemistry. Springer, Berlin. https://doi.org/10.1007/978-3-642-70982-1
10. Graovac, A., Gutman, I. and Trinajstić, N. (1977) Topological Approach to the Chemistry of Conjugated Molecules. Springer, Berlin. https://doi.org/10.1007/978-3-642-93069-0
11. Randić, M. (1975) On Characterization of Molecular Branching, Journal of the American Chemical Society, 97, 6609-6615. https://doi.org/ 10.1021/ja00856a001
12. Bozkurt, S.B. and Bozkurt, D. (2013) Sharp Upper Bounds for Energy and Randić Energy. Communications in Mathematical and in Computer Chemistry, 70, 669-680
13. Dilek Maden, A. (2015) New Bounds on the Incidence Energy, Randić Energy and Randić Estrada Index. Communications in Mathematical and in Computer Chemistry, 74, 367-387
14. Das, K.Ch., Sorgun, S. and Gutman, I. (2015) On Randić Energy. Communications in Mathematical and in Computer Chemistry, 73, 81-92
15. Adiga, C., Bayad, A., Gutman, I. and Srinivas, S.A. (2012) The Minimum Covering Energy of a Graph. Kragujevac Journal of Science, 34, 39-56
16. Bapat, R.B. (2011) Graphs and Matrices. Hindustan Book Agency, Gurgaon, No.32
17. Koolen, J.H. and Moulton, V. (2001) Maximal Energy Graphs. Advances in Applied Mathematics, 26, 47-52. https://doi.org/10.1006/aama.2000.0705
18. Milovanović, I.Z., Milovanović, E.I. and Zakić, A. (2014) A Short Note on Graph Energy. MATCH Communications in Mathematical and in Computer Chemistry, 72, 179-182
19. Biernacki, M., Pidek, H. and Ryll-Nadzewski, C. (1950) Sur une inequalite entre des inegrales defnies. Annales Universitatis Mariae Curie-Sklodowska Sectio A, 4, 1-4
20. Diaz, J.B. and Matcalf, F.T. (1963) Stronger Forms of a Class Inequalities of G. Polja-G. Szego and L. V. Kantorovich. Bulletin of the American Mathematical Society, 69, 415-418. https://doi.org/10.1090/S0002-9904-1963-10953-2
21. Bapat, R.B. and Pati, S. (2011) Energy of a Graph Is Never an Odd Integer. Bulletin of Kerala Mathematics Association, 1, 129-132

Index